Physical Metallurgy of High Manganese Steels

Physical Metallurgy of High Manganese Steels

Special Issue Editors

Wolfgang Bleck
Christian Haase

MDPI • Basel • Beijing • Wuhan • Barcelona • Belgrade

MDPI

Special Issue Editors

Wolfgang Bleck
RWTH Aachen University
Germany

Christian Haase
RWTH Aachen University
Germany

Editorial Office
MDPI
St. Alban-Anlage 66
4052 Basel, Switzerland

This is a reprint of articles from the Special Issue published online in the open access journal *Metals* (ISSN 2075-4701) in 2019 (available at: https://www.mdpi.com/journal/metals/special_issues/ high_manganese_steel).

For citation purposes, cite each article independently as indicated on the article page online and as indicated below:

LastName, A.A.; LastName, B.B.; LastName, C.C. Article Title. *Journal Name* **Year**, *Article Number, Page Range.*

ISBN 978-3-03921-856-1 (Pbk)
ISBN 978-3-03921-857-8 (PDF)

Contents

About the Special Issue Editors

Wolfgang Bleck (Prof. Dr.- Ing.) has been the head of the Steel Institute at RWTH Aachen University in Germany for 25 years. He received his Dipl.-Ing. degree and, subsequently, Dr.-Ing. degree in physical metallurgy from Clausthal University, Germany. He was affiliated with the research department at Thyssen Stahl AG in Duisburg, Germany from 1980 to 1993, where he became Head of the department for process and steel development of flat-rolled products. Since 1994, he has been Professor of ferrous metallurgy at RWTH Aachen University, where he teaches materials science at undergraduate and graduate levels, and participates in research activities on both a national and international level. He has supervised more than 100 PhD students, authored more than 250 publications, and holds several patents. Wolfgang Bleck also served as a member of the senate in 2002–2005, as Vice-Rector of RWTH Aachen University, and in 2014–2016, as Dean of the Faculty of Georesources and Materials Science and Engineering. Since 2007, he has been the spokesman of the collaborative research center SFB 761 "steel—ab initio", dealing with the development of high and medium Mn steels. He belongs to the steering committee of the research cluster "production in high-wage countries". He has been Adjunct Professor at Postech, Pohang in Korea, Honorary Professor at the Northeastern University in Shenyang, and an appointed Baosteel Professor in China. He is a member of the scientific councils of the National University of Science and Technology/MISiS in Moscow and of the King Mongkut University of Technology Thonburi, Bangkok. He is a member of the extended board of the German Steel Institute (VDEh), and Head of the editorial board of the journal Steel Research International. Wolfgang Bleck's research activities are the development and characterization of advanced high-strength steels, new processes for steel products, lightweight structures, principles of steel design, and numerical modeling of material and component properties.

Christian Haase (Dr.-Ing.) received his Dipl.-Ing. degree in materials science and engineering from Otto von Guericke University Magdeburg, Germany, and Dr.-Ing. degree in metal physics in 2016 from RWTH University, Germany. He leads the research group Integrative Computational Materials Engineering at the Steel Institute of RWTH Aachen University. Dr. Haase is involved in and holds or has held leading positions in several large-scale research projects, such as the Collaborative Research Centre (SFB) 761 "steel—ab initio", Cluster of Excellence "production in high-wage countries", and "internet of production" founded by the Deutsche Forschungsgemeinschaft (DFG). He authored more than 50 publications and serves as reviewer for numerous leading journals in the field of materials science. His research work has been honored with several early career researcher awards, such as the DGM Young Researcher Award and establishment of the NanoMatFutur research group by the German Federal Ministry of Education and Research. Christian Haase's research activities focus on alloy design, material characterization, materials processing and application, advanced high-strength steels, high-entropy alloys, integrated computational materials engineering, severe plastic deformation, and additive manufacturing.

metals

MDPI

Editorial

Physical Metallurgy of High Manganese Steels

Wolfgang Bleck * and Christian Haase *

Steel Institute, RWTH Aachen University, D-52072 Aachen, Germany
* Correspondence: bleck@iehk.rwth-aachen.de (W.B.); christian.haase@iehk.rwth-aachen.de (C.H.)

Received: 23 September 2019; Accepted: 26 September 2019; Published: 28 September 2019

1. Introduction and Scope

The development of materials with advanced or new properties has been the primary aim of materials scientists for past centuries. In the field of metallic alloys for structural applications, strength, formability, and toughness are key parameters to achieve desired performance. High manganese steels (HMnS) are characterized by an extraordinary combination of these key parameters, which has aroused the fascination of researchers worldwide.

Although austenitic steels with high manganese content have been known since the original works by Sir Robert A. Hadfield in the 19th century [1], it took until the late 1990s when research into these alloys experienced a resurrection. The present hype in the research of HMnS was initiated by the work of Grässel et al. [2], followed by numerous national and international research activities, such as the Collaborative Research Centre 761 "Steel—ab initio" funded by the German Research Foundation (DFG) [3]. HMnS represent a highly fascinating class of alloys within the field of advanced high strength steels (AHSS). The high interest in HMnS in both academic and industrial research originates from their outstanding mechanical properties. Therefore, potential fields of industrial application supposedly extend from chassis components in the automotive industry over equipment for low-temperature applications to forgings with alternative process routes. Usually, these steels contain a manganese content well above 3% mass, along with significant alloying with carbon and aluminium.

The plasticity of HMnS is strongly influenced by their low stacking fault energy (SFE). Consequently, the low dynamic recovery rate in combination with the activation of additional deformation mechanisms, i.e., twinning-induced plasticity (TWIP), transformation-induced plasticity (TRIP), and microband-induced plasticity (MBIP), promote high work-hardenability. That results in a combination of high ultimate tensile strength (often above 1 GPa) and high uniform elongation (often above 50%). In order to take full advantage of the potential of HMnS, a description of these mechanisms in predictive, physics-based models is required. However, such descriptions constitute a formidable scientific challenge due to the microstructural modifications at various length scales, as well as complex chemical interactions.

The processing of HMnS requires careful consideration of solidification conditions in order to minimize segregation and control precipitation and microstructure development. The further fabrication via rolling, annealing, cutting, and machining needs to be adopted to the specific material behaviour.

Careful review of the related literature at present revealed that there is still a severe need to better understand the physical metallurgical mechanisms of HMnS. Therefore, this Special Issue focuses on fundamental aspects of HMnS including amongst others microstructure evolution, phase transformation, plasticity, hydrogen embrittlement, and fatigue investigated by advanced experimental as well as computational approaches.

2. Contributions

This Special Issue gathers manuscripts from internationally recognized researchers with stimulating new ideas and original results. It consists of fifteen original research papers, seven contributions focus on steels with manganese content above 12% mass [4–10], whereas eight deal with alloys having less manganese [11–18].

The most probable application of HMnS is anticipated to be as sheet products. Therefore, profound understanding of the material behaviour during thermo-mechanical processing is of eminent importance and has been addressed in the contributions by Torganchuk et al. [4], Haupt et al. [5], Oevermann et al. [6], and Quadfasel et al. [7]. As has been shown in [4], the combination of severe cold rolling (86% thickness reduction) and annealing promotes very fine-grained HMnS. The combination of fine recrystallized grains, high carbon content and minor fraction of non-recrystallized grains resulted in a remarkable combination of mechanical properties, i.e., yield strength (YS) of 1 GPa, ultimate tensile strength (UTS) of 1.65 GPa and a total elongation (ε_{tot}) of 40%. Haupt et al. [5] took advantage of the dependence of the SFE on temperature. During rolling at elevated temperatures (up to 500 °C), the contributions of mechanically induced twinning and dislocation slip were adjusted in order to tailor the property profile at room temperature. In contrast, Oevermann et al. [6] applied deep rolling at −196 °C to 200 °C to influence the near surface properties of a HMnS. It was found that deep rolling improved the monotonic mechanical properties, whereas the fatigue performance decreased after cryogenic rolling due to the formation of ε-martensite. Finally, Quadfasel et al. [7] present a computer-aided design approach for the application of HMnS sheets in automotive crash-boxes. Optimum crash behaviour is evaluated based on a multiscale simulation chain with ab initio calculation of the SFE, crystal-plasticity simulation of the strain-hardening behaviour and finite-element simulation of the crash behaviour.

The specific microstructural features that appear in HMnS during plastic deformation strongly influence their fatigue and fracture behaviour. Fluch et al. [8] compared cold worked austenitic CrNi and CrMnN steels during cyclic loading. The higher strength of the CrMnN grade due to the high nitrogen content resulted in superior fatigue behaviour. Contrarily, the CrMnN steel also revealed a higher reduction of fatigue strength with respect to $R_{P0,2}$ as compared to the CrNi counterpart, which has mainly been attributed to the dislocation pattern, i.e., planar in CrMnNi and wavy in CiNi, by the authors. The damage and fracture behaviour of Al-added HMnS was investigated by Madivala et al. [9]. High stress concentration at grain boundaries was observed due to the interception of deformation twins and slip band extrusions and resulted in micro-cracks formation at grain boundaries and triple junctions. Additionally, decreased carbon diffusivity and reduced tendency for Mn-C short-range ordering due to Al-addition caused suppression of serrated flow by dynamic strain aging, which prevents initiation of macro-cracks.

A substantial contribution of the research community during the last two decades was a better understanding of the TWIP effect and its implication for strain hardening. Consequently, this understanding may also serve as a basis for alloy design from a more general perspective. This is addressed in the contribution by Haase and Barrales-Mora [10], who detailed the similarities between HMnS and face-centered cubic high-entropy alloys, with a prospect on mechanism-oriented alloy design.

During the past decade, manganese-alloyed steels with reduced manganese content (mainly with 3–12% of mass) moved into the focus of world-wide steel research. These steels are often referred to as 3rd generation AHSS, MMnS or quenching and partitioning (Q&P/Q+P) steels. Due to the importance of elemental partitioning during annealing, intensive research has been devoted to the microstructure formation during hot deformation, cooling and annealing, especially intercritical annealing. This has also been addressed in the contributions by Speer et al. [11], Mueller et al. [12], Liu et al. [13] and Gramlich et al. [14]. Some novel processing scenarios are presented in [11], namely MMnS for hot-stamping, double-soaked MMnS as well as processing by Q&P. The authors put a focus on steels with increased strength level in order to widen the field of potential applications. Mueller et al. [12] and Liu et al. [13] investigated the influence of pre-deformation on annealing behavior. According to [12], prior cold deformation accelerates the ferrite-to-austenite transformation and decreases the A_{c1} temperatures. This behavior may be attributed to an increased number of austenite nucleation sites as well as an enhanced diffusivity of manganese in ferrite due to higher pre-deformation. In addition, a multi-step deformation and annealing procedure is introduced in [13] and results in ultra-strong (UTS > 2 GPa) and ductile (ε_{tot} > 15%) steel. The authors explain this behavior by a combination of dislocation formation (warm rolling), partial recovery (intercritical annealing), deformation-induced martensitic transformation (cold rolling), austenite reversion (partitioning), and bake hardening. Gramlich et al. designed new MMnS that are suitable for a new annealing process consisting of air cooling after forging followed by austenite reversion tempering (ART). An optimum austenite fraction of about 10% vol. was identified to facilitate improved impact toughness.

As substantiated in the previous section, the multi-phase microstructure formed during annealing determines the mechanical properties. The contributions by Sevsek et al. [15], Glover et al. [16], and Allam et al. [17] were intended to shed more light on the deformation mechanisms in these alloys. A detailed analysis of the strain-rate-dependent deformation behavior in ultrafine-grained austenitic-ferritic MMnS is presented in [15]. Varying mechanically induced transformation behavior was found to be responsible for high strain-rate sensitivity. Glover et al. [16] studied the effects of athermal martensite on yielding behavior and strain partitioning during deformation using in situ neutron diffraction. It was found that athermal martensite, both as-quenched and tempered, led to an improvement in mechanical properties including promotion of continuous yielding and increased work-hardening rate. In addition to mechanical properties, the corrosion behavior of a novel MMnS was studied in [17]. The contribution nicely presents a computational alloy design approach that results in a steel with ultrafine-grained austenite and nano-sized precipitates promoting high strength combined with enhanced corrosion resistance due to chromium and nitrogen additions.

Finally, the scientifically very challenging and industrially relevant topic of hydrogen embrittlement is the focus of the contribution by Shen et al. [18]. Distinctly different microstructures were formed in the same alloy as a consequence of varied annealing treatment after cold rolling, i.e., only ART and austenitization followed by ART. The influence of ultrafine-grained martensite on the contribution of hydrogen-enhanced decohesion and hydrogen-enhanced localized plasticity mechanisms is discussed.

That being said, this Special Issue includes interdisciplinary research works that address current open questions in the field of the physical metallurgy of high manganese steels. The topics are manifold, fundamental-science oriented and, at the same time, relevant to industrial application. We wish an enjoyable and illuminative reading that stimulates future scientific ideas.

Conflicts of Interest: The authors declare no conflict of interest.

References

1. Hadfield, R.A. Hadfield's Manganese Steel. *Science* **1888**, *12*, 284–286.
2. Grässel, O.; Frommeyer, G.; Derder, C.; Hofmann, H. Phase Transformations and Mechanical Properties of Fe-Mn-Si-Al TRIP-Steels. *J. Phys. IV* **1997**, *7*, 383–388. [CrossRef]
3. Sonderforschungsbereich 761. Available online: http://abinitio.iehk.rwth-aachen.de/ (accessed on 20 September 2019).
4. Torganchuk, V.; Belyakov, A.; Kaibyshev, R. Improving Mechanical Properties of 18%Mn TWIP Steels by Cold Rolling and Annealing. *Metals* **2019**, *9*, 776. [CrossRef]
5. Haupt, M.; Müller, M.; Haase, C.; Sevsek, S.; Brasche, F.; Schwedt, A.; Hirt, G. The Influence of Warm Rolling on Microstructure and Deformation Behavior of High Manganese Steels. *Metals* **2019**, *9*, 797. [CrossRef]
6. Oevermann, T.; Wegener, T.; Niendorf, T. On the Evolution of Residual Stresses, Microstructure and Cyclic Performance of High-Manganese Austenitic TWIP-Steel after Deep Rolling. *Metals* **2019**, *9*, 825. [CrossRef]
7. Quadfasel, A.; Teller, M.; Madivala, M.; Haase, C.; Roters, F.; Hirt, G. Computer-Aided Material Design for Crash Boxes Made of High Manganese Steels. *Metals* **2019**, *9*, 772. [CrossRef]
8. Fluch, R.; Kapp, M.; Spiradek-Hahn, K.; Brabetz, M.; Holzer, H.; Pippan, R. Comparison of the Dislocation Structure of a CrMnN and a CrNi Austenite after Cyclic Deformation. *Metals* **2019**, *9*, 784. [CrossRef]
9. Madivala, M.; Schwedt, A.; Prahl, U.; Bleck, W. Strain Hardening, Damage and Fracture Behavior of Al-Added High Mn TWIP Steels. *Metals* **2019**, *9*, 367. [CrossRef]

10. Haase, C.; Barrales-Mora, L.A. From High-Manganese Steels to Advanced High-Entropy Alloys. *Metals* **2019**, *9*, 726. [CrossRef]

11. Speer, J.; Rana, R.; Matlock, D.; Glover, A.; Thomas, G.; De Moor, E. Processing Variants in Medium-Mn Steels. *Metals* **2019**, *9*, 771. [CrossRef]

12. Mueller, J.J.; Matlock, D.K.; Speer, J.G.; De Moor, E. Accelerated Ferrite-to-Austenite Transformation During Intercritical Annealing of Medium-Manganese Steels Due to Cold-Rolling. *Metals* **2019**, *9*, 926. [CrossRef]

13. Liu, L.; He, B.; Huang, M. Processing–Microstructure Relation of Deformed and Partitioned (D&P) Steels. *Metals* **2019**, *9*, 695.

14. Gramlich, A.; Emmrich, R.; Bleck, W. Austenite Reversion Tempering-Annealing of 4 wt.% Manganese Steels for Automotive Forging Application. *Metals* **2019**, *9*, 575. [CrossRef]

15. Sevsek, S.; Haase, C.; Bleck, W. Strain-Rate-Dependent Deformation Behavior and Mechanical Properties of a Multi-Phase Medium-Manganese Steel. *Metals* **2019**, *9*, 344. [CrossRef]

16. Glover, A.; Gibbs, P.J.; Liu, C.; Brown, D.W.; Clausen, B.; Speer, J.G.; De Moor, E. Deformation Behavior of a Double Soaked Medium Manganese Steel with Varied Martensite Strength. *Metals* **2019**, *9*, 761. [CrossRef]

17. Allam, T.; Guo, X.; Sevsek, S.; Lipińska-Chwałek, M.; Hamada, A.; Ahmed, E.; Bleck, W. Development of a Cr-Ni-V-N Medium Manganese Steel with Balanced Mechanical and Corrosion Properties. *Metals* **2019**, *9*, 705. [CrossRef]

18. Shen, X.; Song, W.; Sevsek, S.; Ma, Y.; Hüter, C.; Spatschek, R.; Bleck, W. Influence of Microstructural Morphology on Hydrogen Embrittlement in a Medium-Mn Steel Fe-12Mn-3Al-0.05C. *Metals* **2019**, *9*, 929. [CrossRef]

metals

MDPI

Article

Improving Mechanical Properties of 18%Mn TWIP Steels by Cold Rolling and Annealing

Vladimir Torganchuk, Andrey Belyakov * and **Rustam Kaibyshev**

Laboratory of Mechanical Properties of Nanostructured Materials and Superalloys, Belgorod State University, Pobeda 85, Belgorod 308015, Russia
* Correspondence: belyakov@bsu.edu.ru; Tel.: +7-4722-585457

Received: 15 June 2019; Accepted: 10 July 2019; Published: 11 July 2019

Abstract: The microstructures and mechanical properties of Fe-0.4C-18Mn and Fe-0.6C-18Mn steels subjected to large strain cold rolling followed by annealing were studied. Cold rolling with a total reduction of 86% resulted in substantial strengthening at expense of plasticity. The yield strength and the ultimate tensile strength of above 1400 MPa and 1600 MPa, respectively, were achieved in both steels, whereas total elongation decreased below 30%. Subsequent annealing at temperatures above 600 °C was accompanied with the development of recrystallization leading to fine-grained microstructures with an average grain size of about 1 μm in both steels. The fine-grained steels exhibited remarkable improved mechanical properties with a product of ultimate tensile strength by total elongation in the range of 50 to 70 GPa %. The fine-grained steel with relatively high carbon content of 0.6%C was characterized by ultimate tensile strength well above 1400 MPa that was remarkably higher than that of about 1200 MPa in the steel with 0.4%C.

Keywords: high-Mn steels; twinning induced plasticity; cold rolling; recrystallization annealing; grain refinement; strengthening

1. Introduction

High-Mn steels have aroused a great interest among material scientists and metallurgical engineers because of excellent mechanical performance [1]. These steels have a unique ability to strain hardening, which leads to extraordinary plasticity at room temperature [2–4]. The total elongation during standard tensile tests reaches 100%. Such properties are provided by deformation twinning (i.e., twinning induced plasticity, TWIP effect) and/or deformation martensite (transformation induced plasticity, TRIP effect). Both TWIP and TRIP effects contribute to the hardening of the material during plastic flow, prevent the localization of deformation and increase plasticity. The main consumers of high-Mn steels with TRIP and TWIP effects are car manufacturers such as BMW, Porsche, etc. [5]. These materials are designed to provide a higher level of safety for drivers and passengers and to increase the overall efficiency of road transport. In addition, practical studies of high-Mn steels have recently launched in order to develop technologies for the production and use of such steels as damping elements for seismic resistant structures [6].

A combination of mechanical properties of high-Mn TWIP/TRIP steels depends on their alloying extent and microstructures [7,8]. Specific chemical composition including mainly Mn, C, Al and Si stabilizes austenite and provides appropriate stacking fault energy (SFE), which, in turn, results in TRIP (at SFE below about 20 mJ/m^2) or TWIP (at SFE of 20 to 50 mJ/m^2) effects [1,2]. Regarding the microstructure, it can be controlled by thermo-mechanical treatment involving warm to hot working [9,10]. Depending on application, desired level of strength and ductility of the steels can be obtained by rolling under appropriate conditions. A decrease in rolling temperature commonly promotes the strain hardening of steels with dynamically recrystallized and/or recovered microstructures [11]. A decrease in rolling temperature from 1100 to 500 °C has been shown to result in a significant increase in the yield strength of 18%Mn steels from about 300–400 MPa to 850–950 MPa, while ultimate tensile strength increased from 1000–1100 MPa to 1200–1300 MPa, whereas total elongation decreased to 30% [10]. The development of ultrafine grained microstructure in high-Mn TWIP steels through multiple primary recrystallization has been suggested as another promising method of steel processing [12]. A decrease in the recrystallized grain size provides strengthening without significant degradation of plasticity.

The aim of the present paper is to report our current studies on the microstructure and properties of advanced Fe-0.4C-18Mn and Fe-0.6C-18Mn steels processed by cold rolling followed by recrystallization annealing. It places particular emphasis on a comparison of the microstructures and properties obtained by dynamic recovery/recrystallization and static primary recrystallization. The properties of these steels with dynamically recovered/recrystallized microstructures depended remarkably on the carbon content [10]. Therefore, two steels with different carbon content were studied to reveal a possible solute effect on the mechanical properties of statically recrystallized steels.

2. Materials and Methods

Two high-Mn steels with different carbon content, i.e., Fe-18Mn-0.4C and Fe-18Mn-0.6C, were studied. The steels were produced by an induction melting. Then steel melts were hot rolled at 1150 °C with 60% reduction. The starting materials were characterized by uniform microstructures consisting of equiaxed grains with average sizes of 60 μm and 50 μm in Fe-18Mn-0.4C and Fe-18Mn-0.6C steels, respectively. The steel plates were subjected to rolling at ambient temperature to a total rolling reduction of 86%. After each 15–20% reduction, the samples were subjected to intermediate recrystallization annealing at 700 °C for 30 min. Following the last intermediate annealing, the final rolling reduction was 25% for both steels. Then, the rolled samples were annealed at temperatures of 500–800 °C for 30 min.

The structural investigations were carried out on the sample sections normal to transverse direction (TD) using a Quanta 600 scanning electron microscope (SEM) (FEI, Hillsboro, OR, USA) equipped with an electron back scattering diffraction pattern (EBSD) analyzer incorporating an orientation imaging microscopy (OIM) system. The SEM specimens were electro-polished at a voltage of 20 V at room temperature using an electrolyte containing 10% perchloric acid and 90% acetic acid. The OIM images were subjected to clean up procedure, setting the minimal confidence index of 0.1, except cold rolled sample. In the latter case, the EBSD patterns with confidence index below 0.1 were omitted from the OIM analysis (these data-points appear as black spots in the OIM images). The OIM software (TSL OIM Analysis 6.2) (EDAX, Inc., Mahwah, NJ, USA) was used for evaluation of the mean grain size (D). The grain size was evaluated, counting all boundaries with misorientation of $\theta \geq 15°$, including twin boundaries. The tensile tests were performed along the rolling direction at ambient temperature under a strain rate of 10^{-3} s^{-1} using an INSTRON 5882 on specimens with a gauge length of 12 mm and a cross section of 1.5 mm × 3 mm.

3. Results and Discussion

3.1. Annealed Microstructures

An example of cold rolled microstructure in Fe-0.4C-18Mn steel is presented in Figure 1a. The cold rolling results in significant strain hardening and makes the structural observation difficult, although highly elongated grains along the rolling direction (RD) can be recognized in Figure 1a. The cold rolled microstructure is commonly characterized by rather strong texture components close to brass and copper components (Figure 1b). Similar textures have been frequently observed in various face centered cubic (fcc) metals and alloys subjected to cold rolling [13].

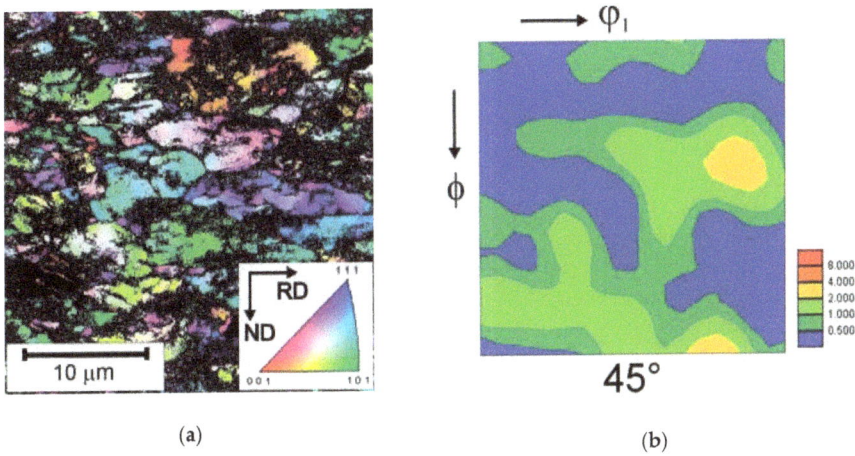

(a)   (b)

Figure 1. Microstructure (a) and orientation distribution function at $\phi_2 = 45°$ (b) of an Fe-0.4C-18Mn steel subjected to cold rolling. Colors in (a) correspond to crystallographic direction along the normal direction (ND).

Annealing softening is shown in Figure 2 as a temperature dependence of hardness. Both steels are characterized by almost the same change in the hardness during annealing. Namely, an increase in annealing temperature to 550 °C leads to gradual decrease in the hardness. The hardness decrease after annealing at 550 °C is about 10% and can be attributed to static recovery leading to a sluggish softening. A drastic decrease in the hardness takes place as temperature increases to 650 °C followed by slow softening upon further increase in temperature. It can be concluded, therefore, that temperature of around 600 °C corresponds to recrystallization temperature of the present steels much similar to other studies on primary recrystallization in high-Mn TWIP steels [14].

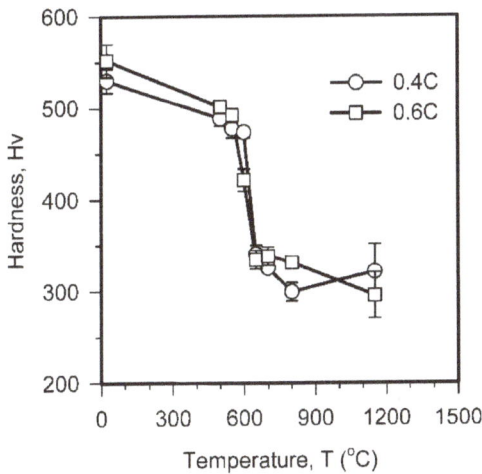

Figure 2. Effect of annealing temperature on hardness of Fe-0.4C-18Mn and Fe-0.6C-18Mn steels (0.4C and 0.6C, respectively) subjected to cold rolling.

Typical annealed microstructures evolved in Fe-0.4C-18Mn and Fe-0.6C-18Mn steel samples after annealing at 600 °C or 650 °C are shown in Figure 3. Some parameters of the annealed microstructures are listed in Table 1. The annealed microstructures consist of almost equiaxed grains with a grain size of about 1 µm irrespective of carbon content and annealing temperature. Numerous Σ3 CSL (coincident site lattice) boundaries corresponding to annealing twins testify to discontinuous recrystallization involving grain nucleation and growth as the main mechanism of microstructure evolution during the present treatment. On the other hand, frequently serrated grain boundaries suggest that the recrystallization processes have not completed. Hence, relatively high strength owing to residual stresses can be expected in these steel samples.

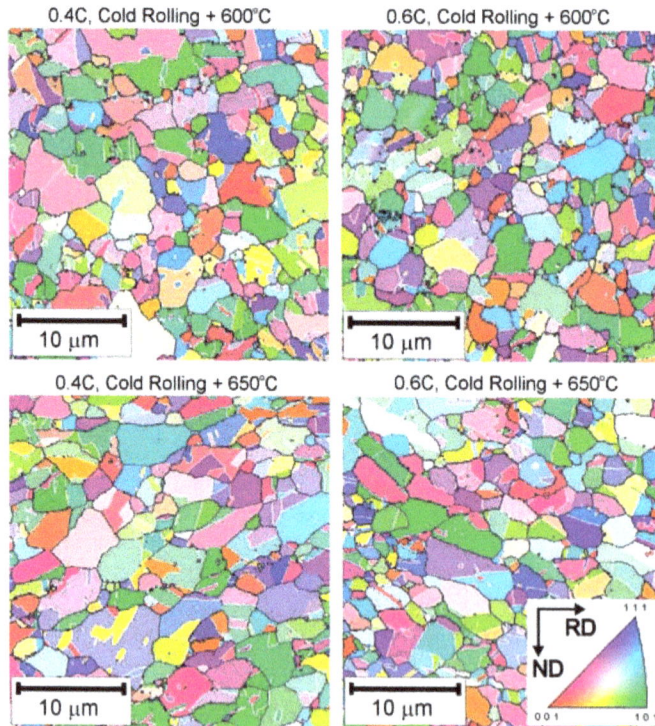

Figure 3. Typical microstructures in Fe-0.4C-18Mn and Fe-0.6C-18Mn steels (0.4C and 0.6C, respectively) subjected to cold rolling and annealing at indicated temperatures. High-angle grain boundaries and Σ3 CSL boundaries are indicated by the black and white lines, respectively. The colors correspond to crystallographic direction along ND.

Table 1. Some parameters of the annealed microstructures.

Steel Processing	Grain Size, µm	Fraction of Σ3 CSL Boundaries	Fraction of Low-Angle Boundaries
Fe-0.4C-18Mn Cold Rolling + 600 °C	1.07	0.40	0.06
Fe-0.4C-18Mn Cold Rolling + 650 °C	1.10	0.39	0.07
Fe-0.6C-18Mn Cold Rolling + 600 °C	1.03	0.29	0.10
Fe-0.6C-18Mn Cold Rolling + 650 °C	1.09	0.37	0.05

Corresponding grain boundary misorientation distributions are shown in Figure 4. Commonly, the misorientation distributions are characterized by a high peak against large misorientations of around 60°. This maximum corresponds to annealing twin boundaries, which frequently develop in fcc-metallic materials with low SFE during static recrystallization [15]. The misorientation distribution of other high-angle boundaries looks like a random one (indicated by the dotted line in Figure 4) with a broad peak at 45° [16]. The most interesting feature of the obtained microstructures is a relatively large fraction of low-angle subboundaries. Besides the sharp peak of twin boundaries, all misorientation distributions in Figure 4 exhibit small peaks corresponding to low-angle subboundaries. This is unusual for discontinuous static (primary) recrystallization [17]. The low-angle subboundaries in the annealed samples might remain from dislocation substructures produced by cold rolling. Again, the presence of dislocation subboundaries in the annealed samples implies incomplete softening. The largest fraction of low-angle subboundaries is observed in the Fe-0.6C-18Mn steel samples after annealing at 600 °C, suggesting a relatively high strength for this condition.

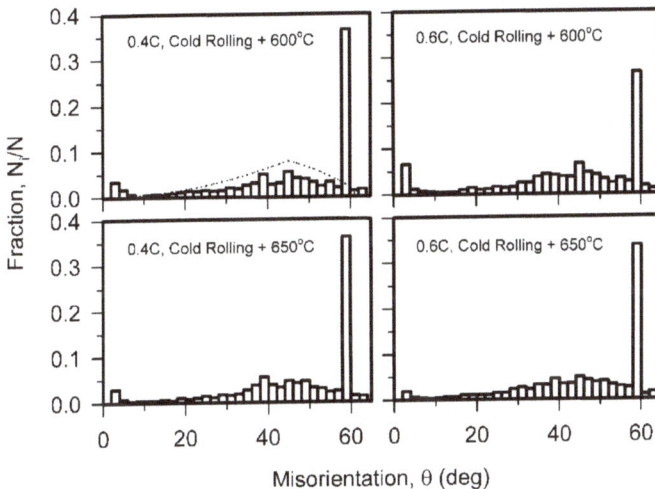

Figure 4. Grain/subgrain boundary misorientation distribution in Fe-0.4C-18Mn and Fe-0.6C-18Mn steels (0.4C and 0.6C, respectively) subjected to cold rolling and annealing at indicated temperatures. The dotted line indicates random misorientation distribution.

The development of discontinuous static recrystallization usually weakens the textures caused by previous cold rolling. Figure 5 shows orientation distribution functions at sections of $\phi_2 = 45°$ for the studied annealed steel samples. As could be expected, the annealed samples do not exhibit any strong textures. The annealed textures include orientations close to Brass and Copper components. The latter is more pronounced in the Fe-0.4C-18Mn steel samples, especially, after annealing at 650 °C. It should be noted that similar texture components were observed in the cold rolled samples (Figure 1b). The annealed textures, therefore, may correspond to early stage of recrystallization, when the deformation microstructures have not been completely replaced by the annealed ones.

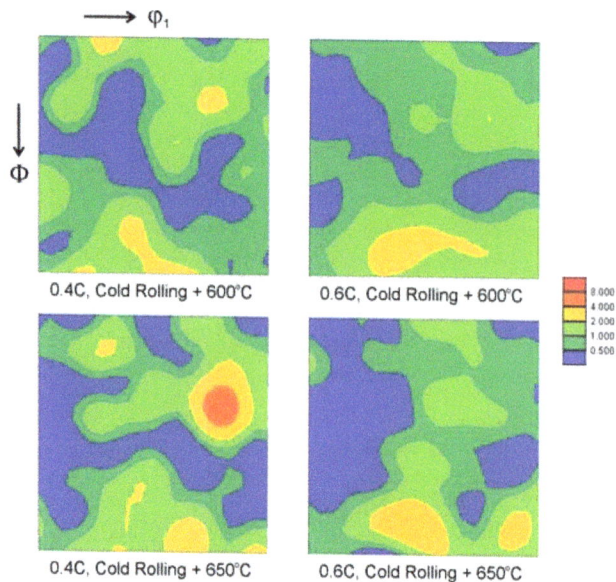

Figure 5. Orientation distribution functions at $\phi_2 = 45°$ for Fe-0.4C-18Mn and Fe-0.6C-18Mn steels (0.4C and 0.6C, respectively) subjected to cold rolling and annealing at indicated temperatures.

3.2. Tensile Behaviour

A series of engineering stress–elongation curves obtained during tensile tests of the cold rolled and annealed steel samples with fine-grained microstructures is shown in Figure 6. The values of yield strength ($\sigma_{0.2}$), ultimate tensile strength (UTS) and total elongation (δ) are represented in Table 2. The cold rolling resulted in significant strengthening. The yield strength above 1400 MPa and UTS above 1600 MPa is obtained in both Fe-0.4C-18Mn and Fe-0.6C-18Mn steels after cold rolling. On the other hand, total elongation of the cold rolled samples does not exceed 30%. Recrystallization annealing at 600–650 °C substantially improves plasticity. Total elongation of 40–60% is obtained after annealing. It should be noted that such enhancement of plasticity is not accompanied by a complete softening. The yield strength remains at a level of about 500 MPa in the Fe-0.4C-18Mn steel samples after annealing and that of about 700 MPa and 1000 MPa is obtained in Fe-0.6C-18Mn steel after annealing at 650 °C and 600 °C, respectively. The annealed samples exhibit pronounced strain hardening. Following yielding, the stress gradually increases up to maximum followed by failure, i.e., total and uniform elongations are almost the same, which is typical of high-Mn TWIP steels [1,7–10]. Therefore, the development of fine-grained microstructures by cold rolling and annealing results in beneficial combination of high strength and plasticity in the present steels. Some serrations on the stress-elongation curves testify to dynamic strain aging, which has been frequently observed in high-manganese steels [1,3,7].

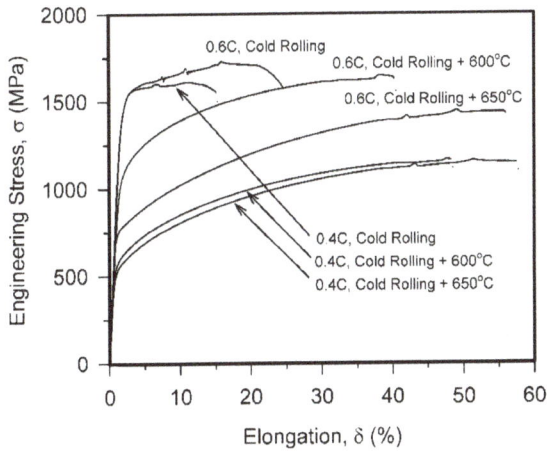

Figure 6. Engineering stress–elongation curves of Fe-0.4C-18Mn and Fe-0.6C-18Mn steels (0.4C and 0.6C, respectively) subjected to cold rolling and annealing at indicated temperatures.

Table 2. The yield strength ($\sigma_{0.2}$), ultimate tensile strength (UTS) and total elongation (δ).

Steel Processing	$\sigma_{0.2}$, MPa	UTS, MPa	δ, %
Fe-0.4C-18Mn Cold Rolling + 600 °C	530	1165	45
Fe-0.4C-18Mn Cold Rolling + 650 °C	465	1155	55
Fe-0.6C-18Mn Cold Rolling + 600 °C	1000	1650	40
Fe-0.6C-18Mn Cold Rolling + 650 °C	730	1445	55

The strengthening by grain refinement is generally discussed in terms of the Hall-Petch relationship [18,19]. The relationship between the grain size and the yield strength of the present steel samples after annealing is shown in Figure 7a along with results obtained by warm to hot rolling of the same steels [10]. The hot rolled samples were characterized by dynamically recrystallized microstructures. Corresponding yield strength can be expressed by Hall-Petch-type relationship. In contrast, warm rolled microstructures were affected by dynamic recovery only and, thus, contained high dislocation densities. In this case, the yield strengths appear well above those predicted by Hall-Petch-type relationship in Figure 7a because of additional strengthening caused by relatively high dislocation density in recovered microstructures. It is interesting to note that the yield strength of the annealed Fe-0.4C-18Mn steel samples roughly corresponds to that predicted by Hall-Petch-type relationship, whereas the Fe-0.6C-18Mn steel samples exhibit remarkably higher yield strength, which is comparable with that of warm rolled steel samples. This strengthening can be attributed to remained work hardening in the steel samples with relatively high carbon content as suggested by microstructural investigations, e.g., large fraction of low-angle dislocation subboundaries in Figure 4. The low-angle dislocation subboundaries resulted from recovery, namely, polygonization. In contrast to recrystallization, which requires incubation period, recovery processes develop just upon heating and lead to partial release of deformation stored energy owing to dislocation rearrangement and annihilation [17]. Partial softening by recovery reduces the driving pressure for recrystallization development. Therefore, recovered portions of microstructure may coexist with recrystallized ones for rather long time, especially during annealing at relatively low temperatures, when diffusion processes are slowed down. The remained recovered portions are characterized by higher dislocation densities as compared to perfect recrystallized grains and, thus, may provide an additional strengthening.

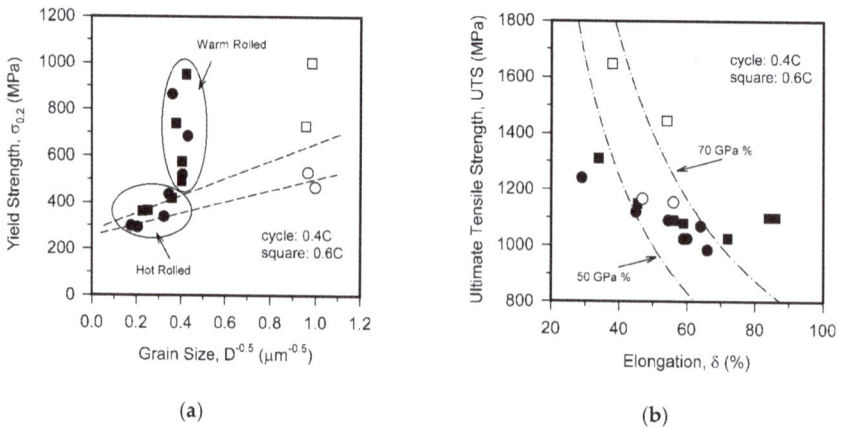

(a)   (b)

Figure 7. Relationship between the grain size and the yield strength (**a**) and the ultimate tensile strength and elongation (**b**) for Fe-0.4C-18Mn and Fe-0.6C-18Mn steels (0.4C and 0.6C, respectively) processed by warm to hot rolling (filled symbols [10]) and cold rolling followed by annealing (open symbols, present study).

The present annealed samples demonstrate improved combination of strength and plasticity as represented by a product of UTS by total elongation in Figure 7b. The values of UTS × δ in the range of 50 GPa to 70 GPa are obtained in the present samples. The larger values correspond to the samples annealed at higher temperature, when annealing softening is compensated by increasing ductility. It should be noted that the Fe-0.4C-18Mn steel samples exhibit almost the same level of mechanical properties as obtained by warm to hot rolling [10]. In contrast, the steel with relatively high carbon content of 0.6% displays remarkably higher strength for the same plasticity after cold rolling followed by annealing.

4. Conclusions

The microstructures and mechanical properties of Fe-0.4C-18Mn and Fe-0.6C-18Mn steels subjected to large strain cold rolling followed by annealing were studied. The main results can be summarized as follows.

Cold rolling with a total reduction of 86% resulted in substantial strengthening. The yield strength above 1400 MPa and ultimate tensile strength above 1600 MPa were achieved in the both steels after cold rolling. On the other hand, corresponding total elongation did not exceed 30%.

Subsequent annealing at temperatures above 600 °C was accompanied with the recrystallization development resulting in the fine-grained microstructures with an average grain size of about 1 μm in the both steels.

The fine-grained steels processed by cold rolling and annealing exhibited beneficial combination of high strength and plasticity. The product of ultimate tensile strength by total elongation comprised 50 to 70 GPa % depending on carbon content and processing conditions.

Author Contributions: Conceptualization, V.T. and R.K.; methodology, V.T.; validation, V.T., formal analysis, A.B.; investigation, V.T.; writing—original draft preparation, V.T.; writing—review and editing, A.B.; supervision, R.K.; funding acquisition, A.B.

Funding: This research was funded by the Ministry of Education and Science, Russia, under Grant No. 11.3719.2017/PCh (11.3719.2017/4.6).

Metals **2019**, *9*, 776

Conflicts of Interest: The authors declare no conflict of interest.

References

1. Bouaziz, O.; Allain, S.; Scott, C.P.; Cugy, P.; Barbier, D. High Manganese Austenitic Twinning Induced Plasticity Steels: A Review of the Microstructure Properties Relationships. *Curr. Opin. Solid State Mater. Sci.* **2011**, *15*, 141–168. [CrossRef]
2. Saeed-Akbari, A.; Mosecker, L.; Schwedt, A.; Bleck, W. Characterization and Prediction of Flow Behavior in High-Manganese Twinning Induced Plasticity Steels: Part I. Mechanism Maps and Work-Hardening Behavior. *Metall. Mater. Trans. A* **2012**, *43*, 1688–1704. [CrossRef]
3. Sevsek, S.; Brasche, F.; Haase, C.; Bleck, W. Combined deformation twinning and short-range ordering causes serrated flow in high-manganese steels. *Mater. Sci. Eng. A* **2019**, *746*, 434–442. [CrossRef]
4. Song, W.; Ingendahl, T.; Bleck, W. Control of Strain Hardening Behavior in High-Mn Austenitic Steels. *Acta Metall. Sin. Engl. Lett.* **2014**, *27*, 546–556. [CrossRef]
5. Keeler, S.; Kimchi, M. *Advanced High-Strength Steels Application Guidelines V5*; WorldAutoSteel: Middletown, OH, USA, 2014.
6. Nikulin, I.; Sawaguchi, T.; Kushibe, A.; Inoue, Y.; Otsuka, H.; Tsuzaki, K. Effect of strain amplitude on the low-cycle fatigue behavior of a new Fe-15Mn-10Cr-8Ni-4Si seismic damping alloy. *Int. J. Fatigue* **2016**, *88*, 132–141. [CrossRef]
7. Kusakin, P.S.; Kaibyshev, R.O. High-Mn twinning-induced plasticity steels: Microstructure and mechanical properties. *Rev. Adv. Mater. Sci.* **2016**, *44*, 326–360.
8. De Cooman, B.C.; Estrin, Y.; Kim, S.K. Twinning-induced plasticity (TWIP) steels. *Acta Mater.* **2018**, *142*, 283–362. [CrossRef]
9. Kusakin, P.; Tsuzaki, K.; Molodov, D.A.; Kaibyshev, R.; Belyakov, A. Advanced thermomechanical processing for a high-Mn austenitic steel. *Metall. Mater. Trans. A* **2016**, *47*, 5704–5708. [CrossRef]
10. Torganchuk, V.; Belyakov, A.; Kaibyshev, R. Effect of rolling temperature on microstructure and mechanical properties of 18%Mn TWIP/TRIP steels. *Mater. Sci. Eng. A* **2017**, *A708*, 110–117. [CrossRef]
11. Sakai, T.; Belyakov, A.; Kaibyshev, R.; Miura, H.; Jonas, J.J. Dynamic and post-dynamic recrystallization under hot, cold and severe plastic deformation conditions. *Prog. Mater. Sci.* **2014**, *60*, 130–207. [CrossRef]
12. Saha, R.; Ueji, R.; Tsuji, N. Fully recrystallized nanostructure fabricated without severe plastic deformation in high-Mn austenitic steel. *Scripta Mater.* **2013**, *68*, 813–816. [CrossRef]
13. Hirsch, J.; Lucke, K. Mechanism of deformation and development of rolling texture in polycrystalline f.c.c. metal—I. Description of rolling texture development in homogeneous CuZn alloys. *Acta Metall. Mater.* **1988**, *36*, 2863–2882. [CrossRef]
14. Yanushkevich, Z.; Belyakov, A.; Kaibyshev, R.; Haase, C.; Molodov, D.A. Effect of cold rolling on recrystallization and tensile behavior of a high-Mn steel. *Mater. Charact.* **2016**, *112*, 180–187. [CrossRef]
15. Mahajan, S. Critique of mechanisms of formation of deformation, annealing and growth twins: Face-centered cubic metals and alloys. *Scripta Mater.* **2013**, *68*, 95–99. [CrossRef]
16. Mackenzie, J.K. Second Paper on the Statistics Associated with the Random Disorientation of Cubes. *Biometrika* **1958**, *45*, 229–240. [CrossRef]
17. Humphreys, F.J.; Hatherly, M. *Recrystallization and Related Annealing Phenomena*, 2nd ed.; Elsevier: Oxford, UK, 2004; pp. 215–268. ISBN 0-08-044164-5.
18. Hall, E.O. The deformation and ageing of mild steel: III discussion of results. *Proc. R. Soc. Lond. Ser. B* **1951**, *64*, 747–753. [CrossRef]
19. Petch, N.J. The cleavage strength of polycrystals. *J. Iron Steel Inst.* **1953**, *174*, 25–28.

metals

MDPI

Article

The Influence of Warm Rolling on Microstructure and Deformation Behavior of High Manganese Steels

Marco Haupt [1,*], **Max Müller** [1], **Christian Haase** [2], **Simon Sevsek** [2], **Frederike Brasche** [3], **Alexander Schwedt** [4] **and Gerhard Hirt** [1]

[1] Institute of Metal Forming, RWTH Aachen University, 52072 Aachen, Germany
[2] Steel Institute, RWTH Aachen University, 52072 Aachen, Germany
[3] Institute of Physical Metallurgy and Metal Physics, RWTH Aachen University, 52074 Aachen, Germany
[4] Central Facility for Electron Microscopy, RWTH Aachen University, 52074 Aachen, Germany
*   Correspondence: marco.haupt@ibf.rwth-aachen.de; Tel.: +49-241-80-93527

Received: 17 June 2019; Accepted: 17 July 2019; Published: 18 July 2019

Abstract: In this work, a Fe-23Mn-0.3C-1Al high manganese twinning-induced plasticity (TWIP) steel is subjected to varying warm rolling procedures in order to increase the yield strength and maintain a notable ductility. A comprehensive material characterization allows for the understanding of the activated deformation mechanisms and their impact on the resulting microstructure, texture, and mechanical properties. The results show a significant enhancement of the yield strength compared to a fully recrystallized Fe-23Mn-0.3C-1Al steel. This behavior is mainly dominated by the change of the active deformation mechanisms during rolling. Deformation twinning is very pronounced at lower temperatures, whereas this mechanism is suppressed at 500 °C and a thickness reduction of up to 50%. The mechanical properties can be tailored by adjusting rolling temperature and thickness reduction to desired applications.

Keywords: high manganese steel; warm rolling; processing; microstructure; texture; mechanical properties; deformation behavior

---

1. Introduction

High manganese twinning-induced plasticity (TWIP) steels are well known for their excellent mechanical properties in terms of exceptional ductility and high tensile strength, which is of high relevance for crash relevant automotive applications. Compared to industrially applied advanced high strength steels, e.g., dual phase steels, fully recrystallized high manganese steels exhibit a comparatively low yield strength, which represents a decisive disadvantage of this class of steel. In order to substantially increase the yield strength of high manganese TWIP steels, numerous approaches have been discussed in the literature such as prestraining [1], grain refinement [2], micro alloying [3], partial recrystallization [4], severe plastic deformation [5], reversion annealing [6] and recovery annealing [7]. Present limitations of these approaches, such as precise temperature control, pronounced anisotropy, or extraordinary degrees of deformation, prevent the use for industrial production routes.

Warm rolling represents a promising alternative processing method to substantially increase the yield strength of high manganese steels [8]. Hence, the impact of warm rolling parameters on microstructure, texture and mechanical properties of a high manganese TWIP steel is investigated in the present study.

The concept of warm rolling takes advantage of the temperature dependence of the stacking fault energy (SFE), which determines the activated deformation mechanisms in high manganese steels. In addition to temperature, the SFE is mainly dependent on chemical composition and grain size. Deformation twinning is the predominant deformation mechanism in the SFE range of 20 to 60 mJ/m^2, in addition to dislocation glide. Above 60 mJ/m^2 the deformation mechanism changes mainly to slip [9]. This dependency in behavior on temperature and therefore the SFE of the deformation mechanisms in high manganese TWIP steels was also reported by [10–12]. The presented results indicate a change of the predominant deformation mechanism from mechanical twinning to dislocation glide with increasing temperature. By rolling at elevated temperatures (higher SFE) and applying the warm-rolled material at room temperature (lower SFE), the contribution of mechanical twinning and dislocation slip can be tailored in both temperature-deformation regimes in order to achieve superior mechanical properties. In order to accomplish this, it is necessary to prevent dynamic recrystallization during rolling, which is typically observed during hot deformation of austenitic TWIP steels at temperatures above ~750 °C [8,13].

Texture evolution in austenitic steels such as TWIP steels strongly depends on the SFE. Alloys with low SFE values, e.g., α-brass, develop a so-called brass-type rolling texture, which is characterized by a strong {110}<112> brass and weak {552}<115> copper twin (CuT), {110}<100> goss, and {123}<634> S texture components, as well as a weak γ-fiber (<111>//ND) [14,15]. A brass-type texture is caused by latent hardening due to planar dislocation glide and the activation of mechanical twinning [16]. In contrast, medium to high SFE materials like copper tend to form a texture consisting of pronounced {112}<111> copper (Cu), brass and S texture components. Therefore, such type of texture is called copper-type texture [17–19]. In previous works the texture evolution during cold rolling of different TWIP steels was investigated. In general, TWIP steels have the tendency to develop a brass-type rolling texture [18–22].

In order to investigate the impact of warm rolling parameters, such as temperature and thickness reduction on microstructure, texture and deformation behavior, a Fe-23Mn-0.3C-1Al TWIP steel was chosen for this study. The steel was processed at warm rolling temperatures ranging from 200 °C to 500 °C with thickness reductions ranging from 50% to 80%. Mechanical properties were analyzed by uniaxial tensile tests, whereas the resulting microstructure and texture evolution was characterized by electron backscatter diffraction (EBSD) and X-ray diffraction (XRD) measurements.

2. Materials and Methods

2.1. Material Processing

The chemical composition of the investigated Fe-23Mn-0.3C-1Al high manganese steel is given in Table 1. The material was ingot-cast in a vacuum induction furnace. The 100 kg ingot was then forged at 1150 °C to reduce the thickness from 140 mm to 50 mm. A subsequent homogenization annealing at 1150 °C for 5 h in an argon atmosphere was applied in order to reduce micro-segregations. The homogenized ingot was subsequently hot-rolled at 1150 °C to further reduce the sheet thickness to 4.0 mm, followed by air cooling.

Table 1. Chemical composition of the investigated Fe-23Mn-0.3C-1Al high manganese steel.

Element	Fe	C	Si	Mn	P	S	Al	Ni	Mo	Cr
wt.%	Bal.	0.322	0.053	22.45	0.008	0.008	0.995	0.027	0.008	0.020

Afterwards, the hot strips were warm-rolled in 7–10 passes in the temperature range between 200 °C and 500 °C on a four-high rolling mill. The rolling degree was varied from 50% to 80%. In addition, a cold rolled strip with 2.0 mm thickness and 50% rolling degree achieved in seven rolling passes was used for comparison of microstructure and mechanical properties. Table 2 gives an overview of the process parameter combinations and the process scheme of the warm rolling process. In order to compensate the temperature loss due to transport of the strip from furnace to rolling mill, the furnace temperature (T_F) was set 20–70 °C higher than the designated rolling temperature (T_R). Consistent rolling temperatures were achieved by reheating of the strips for 5 min in between consecutive rolling passes.

Table 2. Process parameter combinations (marked with an x) and process scheme of warm rolling.

Temperature (°C)		Thickness Reduction (%)				Process Scheme
T_R	T_F	50	60	70	80	
25	-	x				
200	220	x				
300	340	x				
400	460	x				
450	510	x				
500	570	x	x	x	x	

2.2. Sample Preparation and Characterization

EBSD and XRD specimens with the dimensions 10×12 mm^2 (transverse direction (TD) and rolling direction (RD)) were waterjet cut from the cold-rolled and warm-rolled sheets. Both EBSD and XRD specimens were ground utilizing SiC-paper up to 4000 grit followed by mechanical polishing using diamond suspension up to 1 μm. Finally, the XRD samples were electropolished at room temperature. EBSD samples were mechanically polished to 0.25 μm and subsequently electropolished.

The resulting microstructure in the RD-TD section of the cold-rolled and warm-rolled strips was examined by EBSD using a JEOL JSM 7000F scanning electron microscope (JEOL Ltd., Tokyo, Japan) equipped with an EDAX Hikari EBSD detector (EDAX Inc., Mahwah, NJ, USA). The measurements were performed with an acceleration voltage of 20 kV and a step size of 200 nm. EBSD data was analyzed using the EDAX OIM Analysis 8 software (EDAX Inc., Mahwah, NJ, USA).

Macro-texture measurements were conducted in a Bruker D8 Advance diffractometer (Bruker Corporation, Billerica, MA, USA), which was equipped with a HI-STAR area detector and operated with iron-radiation at 30 kV and 25 mA. Three incomplete (0–85°) {111}-, {200}-, and {220}-pole figures were measured and the corresponding orientation distribution functions (ODF) were calculated using the MATLAB®-based MTEX toolbox [23,24]. Besides, volume fractions of selected texture components were computed applying a spread of 15° around the ideal orientation. XRD analysis was performed on the mid-layer of the sheets in the RD-TD section.

For characterization of mechanical properties flat tensile specimens with a gauge length of 30 mm, a gauge width of 6 mm and a fillet radius of 20 mm were waterjet cut from the rolled sheets with tensile direction parallel to RD. After polishing of the edges, quasi-static tensile tests at room temperature were performed utilizing a Zwick Z250 universal testing machine (ZwickRoell, Ulm, Germany) at a constant strain rate of 0.001 s^{-1}.

3. Results

3.1. Microstructure

An EBSD inverse pole figure (IPF) map of the initial hot-rolled Fe-23Mn-0.3C-1Al steel is displayed in Figure 1. Prior to cold or warm rolling the material exhibits an equiaxial grain structure with a relatively high mean grain size of ~30 μm. Twin boundaries are not present in the hot-rolled samples.

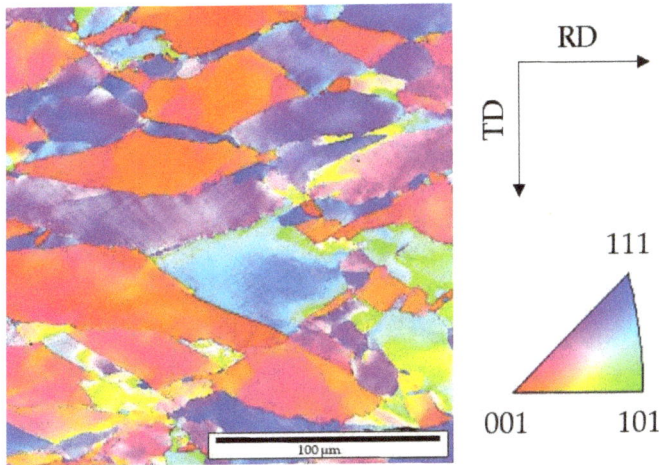

Figure 1. Electron backscatter diffraction inverse pole figure (EBSD-IPF) map of hot-rolled Fe-23Mn-0.3C-1Al steel. The IPF takes the rolling direction (RD) as reference axis.

Figure 2 shows EBSD image quality (IQ) maps of the resulting microstructure of the Fe-23Mn-0.3C-1Al high manganese steel after cold rolling and warm rolling at temperatures ranging from 200 °C to 500 °C, respectively. Detected deformation twin boundaries (60° <111>) are indicated with the color blue. For all samples, a severely deformed austenitic grain structure can be observed. However, the fraction of detected deformation twin boundaries decreases significantly at higher warm rolling temperatures. The material that was warm-rolled at 200 °C and 300 °C shows a high fraction of deformation twin boundaries and is therefore comparable to the cold-rolled sample. In the material that was warm-rolled at 400 °C the deformation twin boundaries decrease drastically compared to the material rolled at lower temperatures. Despite the drastic decrease, the material still shows a significant fraction of twin boundaries. The material which was warm-rolled at 450 °C exhibits an even lower amount of deformation twin boundaries while practically no twin boundaries are present in the material that was warm-rolled at 500 °C. This correlation between warm rolling temperature and detected deformation twin boundaries indicates a change of the predominant deformation mechanism during warm rolling from twinning to slip in the temperature range of 400 to 500 °C. By warm rolling the Fe-23Mn-0.3C-1Al high manganese steel at 500 °C it is possible to achieve a thickness reduction of 50% primarily based on the slip mechanism.

At even higher thickness reductions at a comparably high warm rolling temperature of 500 °C deformation twin boundaries can still be observed, as shown in Figure 3. The material with a thickness reduction of 60% exhibits an amount of deformation twin boundaries comparable to the material that was warm-rolled at 450 °C and 50% thickness reduction. At a higher deformation degree of 70%, the amount of deformation twin boundaries only increases slightly. The overall observed grain structure is strongly elongated parallel to RD due to the high plastic deformation applied during rolling.

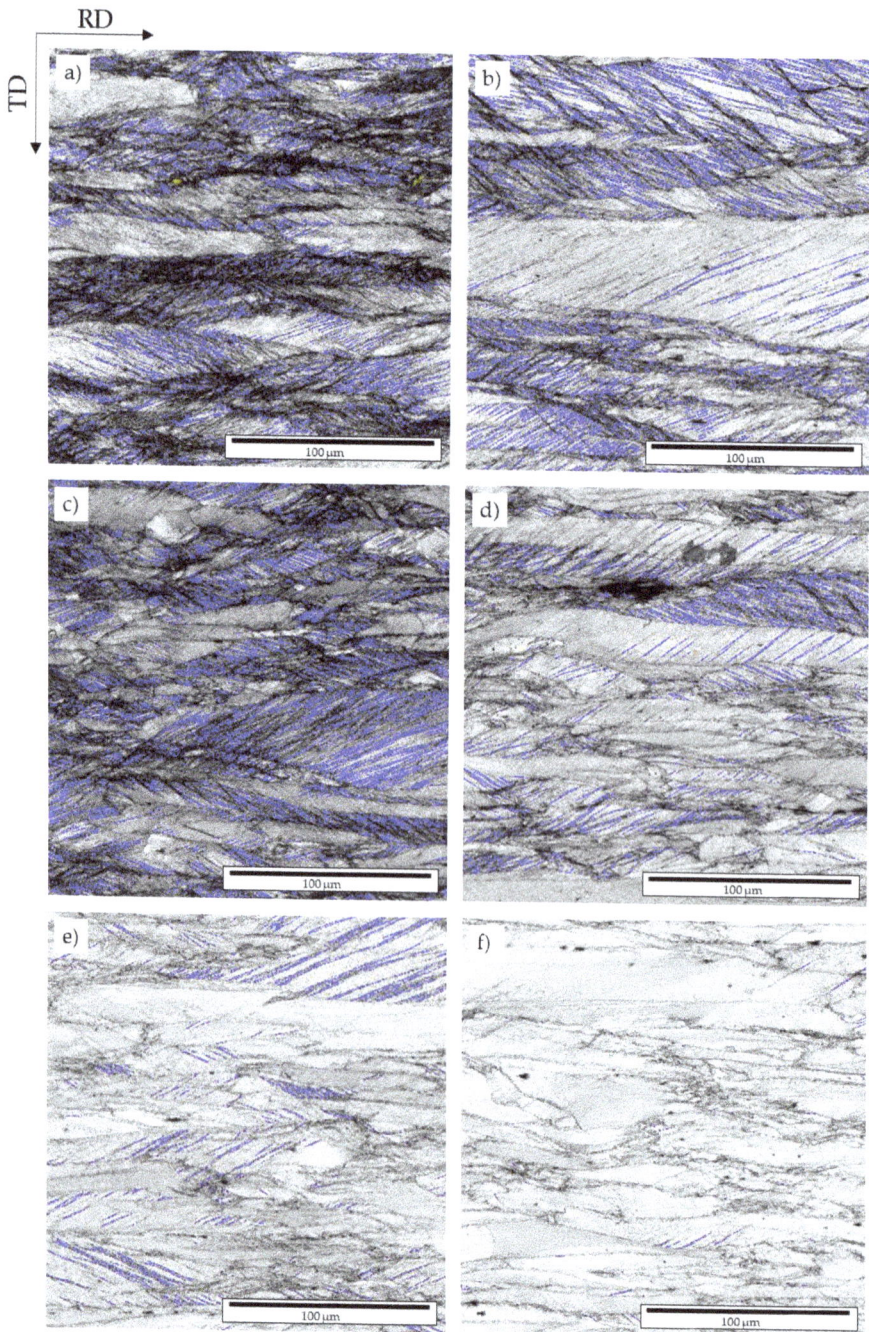

Figure 2. EBSD-IQ maps of the investigated steel deformed to 50% thickness reduction by (**a**) cold rolling and warm rolling at (**b**) 200 °C, (**c**) 300 °C, (**d**) 400 °C, (**e**) 450 °C and (**f**) 500 °C. Blue lines indicate deformation twin boundaries (60° <111>).

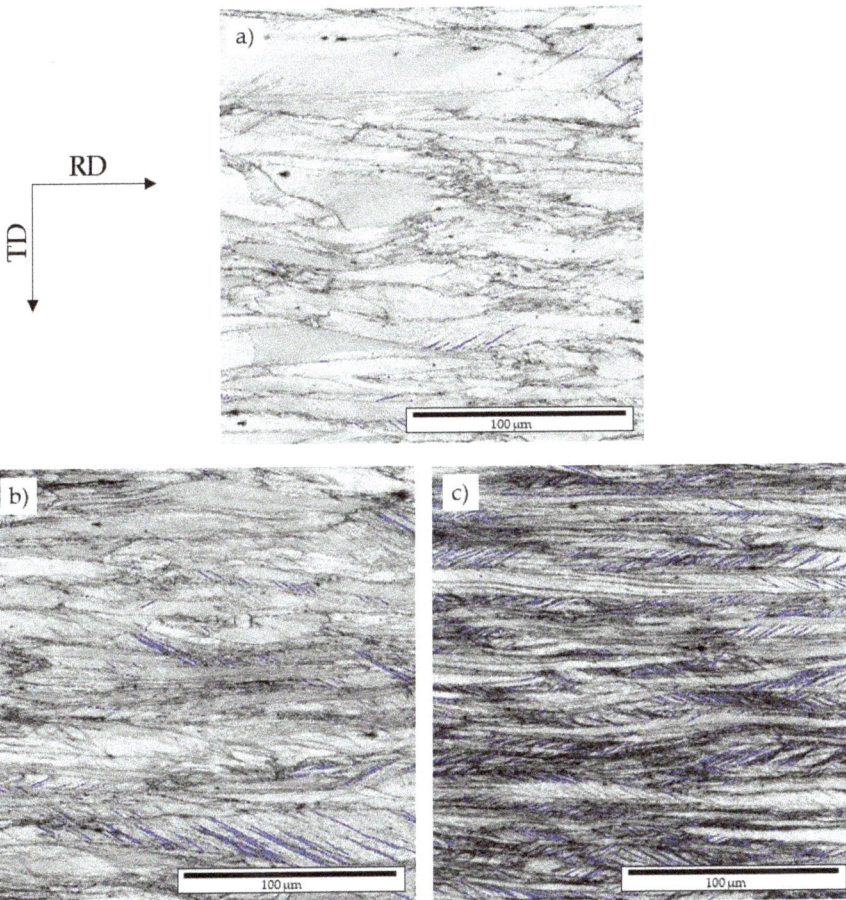

RD

TD

100 µm

100 µm

100 µm

Figure 3. EBSD-IQ maps of the investigated steel warm-rolled at 500 °C to thickness reductions of (**a**) 50%, (**b**) 60%, and (**c**) 70%. Blue lines indicate deformation twin boundaries (60° <111>).

3.2. Influence of Rolling Temperature on Texture Evolution During Warm Rolling

The texture formation of the investigated material after rolling in the temperature range between 25 °C and 500 °C to a rolling reduction of 50% is illustrated in Figure 4 by means of $\varphi_2 = 45°$ sections of the ODF. The ideal location of the relevant texture components is schematically depicted (Figure 4, top left corner) and the corresponding definition of the texture components can be found in Table 3. In addition to the ODF sections, Figure 4h displays the calculated volume fractions of selected texture components. The temperature dependent texture formation in the present investigation can be divided into three stages: 25–300 °C (Stage I), 400 °C (stage II), and 450–500 °C (stage III). Rolling at 25 °C to a deformation degree of 50% results in the development of a relatively strong brass texture component along with weaker goss, CuT, E + F, and Cu texture components. With increasing rolling temperature, the intensity of the brass texture component slightly decreases. This decrease is accompanied by a stagnation in volume fractions of the Cu and CuT texture component and a steady decrease in E + F texture components. Stage II (400 °C) represents an intermediate state. The brass texture component reaches a minimum and a remarkable drop in volume fraction of the CuT texture component is detected. Nevertheless, the volume fraction of the Cu texture component remains at a similar level. Additionally, a small volume fraction of the cube texture component is observed. The third stage comprises the material states at 450 °C and 500 °C. A high intensity of Cu-oriented grains is present and the α-fiber components (brass and goss) intensify, albeit to a smaller extent. In addition, the CuT and γ-fiber components (E + F) weaken strongly.

Figure 5 shows the texture evolution during rolling at 500 °C with a thickness reduction ranging from 50% to 80% as $\varphi_2 = 45°$ ODF sections and the evolution of the volume fraction of the main texture components. With an increasing rolling degree, the volume fractions of the CuT and E + F texture components increase, whereas the amount of Cu-oriented grains decreases. The texture evolution of the present material during cold rolling in the deformation range between 50% and 80% has already been investigated in [19,25,26]. Haase et al. reported about the formation of a γ-fiber accompanied by a strengthening of the α-fiber and a constant low level of Cu-texture component [19]. In the present work, the α-fiber intensifies with increasing rolling degree as well. However, both a stronger Cu-texture component and an absence of the γ-fiber is observed. Therefore, it can be concluded that a temperature increase from 25 °C to 500 °C also effects texture evolution.

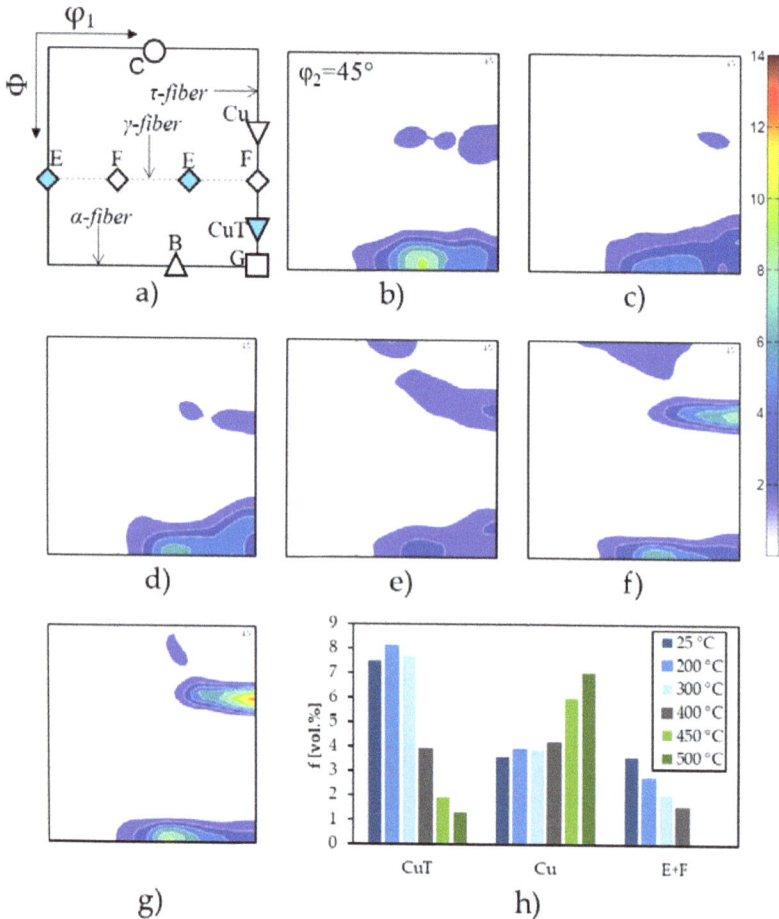

Figure 4. Texture formation of the material rolled to 50% thickness reduction at different temperatures. (a) Schematic presentation of the ideal texture component in the orientation distribution functions (ODF) section at $\varphi_2 = 45°$. ODF sections at $\varphi_2 = 45°$ for (b) 25 °C, (c) 200 °C, (d) 300 °C, (e) 400 °C, (f) 450 °C, (g) 500 °C, and (h) presentation of the volume fractions of selected texture components.

Table 3. Definition of texture components illustrated in Figures 4 and 5.

Component	Symbol	Miller Indices	Euler Angles (φ_1, Φ, φ_2)	Fiber
Brass (B)	△	{110}<112>	(55, 90, 45)	α, β
Goss (G)	□	{110}<100>	(90, 90, 45)	α, τ
Cube (C)	○	{001}<100>	(45, 0, 45)	-
E	◆	{111}<110>	(0/ 60, 55, 45)	γ
F	◇	{111}<112>	(30/ 90, 55, 45)	γ
Copper (Cu)	▽	{112}<111>	(90, 35, 45)	β, τ
Copper Twin (CuT)	▼	{552}<115>	(90, 74, 45)	τ
α-fiber		<110> parallel to ND		
β-fiber		<110> tilted 60° from ND towards RD		
τ-fiber		<110> parallel TD		
γ-fiber		<111> parallel ND		

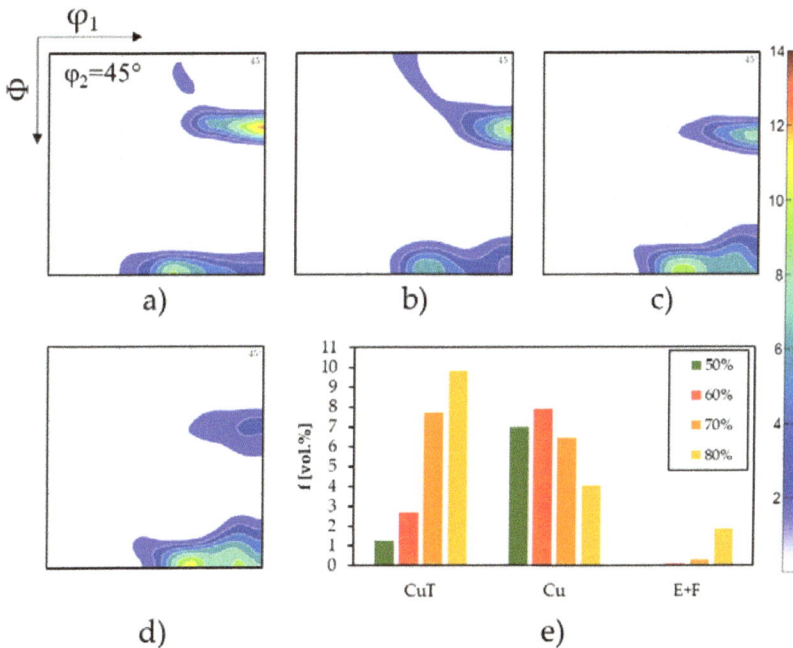

Figure 5. Texture evolution of the material rolled at 500 °C to different rolling reductions. ODF sections at $\varphi_2 = 45°$ of (**a**) 50%, (**b**) 60%, (**c**) 70%, and (**d**) 80% and (**e**) presentation of the volume fractions of selected texture components.

3.3. Mechanical Properties

Figure 6 shows the engineering stress-strain curves derived from uniaxial tensile tests at room temperature of the cold-rolled and warm-rolled material. The corresponding mechanical parameters are summarized in Table 4. The cold-rolled high manganese steel possesses the highest strength but has limited ductility. At a constant thickness reduction of 50% the strength level decreases with increasing rolling temperature. A warm-rolling temperature of 200 °C results in a yield strength (YS) of 1092 MPa, whereas a warm rolling temperature of 500 °C leads to a significantly lower yield strength of 788 MPa. The loss of strength is accompanied by a noticeable increase of uniform elongation (UE) and total elongation (TE). Compared to the material warm-rolled at 200 °C the material warm-rolled at 500 °C exhibits an almost four times higher total elongation of 38.5%.

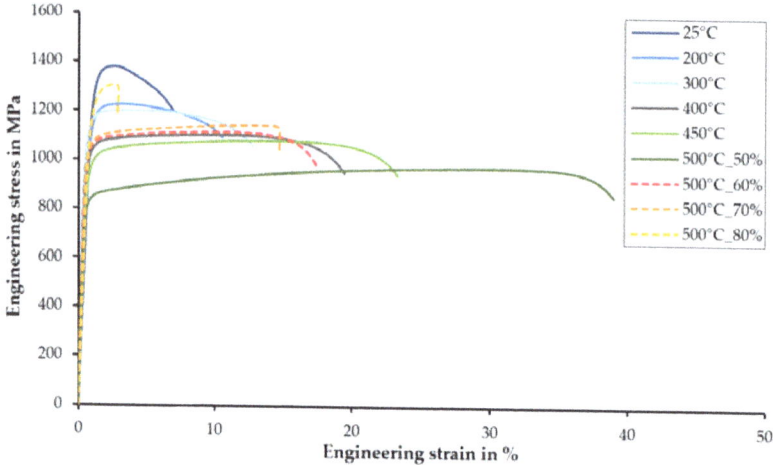

Figure 6. Engineering stress-strain curves of the investigated Fe-23Mn-0.3C-1Al steel subjected to varying rolling procedures.

Table 4. Mechanical properties of the investigated Fe-23Mn-0.3C-1Al steel depending on rolling temperature and thickness reduction.

Rolling Temperature (°C)	Thickness Reduction (%)	Yield Strength (MPa)	Tensile Strength (MPa)	Uniform Elongation (%)	Total Elongation (%)
25	50	1274	1379	1.6	6.1
200	50	1081	1227	2.8	9.8
300	50	1054	1200	3.3	12.3
400	50	959	1103	10.2	18.9
450	50	942	1082	13.6	22.6
500	50	788	975	30.1	38.5
500	60	999	1114	10.0	17.0
500	70	1015	1144	12.7	14.1
500	80	1133	1305	1.6	1.9

By increasing the thickness reduction at a constant warm rolling temperature of 500 °C, the yield and ultimate tensile strength (UTS) also increase. A thickness reduction of 80% results in an ultimate tensile strength of 1305 MPa but poor ductility. By increasing the thickness reduction to 60%, the ultimate tensile strength can be increased to 1114 MPa at a reasonable uniform elongation of 10.0%. The comparison of a thickness reduction of 60% and 70% therefore shows only a slight difference regarding strength and ductility.

In comparison to a fully recrystallized Fe-23Mn-0.3C-1Al steel (YS = 356 MPa, UTS = 787 MPa, UE = 56.8%, TE = 62.1%, mean grain size = 3.1 μm), all warm-rolled materials investigated exhibit a significantly higher yield and ultimate tensile strength.

4. Discussion

The EBSD-IQ maps (Figure 2) confirm the strong temperature dependence of the activated deformation mechanisms during rolling. Twinning is the predominant deformation mechanism in the range of 25 °C to 300 °C, which is indicated by the high fraction of detected deformation twin boundaries. This can also be observed in the according evolution of the crystallographic texture (Figure 4). The CuT texture component is a texture component which is associated with planar dislocation glide and deformation-induced twinning [19,27]. Thus, the volume fractions of the CuT component can be correlated with the densities of deformation twins. The constant high level of CuT volume fraction in the range from 25 °C to 300 °C indicates that twinning is the predominant deformation mechanism during rolling, which has also been reported by Grajcar et al. [10]. The influence of this characteristic microstructure after rolling on the mechanical properties is shown in Figure 6. Due to the strain hardening effect of the already high fraction of deformation twin boundaries, the material exhibits a comparably high yield, and ultimate tensile strength. Therefore, the remaining low work hardening capacity results in a poor ductility and early fracture of the material.

Around the temperature threshold of 400 °C the predominant deformation mechanism shifts from twinning to dislocation glide, which can be concluded by the small volume fractions of the CuT texture component, as shown in Figure 4e. This correlates with the significantly lower amount of deformation twin boundaries detected in the corresponding EBSD-IQ map in Figure 2d. This is in accordance with Barbieri et al. [12], who report a continuous transition of the deformation mechanism with increasing SFE. During subsequent tensile testing at room temperature of the warm-rolled material, the remaining work hardening capacity of the twinning mechanism leads to a significant increase of uniform and total elongation, whereas the overall strength level decreases (Figure 6).

EBSD and XRD measurements of the material 50% warm-rolled at even higher temperatures, e.g., 500 °C, show that the twinning mechanism can almost be suppressed completely. This is indicated by the negligible amount of detected deformation twin boundaries (Figure 2f) and the low fraction of the CuT texture component (Figure 4h). The combination of the slip mechanism, also including cross slip, activated during warm rolling and the twinning mechanism activated during room temperature tensile testing results in excellent mechanical properties (Figure 6). It is important to note that the dynamic recrystallization of the material during rolling, which typically occurs during hot rolling [8,13], was successfully prevented despite the comparably high temperature.

As displayed in Figure 3b,c and Figure 5, a thickness reduction higher than 50% at 500 °C warm rolling temperature without the presence of deformation twinning could not be achieved. This behavior can be explained with the observed texture components present at higher thickness reductions. Deformation twinning proceeds in Cu oriented grains. Consequently, a rise of the CuT texture component effects the elimination of the Cu texture (Figure 5). At high deformation degrees twin-matrix lamellae are rotated into the rolling plane and shear bands occur. This phenomenon is related to the onset of the E + F texture components. By forming the E + F texture components, in turn, the CuT texture component is consumed [19]. Hence, it must be assumed that the formation of shear bands prohibited an even steeper rise of the CuT texture component. The presence of a significant amount of deformation twin boundaries at higher thickness reduction again results in an increased yield and ultimate tensile strength but a drastically decreased ductility.

Despite the high deformation applied during rolling and the different strength-ductility combinations observed, the materials warm-rolled at temperatures higher than 300 °C show a significant work-hardening behavior at room temperature. This is achieved by the activation of the twinning mechanism during the tensile tests at room temperature, which causes the notable work-hardening.

5. Conclusions

In this work, a Fe-23Mn-0.3C-1Al high manganese TWIP steel was processed by warm rolling at various warm rolling temperatures and thickness reductions in order to increase the yield strength and maintain a reasonable ductility, as compared to recrystallized counterparts. The resulting microstructure, texture, and mechanical properties were analyzed, and their correlation was discussed. The following conclusions can be drawn:

- The temperature dependence of the stacking fault energy can be used to control the predominant deformation mechanism of high manganese steels during rolling. The performed EBSD measurements and XRD texture analysis show that the formation of deformation twins can be suppressed at 500 °C and thickness reduction of up to 50%.
- The specimens warm-rolled at temperatures higher than 300 °C exhibit exceptional mechanical properties due to the combination of slip and twinning. The adjustment of rolling temperature and rolling degree allows for tailoring the mechanical properties in a wide range. The resulting uniform and total elongation can be increased by increasing the rolling temperature, whereas yield and ultimate tensile strength decrease accordingly. The material benefits from strain hardening due to the slip effect at elevated temperatures. The twinning mechanism during deformation at room temperature, i.e., after prior warm rolling, accordingly enables a certain ductility and work hardening potential of the warm-rolled high manganese steel.
- The yield strength and overall strength level of a high manganese TWIP steel could be notably improved by warm rolling. A reasonable uniform and ultimate elongation could be maintained. Compared to a fully recrystallized Fe-23Mn-0.3C-1Al steel, the warm-rolled material states exhibit superior mechanical properties, especially in terms of strength.

Author Contributions: Conceptualization, M.H.; formal analysis, M.H.; investigation, M.H., M.M., S.S., F.B. and A.S.; supervision, C.H. and G.H.; visualization, M.H. and M.M.; project administration, G.H.; writing—original draft, M.H. and M.M.

Funding: This research was funded by the Deutsche Forschungsgemeinschaft (DFG) within the Collaborative Research Centre (SFB) 761 "Stahl - ab initio; Quantenmechanisch geführtes Design neuer Eisenbasiswerkstoffe", project number 29898171.

Conflicts of Interest: The authors declare no conflict of interest. The funders had no role in the design of the study; in the collection, analyses, or interpretation of data; in the writing of the manuscript, or in the decision to publish the results.

References

1. Kusakin, P.; Belyakov, A.; Haase, C.; Kaibyshev, R.; Molodov, D.A. Microstructure evolution and strengthening mechanisms of Fe–23Mn–0.3C–1.5Al TWIP steel during cold rolling. *Mater. Sci. Eng. A* **2014**, *617*, 52–60. [CrossRef]
2. Ueji, R.; Tsuchida, N.; Terada, D.; Tsuji, N.; Tanaka, Y.; Takemura, A.; Kunishige, K. Tensile properties and twinning behavior of high manganese austenitic steel with fine-grained structure. *Scr. Mater.* **2008**, *59*, 963–966. [CrossRef]
3. Scott, C.; Remy, B.; Collet, J.L.; Cael, A.; Bao, C.; Danoix, F.; Malard, B.; Curfs, C. Precipitation strengthening in high manganese austenitic TWIP steels. *Int. J. Mater. Res.* **2011**, *102*, 538–549. [CrossRef]
4. Bouaziz, O.; Barbier, D. Benefits of recovery and partial recrystallization of nano-twinned austenitic steels. *Adv. Eng. Mater.* **2013**, *15*, 976–9794. [CrossRef]
5. Haase, C.; Kremer, O.; Hu, W.; Ingendahl, T.; Lapovok, R.; Molodov, D.A. Equal-channel angular pressing and annealing of a twinning-induced plasticity steel: Microstructure, texture, and mechanical properties. *Acta Mater.* **2016**, *107*, 239–253. [CrossRef]
6. Berrenberg, F.; Haase, C.; Barrales-Mora, L.A.; Molodov, D.A. Enhancement of the strength-ductility combination of twinning-induced/transformation-induced plasticity steels by reversion annealing. *Mater. Sci. Eng. A* **2017**, *681*, 56–64. [CrossRef]
7. Haase, C.; Barrales-Mora, L.A.; Molodov, D.A.; Gottstein, G. Tailoring the Mechanical Properties of a Twinning-Induced Plasticity Steel by Retention of Deformation Twins During Heat Treatment. *Metall. Mater. Trans. A* **2013**, *44*, 4445–4449. [CrossRef]
8. Belyakov, A.; Kaibyshev, R.; Torganchuk, V. Microstructure and Mechanical Properties of 18%Mn TWIP/TRIP Steels Processed by Warm or Hot Rolling. *Steel Res. Int.* **2017**, *88*, 1600123. [CrossRef]

9.   Saeed-Akbari, A.; Imlau, J.; Prahl, U.; Bleck, W. Derivation and variation in composition-dependent stacking fault energy maps based on subregular solution model in high-manganese steels. *Metall. Mater. Trans. A* **2009**, *40*, 3076–3090. [CrossRef]

10.   Grajcar, A.; Kozłowska, A.; Topolska, S.; Morawiec, M. Effect of Deformation Temperature on Microstructure Evolution and Mechanical Properties of Low-Carbon High-Mn Steel. *Adv. Mater. Sci. Eng.* **2018**, *2018*, 1–7. [CrossRef]

11.   Zhang, J.; Di, H.; Mao, K.; Wang, X.; Han, Z.; Ma, T. Processing maps for hot deformation of a high-Mn TWIP steel: A comparative study of various criteria based on dynamic materials model. *Mater. Sci. Eng. A* **2013**, *587*, 110–122. [CrossRef]

12.   Barbieri, F.D.; Castro Cerda, F.; Pérez-Ipiña, J.; Artigas, A.; Monsalve, A. Temperature Dependence of the Microstructure and Mechanical Properties of a Twinning-Induced Plasticity Steel. *Metals* **2018**, *8*, 262. [CrossRef]

13.   Dolzhenko, P.; Tikhonova, M.; Kaibyshev, R.; Belyakov, A. Dynamically Recrystallized Microstructures, Textures, and Tensile Properties of a Hot Worked High-Mn Steel. *Metals* **2019**, *9*, 30. [CrossRef]

14.   Hirsch, J.; Lücke, K. Overview no. 76: Mechanism of deformation and development of rolling textures in polycrystalline f.c.c. metals—I. Description of rolling texture development in homogeneous CuZn alloys. *Acta Metall.* **1988**, *36*, 2863–2882. [CrossRef]

15.   Haase, C.; Barrales-Mora, L.A. Influence of deformation and annealing twinning on the microstructure and texture evolution of face-centered cubic high-entropy alloys. *Acta Mater.* **2018**, *150*, 88–103. [CrossRef]

16.   Saleh, A.A.; Haase, C.; Pereloma, E.V.; Molodov, D.A.; Gazder, A.A. On the evolution and modelling of brass-type texture in cold-rolled twinning-induced plasticity steel. *Acta Mater.* **2014**, *70*, 259–271. [CrossRef]

17.   Humphreys, F.J.; Hatherly, M. *Recrystallization and Related Annealing Phenomena*, 2nd ed.; Elsevier: Amsterdam, The Netherlands, 2004.

18.   Bracke, L.; Verbeken, K.; Kestens, L.; Penning, J. Microstructure and texture evolution during cold rolling and annealing of a high Mn TWIP steel. *Acta Mater.* **2009**, *57*, 1512–1524. [CrossRef]

19.   Haase, C.; Barrales-Mora, L.A.; Roters, F.; Molodov, D.A.; Gottstein, G. Applying the texture analysis for optimizing thermomechanical treatment of high manganese twinning-induced plasticity steel. *Acta Mater.* **2014**, *80*, 327–340. [CrossRef]

20.   Lü, Y.; Molodov, D.A.; Gottstein, G. Correlation Between Microstructure and Texture Development in a Cold-rolled TWIP Steel. *ISIJ Int.* **2011**, *51*, 812–817. [CrossRef]

21.   Vercammen, S.; Blanpain, B.; Cooman, B.C.D.; Wollants, P. Cold rolling behaviour of an austenitic Fe-30Mn-3Al-3Si TWIP-steel: The importance of deformation twinning. *Acta Mater.* **2004**, *52*, 2005–2012. [CrossRef]

22.   Haase, C.; Chowdhury, S.G.; Barrales-Mora, L.A.; Molodov, D.A.; Gottstein, G. On the Relation of Microstructure and Texture Evolution in an Austenitic Fe-28Mn-0.28C TWIP Steel During Cold Rolling. *Metall. Mater. Trans. A* **2013**, *44*, 911–922. [CrossRef]

23.   Hielscher, R.; Schaeben, H. A novel pole figure inversion method: Specification of the MTEX algorithm. *J. Appl. Crystallogr.* **2008**, *41*, 1024–1037. [CrossRef]

24.   Bachmann, F.; Hielscher, R.; Schaeben, H. Texture Analysis with MTEX—Free and Open Source Software Toolbox. *SSP* **2010**, *160*, 63–68. [CrossRef]

25.   Haase, C.; Ingendahl, T.; Güvenc, O.; Bambach, M.; Bleck, W.; Molodov, D.A.; Barrales-Mora, L.A. On the applicability of recovery-annealed twinning-induced plasticity steels: Potential and limitations. *Mater. Sci. Eng. A* **2016**, *649*, 74–846. [CrossRef]

26.   Haase, C.; Barrales-Mora, L.A.; Molodov, D.A.; Gottstein, G. Application of Texture Analysis for Optimizing Thermo-Mechanical Treatment of a High Mn TWIP Steel. *AMR* **2014**, *922*, 213–218. [CrossRef]

27.   Haase, C.; Zehnder, C.; Ingendahl, T.; Bikar, A.; Tang, F.; Hallstedt, B.; Hu, W.; Bleck, W.; Molodov, D.A. On the deformation behavior of κ-carbide-free and κ-carbide-containing high-Mn light-weight steel. *Acta Mater.* **2017**, *122*, 332–343. [CrossRef]

metals

MDPI

Article

On the Evolution of Residual Stresses, Microstructure and Cyclic Performance of High-Manganese Austenitic TWIP-Steel after Deep Rolling

Torben Oevermann*, Thomas Wegener and Thomas Niendorf[ORCID]

Institute of Materials Engineering, University of Kassel, Moenchebergstraße 3, 34125 Kassel, Germany
* Correspondence: oevermann@uni-kassel.de; Tel.: +49-0561-804-3701

Received: 14 June 2019; Accepted: 19 July 2019; Published: 25 July 2019

Abstract: The mechanical properties and the near surface microstructure of the high-manganese twinning-induced plasticity (TWIP) steel X40MnCrAl19-2 have been investigated after deep rolling at high (200 °C), room and cryogenic temperature using different deep rolling forces. Uniaxial tensile tests reveal an increase in yield strength from 400 MPa to 550 MPa due to surface treatment. The fatigue behavior of selected conditions was analyzed and correlated to the prevailing microstructure leading to an increased number of cycles to failure after deep rolling. Deep rolling itself leads to high compressive residual stresses with a stress maximum of about 800 MPa in the subsurface volume characterized by the highest Hertzian pressure and increased hardness up to a distance to the surface of approximately 1 mm with a maximum hardness of 475 HV0.1. Due to more pronounced plastic deformation, maximum compressive residual stresses are obtained upon high-temperature deep rolling. In contrast, lowest compressive residual stresses prevail after cryogenic deep rolling. Electron backscatter diffraction (EBSD) measurements reveal the development of twins in the near surface area independently of the deep rolling temperature, indicating that the temperature of the high-temperature deep rolling process was too low to prevent twinning. Furthermore, deep rolling at cryogenic temperature leads to a solid–solid phase transformation promoting martensite. This leads to inferior fatigue behavior especially at higher loads caused by premature crack initiation. At relatively low loads, all tested conditions show marginal differences in terms of number of cycles to failure.

Keywords: high-manganese steel; deep rolling; TWIP; TRIP; near surface properties; residual stresses; fatigue behavior

---

1. Introduction

In order to meet the high demands of many current applications, e.g., in the mobility sector, not only high-strength materials are required, but also appropriate processes to form components with locally tailored properties. Due to their well balanced properties and low costs, steels are still the material of choice for numerous applications. Since most commercially available steels in general are suffering from a "strength-ductility trade-off", i.e., showing either high strength and limited formability or high ductility and low strength, remarkable effort has been spent on the development of materials overcoming current limitations.

Austenitic metastable high-manganese steels are a representative of advanced high-strength steels with superior mechanical properties, showing high strength combined with excellent ductility, opening up great potentials for new designs in the automotive sector, e.g., relevant for crash-boxes and advanced lightweight components. The balanced combination of high strength and high ductility as well as the generally outstanding properties of high-manganese steels can be attributed to the deformation mechanisms being active. Besides dislocation glide, twinning and martensitic transformation can occur leading to two well-known effects: twinning-induced plasticity (TWIP) and transformation-induced plasticity (TRIP). Here, the TRIP-effect is based on a solid–solid phase transformation from austenite to either α'-martensite or ϵ–martensite. Contribution of each single deformation mechanism can be tailored based on the stacking fault energy (SFE) [1,2]. For instance, a relatively low SFE promotes the martensitic phase transformation, whereas higher values lead to a suppression of this mechanism [3] favoring the formation of mechanical twins or even pure dislocation glide instead [4].

The SFE can on the one hand be influenced by alloying elements. Aluminum, for example, increases the SFE [5], whereas silicon decreases the SFE, supporting the martensitic phase transformation [6]. On the other hand, high strain rate deformation increases the SFE in face-centered cubic alloys, as shown by Fiarro et al. [7]. Another way to tailor the SFE, which can be easily integrated into a process, is to decrease or increase the temperature of the material during deformation. As a result, the different deformation mechanisms can be activated as a function of the deformation temperature for a material with a given chemical composition. This was shown, e.g., by Rüsing et al. [8] investigating the effect of pre-deformation temperature on the fatigue behavior of the high-manganese austenitic steel X40MnCrAl19-2. In that study, twinning was observed upon deformation at room temperature, whereas, in addition to twinning, ϵ- as well as α'-martensite could be detected after pre-straining at $-196\,^\circ\text{C}$. Following deformation at a temperature of $400\,^\circ\text{C}$, twinning and phase transformation could not be seen.

The fatigue behavior of high-manganese steels in different regimes, i.e., low-cycle and high-cycle fatigue (LCF/HCF) regimes as well as the crack growth regime were investigated and reported in numerous studies. Schilke et al. [9] studied the LCF behavior of austenitic medium-manganese steels with a manganese content of 12–13 wt% in rolled and as-cast conditions revealing initial cyclic hardening, in line with the monotonic behavior, followed by cyclic softening. This is in good agreement with the findings of Wu et al. [10], who investigated the cyclic deformation response and deformation mechanisms of a TWIP steel in the LCF regime. They observed twinning and the formation of persistent slip bands providing for a strong work hardening effect. Another LCF study for TRIP/TWIP steels with varying carbon contents was conducted by Shao et al. [11]. In addition to the already mentioned initial cyclic hardening, they observed a more pronounced hardening with an increasing strain amplitude. Another study of the same group focusing on the effect of grain size on the cyclic deformation response in the LCF regime revealed three characteristic stages of the stress response [12]. An initial cyclic hardening attributed to intense dislocation interaction was followed by a stage of cyclic softening caused by a rearrangement of dislocations. The softening was followed by a second cyclic hardening stage related to the formation of a particular dislocation structure. Moreover, the results showed superior LCF properties upon grain refinement.

Similar results on the effect of grain size on the cyclic deformation response of the X-IP™1000 TWIP steel were reported by Rüsing et al. [13]. Compared to fine grained material, lower stress amplitudes for given strain amplitudes as well as a more pronounced cyclic softening at medium to high strain amplitudes were found for the coarse grained condition and attributed to a rearrangement of dislocations. The effects of phase transformation on the LCF behavior of austenitic high-manganese steels have been analyzed by Ju et al. [14] and Nikulin et al. [15]. The first study reported on fatigue crack propagation along the γ/ϵ interface focusing on short cracks. Dependent on the actual alloying composition, the formation of ϵ–martensite either contributed to a decrease of fatigue crack growth rates due to crack tip distortion or increased crack growth rates due to phase transformation promoted embrittlement. These results are in good agreement with the investigations of Nikulin et al. focusing on the effect of $\gamma \rightarrow \epsilon$–martensitic phase transformation on the LCF behavior and the fatigue induced microstructure of Fe-Mn-Cr-Ni-Si alloys. They showed that, as a function of varying Si contents, the ϵ–martensite formation can either improve or deteriorate fatigue resistance of this alloy. Martensite stability was found to be the key, while fatigue crack advance within the ϵ–martensite was strongly affected. Pathways towards improvement of LCF properties of high-manganese steels were reported by Niendorf et al. [16] and Guo et al. [17], both applying pre-deformation. Niendorf et al. showed that monotonic pre-straining of the X-IP^{TM}1000 TWIP steel up to 20% led to impeded dislocation mobility due to an increased density of twins acting as effective barriers against dislocation glide. Eventually, this led to a stable cyclic deformation response (CDR) with significantly improved fatigue lives. These results are in excellent agreement with the investigations of Guo et al. on a Fe-17Mn-0.8C steel. In addition to pre-deformation in tension, drawing was carried out. In case of the stretched samples, the CDR was stabilized, resulting in increased fatigue lives. For the pre-drawn condition, the fatigue life in the LCF was further enhanced. This was attributed to a more effective grain refinement as well as multi-variant twins formed by drawing.

Hamada et al. [18,19] conducted HCF experiments on four different high-manganese TWIP steels in reverse plane bending. Despite the different chemical compositions, a fatigue limit of 400 MPa was reported. Ratios of fatigue limit and tensile strength were varying between 0.42 and 0.48. Mechanical twins were not formed during cyclic loading; however, planar slip bands formed during the early stages of deformation. These were found to induce micro-cracks by interactions with grain boundaries and annealing twins. Another study of the same group [20] focused on enhancing HCF properties by grain refinement. Imposed by cold-rolling and recrystallization annealing, for the high-manganese Fe–22Mn–0.6C steel an average grain size of 1.8 μm could be obtained, while the coarse grained (CG) material was characterized by an average grain size of 35 μm. Upon grain refinement, the fatigue limit as well as the ratio of the fatigue limit to the ultimate tensile strength were increased from 400 MPa to 560 MPa and from 0.46 to 0.50, respectively. Despite the ultra-fine grained microstructure, the activity of deformation mechanisms during cyclic deformation was found to be similar to the CG counterpart. Likewise, Song et al. [21] reported on enhanced HCF properties for cold-drawn Fe–Mn–C TWIP steels. Cold-drawing up to a strain of $\epsilon = 0.45$ resulted in a maximum fatigue strength of 875 MPa, a subsequent stress relieving annealing for two hours at 400 °C increased the fatigue strength to even 975 MPa. The improvement upon heat-treatment was related to a relief of dislocation pile-ups alongside grain boundaries. In line with the results reported for the X-IP^{TM}1000 TWIP steel in the LCF regime, Niendorf et al. [22] showed that, in case of loading in the HCF regime, superior properties are obtained upon relatively high pre-strain levels of 40%. Most recent results on improvement of HCF properties of high-manganese steels were reported by Shao et al. [23,24]. In one study, a gradient microstructure in the surface layer of a high-manganese TWIP steel was achieved by a novel surface treatment process named *surface spinning strengthening* (3S). Due to this treatment, a hardened layer with a thickness of about 1 mm, accompanied by a grain size gradient, was established. As a result, the fatigue limit of the 3S samples increased by about 26% compared to their as-received counterparts. This enhancement was mainly attributed to the suppression of fatigue crack initiation. In a second study of the same group, a linear gradient in grain size was established in an Fe-30Mn-0.6C TWIP steel sample by pre-torsion and annealing (*T & A treatment*). The graded sample exceeded the fatigue strength of both the coarse-grained as well as the fine-grained structure. The superior properties of the microstructurally graded material in the HCF regime were ascribed to a pronounced generation of dislocations and the formation of a hard core and soft shell during cyclic loading.

Generally, the fatigue behavior of a material is significantly influenced by its surface condition, i.e., the near surface microstructure, roughness, phase fractions and the distribution of residual stresses [25–27]. The surface condition can be enhanced by different types of surface treatments like attrition, turning, deep rolling or shot-peening. For metastable austenitic stainless steels, numerous studies are available in a literature reporting on the influence of the aforementioned treatments. These were applied to modify the surface layer by introducing mechanical twins, martensite, compressive residual stresses and/or increased dislocation density. The improvement of fatigue properties by the generation of compressive residual stresses in the surface layer has been discussed in detail, e.g., in [28]. Nevertheless, currently only limited data reporting on the influence of residual stresses on the fatigue behavior of high-manganese TWIP/TRIP steel are available. Teichmann et al. [29] investigated the effect of shot-peening on the mechanical properties of a high-manganese TWIP steel. As a result of shot-peening with an intensity of 0.30 mmA, compressive residual stresses reached a maximum of 1400 MPa. Moreover, the fatigue strength in rotation bending testing was increased from 400 to 600 MPa for the shot-peened condition. Another study from Klein et al. [30] focused on the surface morphology and its influence on the cyclic deformation behavior of a commercial TWIP steel. Phase fractions as well as residual stresses were determined for the as-received, an up- and down-milled and a polished reference condition. Compressive residual stresses were found for each condition investigated, revealing a maximum value of 800 MPa for the up-milled surface. In the LCF regime, fatigue life was mainly influenced by the surface topography resulting in the lowest number of cycles to failure for the as-received condition due to its high roughness. Improved fatigue life was found for the polished condition. However, the cyclic deformation response in the HCF regime was significantly influenced by other surface features, i.e., residual stresses and nano-crystalline sub-surface layers, respectively. Consequently, the best performance in the HCF regime was reported for the up-milled condition being characterized by highest compressive residual stresses and a nano-crystalline surface layer.

Deep rolling at different temperatures is particularly suitable to simultaneously tailor local plastic deformation and the global SFE of the material treated. The process allows for high plastic deformation of the near surface layer and can be conducted at very low or very high temperatures. Moreover, deep rolling is a mechanical surface treatment that is known for smoothing the surface, inducing high compressive residual stresses and providing a significant amount of work hardening, reported in several studies, e.g., in [27,31–33]. The changes within the surface layer strongly improve the fatigue life and, thus, are of highest importance for the material behaviour under cyclic loading. Based on the experiments conducted, the present paper reports on the performance of the high-manganese TWIP steel X40MnCrAl19-2 after deep rolling at various temperatures (−196 °C, room temperature, 200 °C). The following novel insights into process-microstructure-property relationships in TWIP steels will be presented:

- Evolution of the microstructure dependent on the deep rolling temperature affected by the different activated elementary deformation mechanisms upon surface treatment.
- Development of residual stresses dependent on the deep rolling temperature.
- Improvement of the cyclic deformation behavior imposed by microstructure and residual stress state.

2. Materials and Methods

The material studied in the present work is the high-manganese austenitic steel X40MnCrAl19-2 (precidur® H-Mn LY) provided by thyssenkrupp Hohenlimburg GmbH (Hagen, Germany). The chemical composition is given in Table 1. The electron backscatter diffraction (EBSD) micrograph shown in Figure 1 reveals the presence of recrystallization twins in the homogenous as-received microstructure being characterized by an average grain size of about 20 μm. The material was provided as hot-rolled blank with a thickness of 9 mm. Cylindrical specimens were machined by turning. Before, cuboid specimens of about 9 mm × 121 mm × sheet thickness were obtained by water jet cutting with the longitudinal axis of all specimens being perpendicular to the rolling direction of the sheet. The final specimen geometry is shown in Figure 2. Tensile testing of the untreated condition at room temperature revealed a relatively low yield strength of about 400 MPa, an ultimate tensile strength of 880 MPa and an elongation to fracture of 52% (see Figure 3). The mechanical properties are similar to those determined by Rüsing [34], who also performed tensile tests at elevated and cryogenic temperatures, showing that the material is characterized by higher tensile strength at cryogenic temperature, and an increase of the yield strength by a factor of two. The average hardness of the as-received material was found to be 250 HV.

Table 1. Chemical composition of the austenitic TWIP steel X40MnCrAl19-2 (Fe: Bal).

C Wt.%	Si Wt.%	Mn Wt.%	P Wt.%	S Wt.%	Al Wt.%
0.45	0.40	20.00	0.03	0.005	2.50
Cr Wt.%	Cu Wt.%	V Wt.%	Mo Wt.%	Ti Wt.%	Ni Wt.%
2.50	0.20	0.20	0.20	0.20	1.00

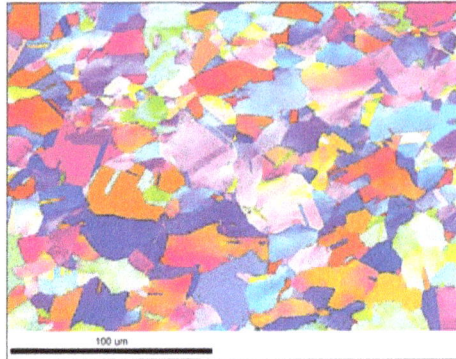

Figure 1. Inverse Pole Figure (IPF) map of the as-received condition showing a homogenous microstructure with recrystallization twins and an average grain size of about 20 µm.

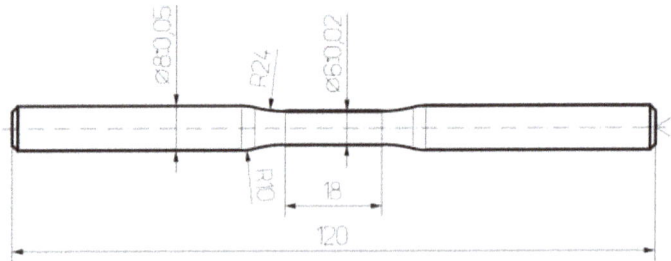

Figure 2. Specimen geometry employed for tensile and fatigue tests, units in millimeters.

To manufacture specimens with different surface properties, the turned specimens were deep rolled using different process parameters under controlled conditions on a servo-conventional lathe of type Weiler C30 (Weiler Werkzeugmaschinen GmbH, Emskirchen, Germany). In preliminary studies, the deep rolling force was varied to analyze its influence on the near surface hardness and residual stress depth profiles to obtain adequate surface properties for further investigations. With respect to the obtained results (see Figure 4) and to avoid bending of the specimens during the deep rolling process at even higher forces, a deep rolling force of 855 N was chosen. The deep rolling force was applied by a one-roller tool made by Ecoroll (Celle, Germany) with a diameter of Ø40 mm. A constant feed rate of 1 mm per revolution at a rotational speed of 80 rev./min was applied. Deep rolling was carried out at room, cryogenic (−196 °C) and high temperature (200 °C). Cryogenic deep rolling was performed using liquid nitrogen employing a custom-built device. High-temperature deep rolling was carried out using an induction heater (TRUMPF Hüttinger GmbH, Freiburg, Germany) controlled by a pyrometer (Sensortherm GmbH, Sulzbach, Germany).

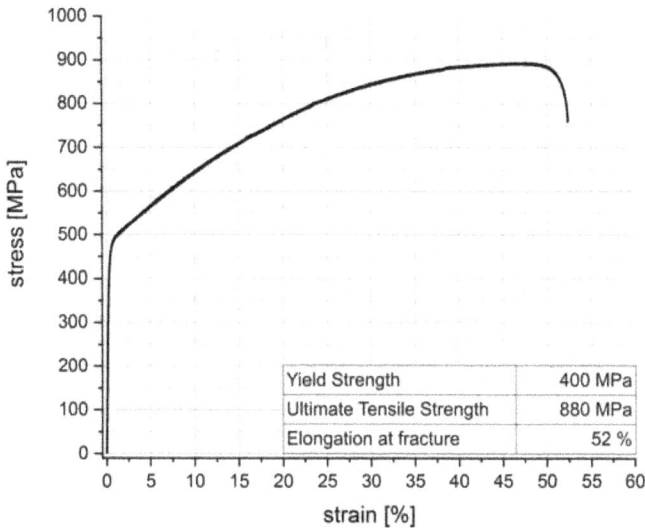

Figure 3. Stress–strain diagram for the as-received twinning-induced plasticity (TWIP) steel tested at room temperature. The inset summarizes the most important characteristic values.

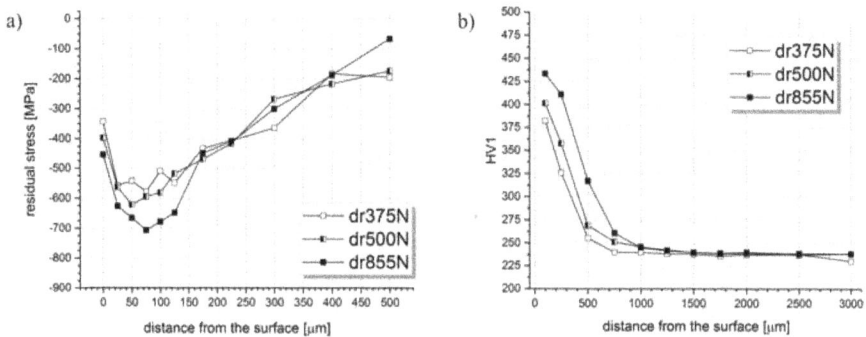

Figure 4. Residual stress depth profiles in longitudinal direction (**a**) and near-surface hardness distribution (**b**) after deep rolling at room temperature applying deep rolling forces between 375 N and 855 N.

After deep rolling, the near surface properties were characterized. Vickers hardness measurements were carried out on polished cross sections to determine the overall hardness distribution and the increase of hardness in the near surface area induced by the surface treatment process. The measurements were conducted using a Struers DuraScan-70 system (Hannover, Germany) employing loads of 9.81 N and 1.96 N, respectively. Microstructure analysis, focusing on mechanical twinning and strain induced phase transformation in the surface layer, was conducted using a Zeiss ULTRA GEMINI high-resolution scanning electron microscope (SEM) (Jena, Germany) at an acceleration voltage of 30 kV. The SEM employed is equipped with an EBSD unit and a back scattered electron (BSE) detector. A second SEM system (CamScan MV 2300, Tescan GmbH, Dortmund, Germany) operated at 20 kV was used to carry out fracture surface investigations for representative fatigued specimens. For EBSD characterization, cross sections of the specimens were mechanically ground down to 5 µm grit size using SiC paper and polished using a diamond suspension to a minimum of 1 μm. The final step of the polishing process was vibration polishing for 16 hours using conventional oxide polishing suspension (OPS) (0.04 μm).

Residual stress depth profiles and integral width values were analyzed by X-ray diffraction (XRD) using a HUBER-diffractometer (Huber Diffraktionstechnik GmbH, Rimsting, Germany) equipped with a 1 mm collimator. The {220}-planes of the austenitic phase were considered at a 2θ angle of 128.78° using CrKα-radiation. To analyze stress profiles in depth, surface layers were removed layerwise by electro-chemical polishing. All values obtained were evaluated utilizing the standard $\sin^2 \Psi$-method ($\frac{1}{2}s_2 = 6,75E^{-6}\mathrm{mm}^2\,\mathrm{N}^{-1}$) not considering any mathematical stress correction. In addition, phase analysis was performed by XRD.

To determine the influence of the different deep rolling parameters on the mechanical behavior, monotonic tensile and stress controlled fatigue tests in the HCF regime were performed using a digitally controlled servo-hydraulic test rig equipped with a 63 kN load cell. Monotonic stress–strain diagrams were plotted based on strains calculated based on crosshead displacement taking into account the nominal specimen gauge length of 18 mm (cf. Figure 5). The uniaxial tensile tests were performed under displacement control with a constant crosshead speed of 2 mm/min. The HCF tests were carried out in stress control under fully reversed push-pull loading with a stress ratio of $R = -1$. For strain measurement, an extensometer featuring a gauge length of 12 mm, directly attached to the surface of the specimens, was used. To avoid heating of the specimens, compressed air cooling was used. Specimen temperature was limited to a maximum of 30 °C. The standard frequency for the HCF tests was 15 Hz, however, had to be reduced to 1 Hz at loads above 400 MPa. For all load levels, an adequate number of specimens was considered; however, no statistical evaluation was carried out. Representative fracture surfaces were chosen for examination using the SEM mentioned above.

3. Results and Discussion

3.1. Characterization of the Near Surface Properties

As expected, the deep rolling process leads to increased hardness in the near surface area. The hardness values presented in Figure 4b highlight the relationship between the applied deep rolling force and the degree of cold hardening. For all conditions, the maximum hardness is always obtained at the surface. The lowest applied deep rolling force of 375 N increases the hardness in a depth of 0.15 mm to a maximum of about 400 HV. Higher deep rolling forces lead to an even more pronounced cold hardening effect with hardness values increased by a factor of 1.65 and 1.75, respectively, compared to the hardness of the as-received state. In line with the relations in terms of the maximum hardness, the in-depth effect imposed by the surface treatment increases with increasing deep rolling force. The lowest deep rolling force applied leads to increased hardness values up to a depth of 0.6 mm compared to 0.9 mm for the highest deep rolling force. This is in good agreement with the results of Teichmann et al. [29] reporting on the influence of shot peening on the mechanical behavior of a different TWIP steel grade. Compared to a hardness of 250 HV0.1 in the as-received condition, a maximum hardness of 550 HV0.1 was observed after shot peening with the highest Almen intensity of 0.3 mmA. Concomitantly, a higher penetration depth was found with increasing peening intensity. The saturation seen for shot peening could not be observed for the deep rolling parameters applied in the present study. Nevertheless, maximum hardness was shown for the highest peening intensity [29]. Comparing the two surface treatment processes in terms of hardness evolution, it can be revealed that the depth effect for deep-rolling is almost 4 to 5 times higher than in the case of shot peening.

Results obtained by residual stress measurements, plotted for longitudinal direction, are presented in Figure 4a. Compared to the hardness distribution (Figure 4b), discussed above, a similar trend can be observed. Regardless of the process parameters, compressive residual stresses are generated. The maximum values are always seen in direct vicinity of the surface, i.e., located at a distance to the surface of about 50 μm to 100 μm. This depth can be related to the area of the highest Hertzian pressure during deep rolling. In general, a higher deep rolling force results in higher residual stress values at the surface. However, for all conditions, the values at distances to the surface above 175 µm are found to be almost identical. In contrast to the results from hardness measurements, the analysis of residual stress distribution presented by Teichmann et al. [29] for the shot peened TWIP steel reveals some differences compared to present results. While penetration depth increases with higher shot intensities, the residual stress maxima reveal a different behavior. After an intensity increase from 0.15 to 0.22 mmA, hardly any changes are seen in residual stress maxima; however, a further increase to 0.30 mmA results in a significant increase of compressive stresses to 1400 MPa. This different behavior can be rationalized based on the characteristics of both surface treatment processes. While for deep rolling the force exerted to the specimen is concentrated in one single contact zone, the effective force imposed by shot peening is distributed in a contact zone area.

Tensile testing reveals that deep rolling at room temperature leads to an increased strength. From Figure 5, it can be seen that the yield strength, in many studies stated to be a roadblock towards use of high-manganese steels [35], is increased by the deep rolling process, i.e., from 400 MPa in the untreated condition to 550 MPa. Furthermore, a minor increase of the ultimate tensile strength to a maximum value of 940 MPa is seen, whereas the elongation to fracture remains stable at about 50%. Using different deep rolling temperatures, the monotonic mechanical properties are affected similarly, cf. Figure 5.

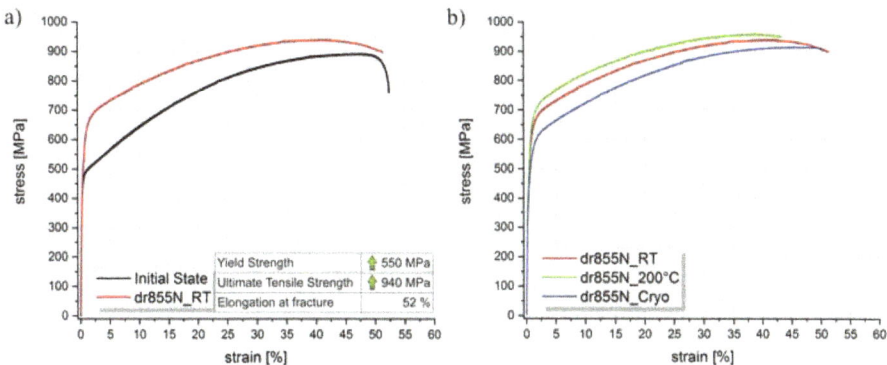

Figure 5. Monotonic stress–strain response of the as-received material compared to a specimen deep rolled at room temperature using a deep rolling force of 855 N (**a**) and direct comparison of the conditions deep rolled at different temperatures applying the same deep rolling force (**b**).

Figure 6 shows representative micrographs obtained by EBSD for the three deep rolled conditions characterized in the present study, i.e., the (a) room temperature, (b) 200 °C and (c) cryogenic temperature deep rolled specimens. The step size for all EBSD maps shown was 1 µm. A very similar homogeneous microstructure, as shown for the as-received condition in Figure 1, can be seen (with respect to grain size and texture) for all conditions. Considering the room temperature deep rolled condition in detail, some grains featuring mechanical twins can be made out. The majority of these twins is located in a distance to the surface of about 100 µm, which can be related to the area of the highest Hertzian pressure during the deep rolling process as already mentioned above. Phase transformation (corresponding phase maps are not shown for sake of brevity) from the austenitic γ-phase to either ϵ- or α'-martensite could not be detected as a result of deep rolling at room temperature. From these findings it can be deduced that, due to the high SFE at room temperature, the austenite grains were mainly deformed by mechanical twinning and dislocation glide. This behavior is consistent with current literature. A very similar microstructure revealing grains with abundant mechanical twins and even formation of secondary and double twins was reported by Song et al. [21] for a Fe-Mn-C TWIP steel upon drawing at room temperature. Furthermore,

Rüsing et al. [8] found numerous twins, stacking faults and a high density of dislocations using transmission electron microscopy (TEM) in the material investigated in the current study as a result of room temperature tensile deformation to an elongation of 50%. The EBSD micrograph of the specimen deep rolled at 200 °C shown in Figure 6b is characterized by a very similar microstructure as compared to the room temperature deep rolled condition displayed in (a). Again, mechanical twins can be observed in several grains in the area being characterized by the highest Hertzian pressure, whereas no phase transformation was detected. Both martensitic transformation and twinning cannot be seen after stretching at a significantly higher deformation temperature of 400 °C in the study of Rüsing et al. [8]. From these findings, it can be deduced that a deep rolling temperature of 200 °C is not high enough to suppress twinning. With respect to the prevailing deformation mechanisms and resulting microstructure evolution, a different behavior can be seen for the condition deep rolled at cryogenic temperature displayed in Figure 6c. Deformation induced phase transformation seems to be the dominant deformation mechanism for this condition. This is confirmed by the phase distribution map displayed in Figure 6d where large areas are indexed as hexagonal ϵ–martensite. In line with the deep rolled conditions at 200 °C and room temperature, the largest fraction of ϵ–martensite is detected in a distance to the surface of approximately 100 µm, again being rationalized by the highest Hertzian pressure. These results are once more consistent with findings of Rüsing et al. [8], showing fractions of ϵ–martensite as well as α'-martensite in addition to twins for a specimen deformed in tension to an elongation of 40% at a temperature of −196 °C. The absence of α'-martensite in the deep rolled condition of the current study is thought to be due to the different loading situation in deep rolling as compared to tensile testing.

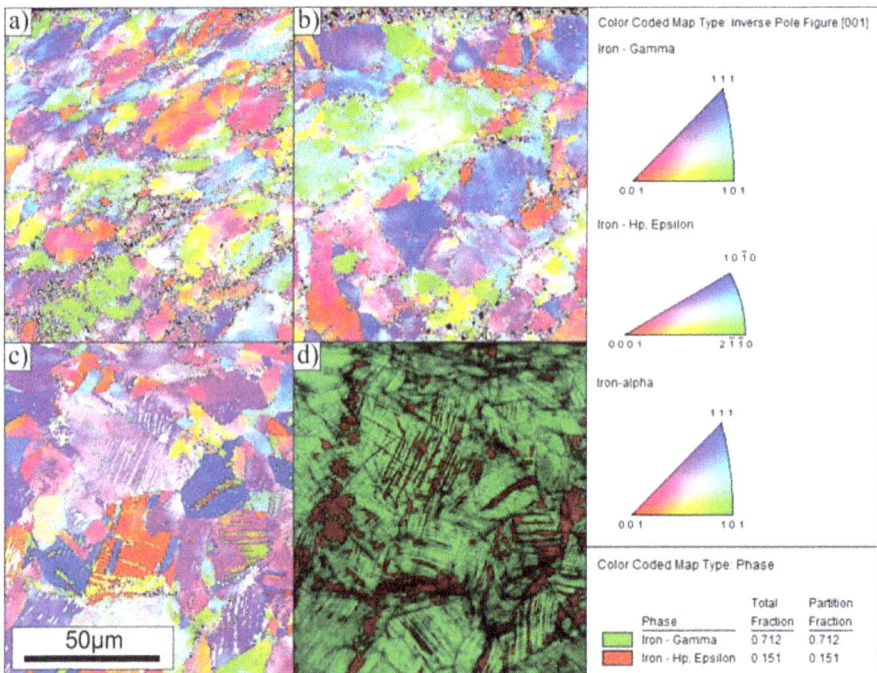

Figure 6. IPF maps (**a–c**) and phase map with superimposed image quality (IQ) (**d**) for differently treated specimens. IPF maps depict the near surface microstructure following deep rolling at room temperature (**a**), 200 °C (**b**) and cryogenic temperature (**c**) using a deep rolling force of 855 N. The IQ-phase map for a surface layer after deep rolling at cryogenic temperature is shown in (**d**). What appears to be twinning in the IPF map clearly is resolved as a $\gamma \rightarrow \epsilon$ phase transformation in (**d**). Color coding in (**a–d**) is in accordance to the standard triangles and the nomenclature shown to the right.

Focusing on the mechanical properties of the surface layer, both the degree of work hardening and the affected depth can be tailored by changing the temperature of the deep rolling process, cf. Figure 7. Increasing the temperature to 200 °C leads to generally increased hardness values throughout the probed volume with a maximum hardness of 460 HV near the surface and increased hardness values up to a depth of 1.0 mm. On the contrary, as a result of decreasing the deep rolling temperature to −196 °C by using liquid nitrogen, local plastic deformation seems to be impeded, leading to a minor increase of the hardness. In order to compare hardness and work hardening of the different conditions, it has to be considered that the deep rolling force was constant, while the temperature varied between −196 °C and 200 °C. Since a decreasing temperature leads to differences in mechanical properties (higher yield strength and ultimate tensile strength) as seen in tensile tests conducted by Rüsing [34], here a constant load results in less pronounced work hardening as the resistance of the material against plastic deformation increases. At high temperatures, the opposite effect is achieved. Even if martensitic transformation is seen in the specimens deep rolled at cryogenic temperature, hardness is lower as compared to the other two conditions. Further work including varying deep rolling forces at given temperatures will have to be conducted for an in-depth evaluation of evolving microstructural features and related local hardness. This, however, is beyond the scope of the current work.

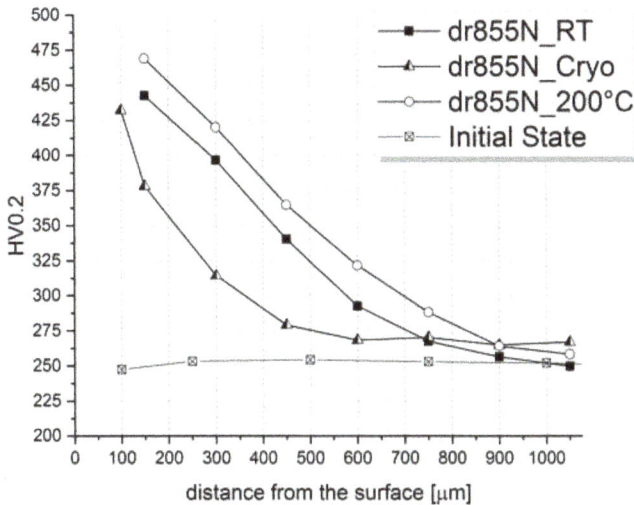

Figure 7. Hardness distribution in the near surface area after deep rolling in a temperature range from −196 °C (Cryo) to 200 °C at a constant deep rolling force of 855 N.

In line with the hardness measurements, the remarkable influence of process temperature on the development of residual stresses is revealed by the results presented in Figure 8. Deep rolling at 200 °C strongly promotes the evolution of compressive residual stresses, leading to a maximum of about 840 MPa at a distance to the surface of 75 µm compared to only 700 MPa after deep rolling at room temperature determined in the same depth (results for longitudinal direction). Clearly, the effect of deep rolling temperature is much more pronounced as compared to the effect of deep rolling force. The residual stress depth profile maxima for the measurements in circumferential direction are slightly shifted towards the core of the specimens and a lower absolute value of the compressive residual stresses can be observed here. The integral width values for austenitic steels after surface treatment are expected to be characterized by maximum values at the surface and a small plateau in the area of the highest Hertzian pressure, as has been shown in [27].

All residual stress measurements were performed solely considering the austenitic phase of the material. Rüsing et al. reported in [8] that a martensitic phase transformation to either α'–martensite or ε–martensite has to be expected after plastic deformation at cryogenic temperature for the TWIP steel considered. This observation was also confirmed by phase analysis using EBSD (cf. Figure 6c,d). In order to provide for better statistics, XRD phase analysis was carried out in parallel to every step of the residual stress profile determination, i.e., on the same spot used for stress determination in longitudinal direction. The qualitative results of the measurement are shown in Figure 9. The 3D-plot depicted clearly reveals that the maximum ε–martensite phase fraction is present at a distance of about 0.075 mm to 0.125 mm from the surface. Compared to the residual stress depth profile (Figure 8) in the longitudinal direction, this is apparently the area undergoing the highest Hertzian pressure during the deep rolling process [36]. After the ε–martensite phase fraction reaches its maximum, the amount decreases constantly. The intensity of both austenitic peaks shows no clear trends. This is attributed to the relatively coarse austenite grains. Consequently, results will not be further interpreted. X-ray diffraction peaks used for determination of residual stresses are shown in Figure 10.

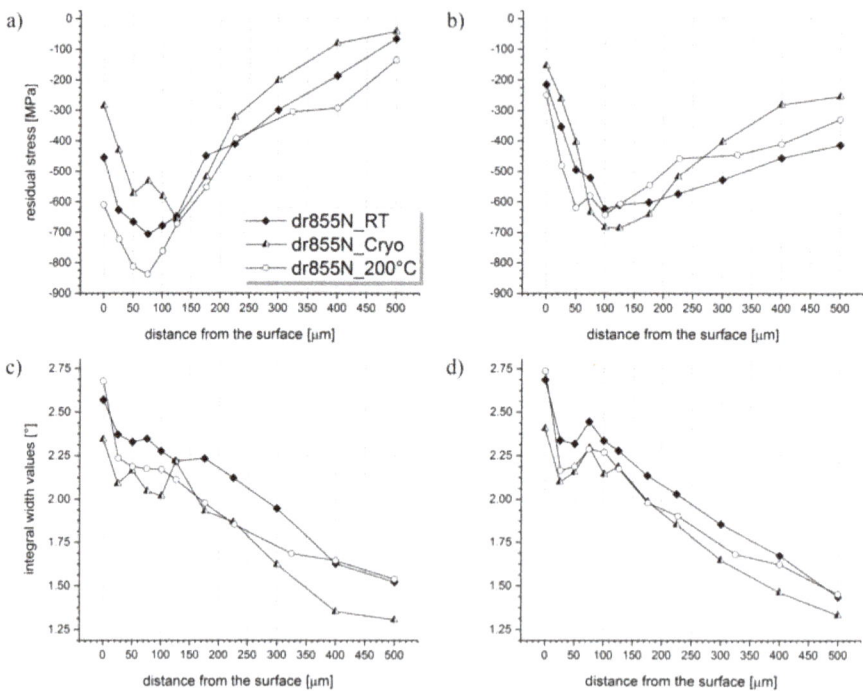

Figure 8. Residual stress profiles and integral width values in longitudinal (**a+c**) and circumferential (**b+d**) direction plotted as a function of deep rolling temperature at constant deep rolling force.

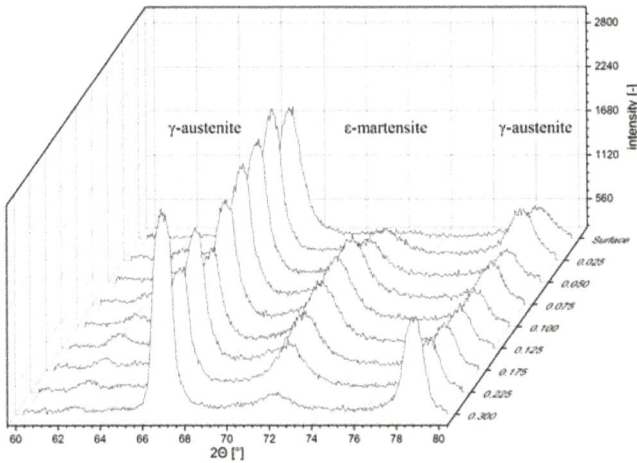

Figure 9. XRD phase analysis using *CrKα*-radiation revealing a significant martensitic phase fraction upon deep rolling at cryogenic temperature. Peak profiles are plotted as a function of distance from the surface.

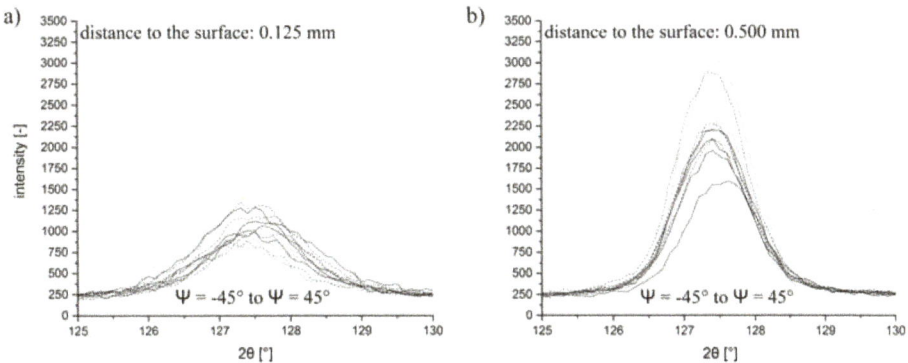

Figure 10. XRD peaks used for determination of residual stresses in a depth of 0.125 mm (**a**) and 0.500 mm (**b**), respectively, plotted for Ψ-angles ranging from −45° to 45°.

3.2. Fatigue Tests

After thorough characterization of the near surface properties, fatigue tests in the HCF regime were performed considering the conditions deep rolled at 200 °C, room temperature and cryogenic temperature. Taking into account the residual stress depth profiles only, the superior specimen condition was expected to be the condition deep rolled at high temperature. Highest residual stress values in the near surface area and the assumedly higher stability of the residual stresses, as reported in literature for other alloys [33,37], were expected to be the major factors in this regard. The behavior of the condition deep rolled at cryogenic temperature could not be predicted based on literature results, as the consequences of the phase transformation to ε−martensite in the near surface area have only been analyzed scarcely so far. However, in light of the lower residual stress level, determined based on the austenitic phase, and the lower hardness level, the number of cycles to failure at a given loading amplitude was expected to be lower. Figure 11 highlights the fatigue lives of all analyzed conditions. Compared to the results of the as-received state, deep rolling at room temperature leads to an increased fatigue life for all load levels, consistent with the current literature reporting on the fatigue properties of room temperature mechanically surface treated TWIP steels [29,30]. At a loading amplitude of

460 MPa, i.e., the highest level considered for the as-received condition, the average difference of the number of cycles to failure amounts to a factor of about 14. This factor decreases with decreasing loading amplitude to a minimum of about 6 at a loading amplitude of 425 MPa. The advantages of deep rolling at elevated temperature known from literature for austenitic steels [27,38] and other steel grades [39] are not yet achieved at the deep rolling temperature of 200 °C in case of the TWIP steel studied. At relatively high loads, deep rolling at 200 °C seems to have a positive effect on the fatigue life; however, this has to be analyzed in future studies based on a statistical basis. An absolutely unexpected behavior can be seen for the specimen deep rolled at cryogenic temperature. Hardly any differences can be seen between the deep rolled specimens being characterized by differences in strain hardening, residual stresses or phase transformation with respect to the fatigue life for loading amplitudes below 460 MPa. A loading amplitude of 460 MPa, however, seems to be a critical loading level, as here significant divergences in terms of trend lines for the different conditions can be derived.

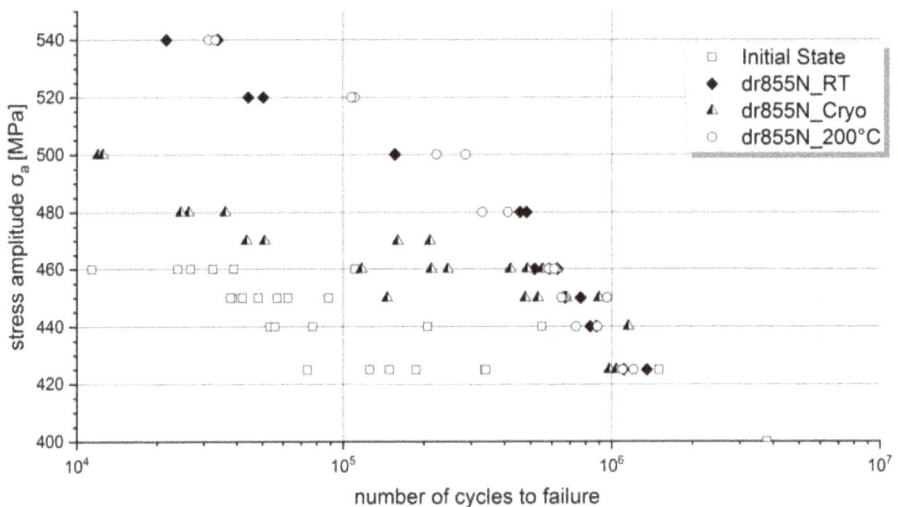

Figure 11. Woehler-type S-N plots for conditions deep rolled at high, room and cryogenic temperature applying a constant deep rolling force of 855 N. See text for details.

Fracture surface analysis reveals that the common advantages of deep rolling known from literature, i.e., the shifting of the point of crack initiation away from the surface to the specimen interior by introducing compressive residual stresses in the near surface layers, can be also seen for the TWIP steel conditions deep rolled at room temperature (cf. Figure 12a)) and 200 °C. (The latter condition is not shown for sake of brevity.) For the cryogenic deep rolled specimens, these well-established relationships seem not to hold true. At a loading amplitude of 480 MPa, all specimens deep rolled at cryogenic temperature show crack initiation at the surface, cf. Figure 12b. Consequently, fatigue lives are significantly reduced. A reduction of fatigue life by a factor as large as twenty can be seen. Another critical aspect in evaluating the cryogenic deep rolled condition is the huge scatter seen in the results (between load levels of 480 MPa and 450 MPa). Specimens characterized by crack initiation below the surface show comparably high numbers of cycles to fracture, approaching the room temperature deep rolled condition, whereas some specimens show premature failure induced by crack initiation close to the surface, Figure 13. These specimens are characterized by numbers of cycles to failure similar to the initial state. This behavior, however, currently seems to be unpredictable and probably is related to the phase fraction of ϵ–martensite in the near surface area and/or the morphology of the martensitic phase (cf. Figure 6d)). How far these factors promote a microstructure notch effect, internal friction, local cyclic plasticity and degradation mechanisms related hereto cannot be deduced from current results and, thus, analysis of relationships has to be subject of future work.

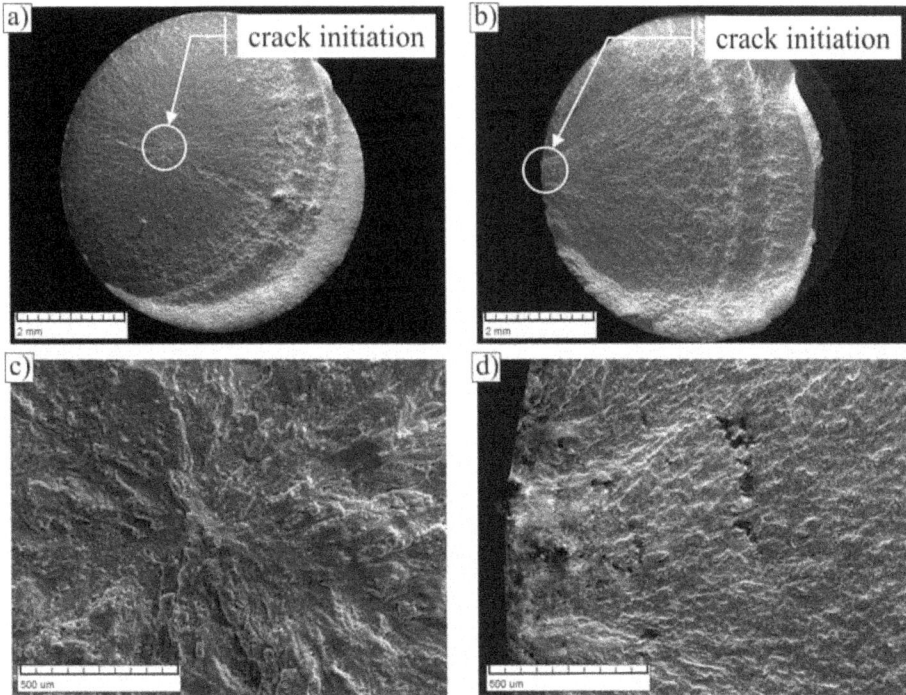

Figure 12. Fracture surfaces of specimen deep rolled at room temperature and cryogenic temperature fatigued at a cyclic loading amplitude of 480 MPa. Overview images of conditions deep rolled at room (**a**) and cryogenic (**b**) temperature using a deep rolling force of 855 N are depicted with encircled crack initiation spots. High resolution micrographs depicting the microstructural features being responsible for crack initiation are shown in (**c**+**d**).

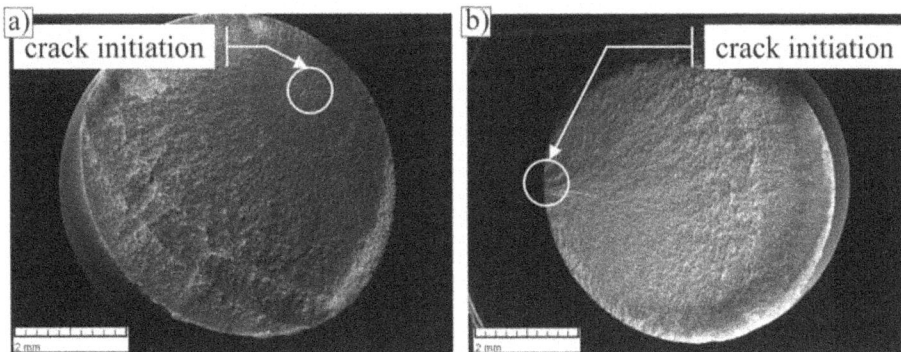

Figure 13. Fracture surfaces of specimens deep rolled at cryogenic temperature after fatigue tests applying a cyclic loading amplitude of 460 MPa. The overview SEM micrographs depict the respective crack initiation points. The specimens are characterized by fundamentally different fatigue lives, i.e., 500,000 cycles (**a**) and 25,000 cycles (**b**).

4. Conclusions

The present study was conducted to evaluate the influence of deep rolling in a temperature range from −196 °C to 200 °C on the near surface microstructure, hardness distribution and residual stresses of the high-manganese twinning-induced plasticity (TWIP) steel X40MnCrAl19-2 and the related fatigue behavior in the high-cycle fatigue (HCF) regime. The following conclusions can be drawn from the results presented:

- Deep rolling of the TWIP steel improves the monotonic mechanical properties, e.g., the yield strength (from 400 MPa to 550 MPa), without having any negative effect on the elongation at fracture. Furthermore, high compressive residual stresses with a maximum of 800 MPa are generated in the near surface area accompanied by high hardness values up to 475 HV0.1.
- The martensitic phase transformation promoting ϵ–martensite in the near surface area induced by cryogenic deep rolling has a negative impact on the fatigue performance at least at relatively high loading amplitudes. In addition, an unpredictable behavior caused by random premature crack initiation is found at an intermediate stress level. Reasons are thought to be linked to increased internal friction going along with increased plastic deformation during the fatigue tests and/or notch effects in the two-phase region established in the near surface area.
- Superior performance is seen for the TWIP steel deep rolled at elevated temperature. However, in light of findings being present in literature for alternative steel grades, pathways towards further property optimization for the TWIP steel considered could be derived.

Author Contributions: Conceptualization, T.O.; Funding acquisition, T.W. and T.N.; Methodology, T.O.; Project administration, T.N.; Supervision, T.N.; Visualization, T.O.; Writing—original draft, T.O. and T.W.; Writing—review and editing, T.O. and T.W.; All authors contributed equally to the interpretation of results and writing the final version of the manuscript.

Funding: This research was funded by "Deutsche Forschungsgemeinschaft" (DFG) (German Research Foundation), Grant No. 406320672, within the research project "Mechanische Oberflächenbehandlung von hochmanganhaltigen TWIP/TRIP Stählen—Mikrostrukturelle Stabilität and Mechanische Eigenschaften".

Acknowledgments: The authors would like to thank Rolf Diederich and Rainer Hunke for technical support regarding the maintenance and operation of the laboratory equipment. The assistance of Leoni Hübner is greatly appreciated. The authors acknowledge thyssenkrupp Hohenlimburg GmbH (Hagen, Germany) for providing the material.

Conflicts of Interest: The authors declare no conflict of interest. The funding agency had no influence on the design of the study, the collection, analysis, or interpretation of data, the writing of the manuscript, or the decision to publish the results.

References

1. Ferreira, P.; Müllner, P. A thermodynamic model for the stacking-fault energy. *Acta Mater.* **1998**, *46*, 4479–4484. [CrossRef]
2. Karaman, I.; Sehitoglu, H.; Gall, K.; Chumlyakov, Y.; Maier, H. Deformation of single crystal Hadfield steel by twinning and slip. *Acta Mater.* **2000**, *48*, 1345–1359. [CrossRef]
3. Sato, K.; Ichinose, M.; Hirotsu, Y.; Inoue, Y. Effects of deformation induced phase transformation and twinning on the mechanical properties of austenitic Fe-Mn-Al alloys. *ISIJ Int.* **1989**, *29*, 868–877. [CrossRef]
4. Grässel, O.; Krüger, L.; Frommeyer, G.; Meyer, L.W. High strength Fe–Mn–(Al, Si) TRIP/TWIP steels development—properties—application. *Int. J. Plast.* **2000**, *16*, 1391–1409. [CrossRef]
5. Ishida, K.; Nishizawa, T. Effect of Alloying Elements on the Stability of Epsilon Iron. *J. Jpn. Inst. Metals* **1972**, *36*, 1238–1245. [CrossRef]
6. Schramm, R.E.; Reed, R.P. Stacking fault energies of seven commercial austenitic stainless steels. *Metall. Trans. A* **1975**, *6*, 1345–1351. [CrossRef]
7. Firrao, D.; Matteis, P.; Scavino, G.; Ubertalli, G.; Pozzi, C.; Ienco, M.G.; Piccardo, P.; Pinasco, M.R.; Costanza, G.; Montanari, R.; et al. Microstructural Effects in Face-Centered-Cubic Alloys after Small Charge Explosions. *Metall. Mater. Trans. A* **2007**, *38*, 2869–2884. [CrossRef]

8.   Rüsing, C.; Lambers, H.G.; Lackmann, J.; Frehn, A.; Nagel, M.; Schaper, M.; Maier, H.; Niendorf, T. Property Optimization for TWIP Steels–Effect of Pre-deformation Temperature on Fatigue Properties. *Mater. Today Proc.* **2015**, *2*, S681–S685. [CrossRef]
9.   Schilke, M.; Ahlström, J.; Karlsson, B. Low cycle fatigue and deformation behaviour of austenitic manganese steel in rolled and in as-cast conditions. *Procedia Eng.* **2010**, *2*, 623–628. [CrossRef]
10.   Wu, Y.X.; Tang, D.; Jiang, H.T.; Mi, Z.L; Xue, Y.; Wu, H.P. Low Cycle Fatigue Behavior and Deformation Mechanism of TWIP Steel. *J. Iron Steel Res. Int.* **2014**, *21*, 352–358. [CrossRef]
11.   Shao, C.; Zhang, P.; Liu, R.; Zhang, Z.; Pang, J.; Zhang, Z. Low-cycle and extremely-low-cycle fatigue behaviors of high-Mn austenitic TRIP/TWIP alloys: Property evaluation, damage mechanisms and life prediction. *Acta Mater.* **2016**, *103*, 781–795. [CrossRef]
12.   Shao, C.; Zhang, P.; Zhu, Y.; Zhang, Z.; Pang, J.; Zhang, Z. Improvement of low-cycle fatigue resistance in TWIP steel by regulating the grain size and distribution. *Acta Mater.* **2017**, *134*, 128–142. [CrossRef]
13.   Rüsing, C.J.; Niendorf, T.; Frehn, A.; Maier, H.J. Low-Cycle Fatigue Behavior of TWIP Steel - Effect of Grain Size. *Adv. Mater. Res.* **2014**, *891–892*, 1603–1608. [CrossRef]
14.   Ju, Y.B.; Koyama, M.; Sawaguchi, T.; Tsuzaki, K.; Noguchi, H. Effects of ϵ-martensitic transformation on crack tip deformation, plastic damage accumulation, and slip plane cracking associated with low-cycle fatigue crack growth. *Int. J. Fatigue* **2017**, *103*, 533–545. [CrossRef]
15.   Nikulin, I.; Sawaguchi, T.; Ogawa, K.; Tsuzaki, K. Effect of γ to ϵ martensitic transformation on low-cycle fatigue behaviour and fatigue microstructure of Fe–15Mn–10Cr–8Ni– x Si austenitic alloys. *Acta Mater.* **2016**, *105*, 207–218. [CrossRef]
16.   Niendorf, T.; Lotze, C.; Canadinc, D.; Frehn, A.; Maier, H.J. The role of monotonic pre-deformation on the fatigue performance of a high-manganese austenitic TWIP steel. *Mater. Sci. Eng. A* **2009**, *499*, 518–524. [CrossRef]
17.   Guo, Q.; Chun, Y.S.; Lee, J.H.; Heo, Y.U.; Lee, C.S. Enhanced low-cycle fatigue life by pre-straining in an Fe-17Mn-0.8C twinning induced plasticity steel. *Metals Mater. Int.* **2014**, *20*, 1043–1051. [CrossRef]
18.   Hamada, A.; Karjalainen, L.; Puustinen, J. Fatigue behavior of high-Mn TWIP steels. *Mater. Sci. Eng. A* **2009**, *517*, 68–77. [CrossRef]
19.   Hamada, A.; Karjalainen, L.; Ferraiuolo, A.; Sevillano, J.G.; de las Cuevas, F.; Pratolongo, G.; Reis, M. Fatigue Behavior of Four High-Mn Twinning Induced Plasticity Effect Steels. *Metall. Mater. Trans. A* **2010**, *41*, 1102–1108. [CrossRef]
20.   Hamada, A.; Karjalainen, L. High-cycle fatigue behavior of ultrafine-grained austenitic stainless and TWIP steels. *Mater. Sci. Eng. A* **2010**, *527*, 5715–5722. [CrossRef]
21.   Song, S.W.; Lee, J.H.; Lee, H.J.; Bae, C.M.; Lee, C.S. Enhancing high-cycle fatigue properties of cold-drawn Fe–Mn–C TWIP steels. *Int. J. Fatigue* **2016**, *85*, 57–64. [CrossRef]
22.   Niendorf, T.; Klimala, P.; Maier, H.J.; Frehn, A. The Role of Notches on Fatigue Life of TWIP Steel in the HCF Regime. *Mater. Sci. Forum* **2012**, *706–709*, 2205–2210. [CrossRef]
23.   Shao, C.; Zhang, P.; Wang, X.; Wang, Q.; Zhang, Z. High-cycle fatigue behavior of TWIP steel with graded grains: breaking the rule of mixture. *Mater. Res. Lett.* **2018**, *7*, 26–32. [CrossRef]
24.   Shao, C.; Wang, Q.; Zhang, P.; Zhu, Y.; Zhao, Z.; Wang, X.; Zhang, Z. Improving the high-cycle fatigue properties of twinning-induced plasticity steel by a novel surface treatment process. *Mater. Sci. Eng. A* **2019**, *740-741*, 28–33. [CrossRef]
25.   Skorupski, R.; Smaga, M.; Eifler, D. Influence of Surface Morphology on the Fatigue Behavior of Metastable Austenitic Steel. *Adv. Mater. Res.* **2014**, *891–892*, 464–469. [CrossRef]
26.   Mughrabi, H. Cyclic Slip Irreversibilities and the Evolution of Fatigue Damage. *Metall. Mater. Trans. B* **2009**, *40*, 431–453. [CrossRef]
27.   Nikitin, I.; Scholtes, B. Deep rolling of austenitic steel AISI 304 at different temperatures–near surface microstructures and fatigue. *HTM J. Heat Treat. Mater.* **2012**, *67*, 188–194. [CrossRef]
28.   Webster, G.; Ezeilo, A. Residual stress distributions and their influence on fatigue lifetimes. *Int. J. Fatigue* **2001**, *23*, 375–383. [CrossRef]
29.   Teichmann, C.; Wagner, L. Shot peening of TWIP steel-influence on mechanical properties. In Proceedings of the ICSP-12, Goslar, Germany, 15–18 September 2014.
30.   Klein, M.; Smaga, M.; Beck, T. Surface Morphology and Its Influence on Cyclic Deformation Behavior of High-Mn TWIP Steel. *Metals* **2018**, *8*, 832. [CrossRef]

31. Abrão, A.; Denkena, B.; Köhler, J.; Breidenstein, B.; Mörke, T.; Rodrigues, P. The influence of heat treatment and deep rolling on the mechanical properties and integrity of AISI 1060 steel. *J. Mater. Proc. Technol.* **2014**, *214*, 3020–3030. [CrossRef]
32. Kloos, K.H.; Adelmann, J. Schwingfestigkeitssteigerung durch Festwalzen. *Materialwissenschaft und Werkstofftechnik* **1988**, *19*, 15–23. [CrossRef]
33. Nikitin, I.; Altenberger, I.; Cherif, M.A.; Juijerm, P.; Maier, H.J.; Scholtes, B. Festwalzen bei erhöhten Temperaturen zur Steigerung der Schwingfestigkeit. *HTM Härtereitechnische Mitteilungen* **2006**, *61*, 289–295. [CrossRef]
34. Rüsing, C.J. Optimierung der monotonen und zyklischen Eigenschaften von hoch manganhaltigen TWIP-Stählen–Einfluss von Temperatur und Vorverformung auf die Mikrostrukturentwicklung. Ph.D. Thesis, Fakultät für Maschinenbau, Universität Paderborn, Paderborn, Germany, 2015.
35. Cooman, B.C.D.; Estrin, Y.; Kim, S.K. Twinning-induced plasticity (TWIP) steels. *Acta Mater.* **2018**, *142*, 283–362. [CrossRef]
36. Kongthep, J.; Timmermann, K.; Scholtes, B.; Niendorf, T. On the impact of deep rolling at different temperatures on the near surface microstructure and residual stress state of steel AISI 304. *Materialwissenschaft und Werkstofftechnik* **2019**, *50*, 788–795. [CrossRef]
37. Nikitin, I.; Altenberger, I. Comparison of the fatigue behavior and residual stress stability of laser-shock peened and deep rolled austenitic stainless steel AISI 304 in the temperature range 25–600 °C. *Mater. Sci. Eng. A* **2007**, *465*, 176–182. [CrossRef]
38. Nikitin, I.; Altenberger, I.; Scholtes, B. Effect of Deep Rolling at Elevated and Low Temperatures on the Isothermal Fatigue Behavior of AISI 304. *Proc. Int. Conf. Shot Peen.* **2005**, *9*, 185–190.
39. Cherif, A.; Scholtes, B. Kombinierte thermische und mechanische Festwalzbehandlungen von gehärtetem und vergütetem Stahl Ck 45. *HTM Härtereitechnische Mitteilungen* **2008**, *63*, 155–161. [CrossRef]

![metals logo] *metals*

MDPI

Article

Computer-Aided Material Design for Crash Boxes Made of High Manganese Steels

Angela Quadfasel [1,*], **Marco Teller** [1], **Manjunatha Madivala** [2] ⓘ, **Christian Haase** [2] ⓘ, **Franz Roters** [3] and **Gerhard Hirt** [1]

1   Institute of Metal Forming (IBF), RWTH Aachen University, 52072 Aachen, Germany
2   Steel Institute (IEHK), RWTH Aachen University, 52072 Aachen, Germany
3   Max-Planck-Institut für Eisenforschung GmbH, 40237 Düsseldorf, Germany
*   Correspondence: angela.quadfasel@ibf.rwth-aachen.de

Received: 14 June 2019; Accepted: 4 July 2019; Published: 10 July 2019

Abstract: During the last decades, high manganese steels (HMnS) were considered as promising materials for crash-relevant automobile components due to their extraordinary energy absorption capability in tensile tests. However, in the case of a crash, the specific energy, absorbed by folding of a crash box, is lower for HMnS as compared to the dual phase steel DP800. This behavior is related to the fact that the crash box hardly takes advantage of the high plastic formability of a recrystallized HMnS during deformation. It was revealed that with the help of an alternative heat treatment after cold rolling, the strength of HMnS could be increased for low strains to achieve a crash behavior comparable to DP800. In this work, a multi-scale finite element simulation approach was used to analyze the crash behavior of different material conditions of an HMnS. The crash behavior was evaluated under consideration of material efficiency and passenger safety criteria to identify the ideal material condition and sheet thickness for crash absorption by folding. The proposed simulation methodology reduces the experimental time and effort for crash box design. As a result of increasing material strength, the simulation exhibits a possible weight reduction of the crash box, due to thickness reduction, up to 35%.

Keywords: high manganese steel; crash box; lightweight; multiscale simulation

---

1. Introduction

For crash-relevant components of the body-in-white design of vehicles, a combination of high strength and high plasticity is necessary to guarantee passenger safety and lightweight construction at the same time. This combination of mechanical material properties is provided by high manganese steels (HMnS). Due to the additional deformation mechanism of twinning, the group of TWIP (twinning-induced plasticity) steels exhibit the possibility of forming complex geometries in combination with a high impact-energy-absorption potential [1]. Strain hardening engineering can be applied on HMnS to adjust the hardening mechanism for ideal hardening behavior for specific applications. Bambach et al. [2] argued that a good crash performance for a crash box, built for frontal impact, is achieved by a high initial strength of HMnS. The works of Haase et al. [3,4] showed that recovery (RV) annealing is a simple, promising processing route to overcome the shortcoming of low yield strength usually associated with TWIP steels in recrystallized (RX) condition. While dislocation density of the cold rolled (CR) condition of a Fe-23Mn-1.5Al-0.3C is reduced during the RV annealing, the twin fraction remains constant, resulting in high yield strength with moderate formability. The material condition after RV annealing of Fe-23Mn-1.5Al-0.3C has been applied to the experimental crash test by drop tower tests and successfully increased the specific energy absorption of the crash box [2,5]. Due to the higher yield strength compared to the fully RX material condition, an increase of the energy absorption per mass was achieved and resulted in a shorter crash distance. In addition to material strength, the crash behavior and especially the buckling stress, which is necessary for initiation of the folds, is depending on the geometry of the crash boxes. Bambach et al. showed in their work [2] that the cross-section geometry of a crash box of the same material has a significant influence on the crashworthiness. While a square cross section needs a higher crash distance, the hexagon shape showed a performance nearly as good as a circular shape with the same cross-sectional area. Therefore, an additional degree of freedom exists in the design of a crash box in the form of the geometry. In this work, only a hexagonal cross section with constant diameter is used and the influence of geometry is only considered by a variation of the sheet thickness.

Experimental crash tests are very time consuming and expensive. In order to save costs, a computer-aided design of the ideal material condition and geometry design can be realized by finite element (FE) simulation in combination with sophisticated material models. In this work, an ICME (integrated computational material engineering) approach, depicted in Figure 1, is used to reduce the experimental effort of the crash box design. Material parameters of HMnS were calculated at different scales and used to predict the hardening behavior of different material conditions, which is finally used to simulate the crash behavior of a crash box. On the atomistic scale, the stacking fault energy of the HMnS was calculated by a combination of thermodynamic and ab initio calculations and was taken from [6]. The stacking fault energy (SFE) influences the activation of different deformation mechanisms, which are considered during the simulation of hardening behavior with a physics-based crystal plasticity (CP) model, which is described in detail in [7]. The flow curves of different material conditions are then used in FE simulation to predict the crash behavior. The consideration of the microstructure during FE simulation is necessary because the hardening behavior is significantly correlated to the microstructure and deformation mechanisms. Due to high deformation rates and high dissipation during the crash, the deformation mechanism can change. With the help of the ICME approach, different initial microstructures can be analyzed with respect to their influence on the crash behavior and give a direct insight into promising microstructure-design strategies.

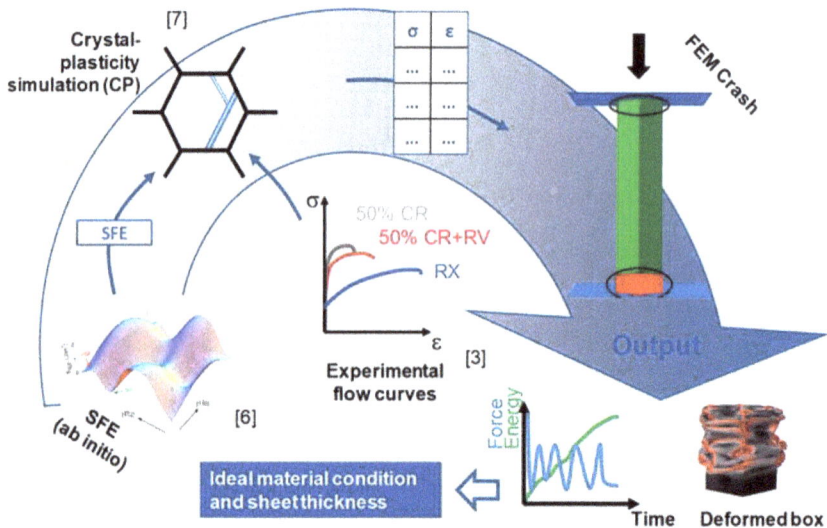

Figure 1. Multiscale simulation chain with ab initio calculation of stacking fault energy (SFE), crystal plasticity simulation for hardening behavior and finite element (FE) simulation for crash behavior. The references in the figure indicate the sources used for data acquisition.

From an economic point of view, a successfully validated simulation chain enables the reduction of experimental effort. Figure 2 compares the steps for an experimental and a more simulation-based design of a crash box. The traditional experimental-based design process is coupled to a high number of crash experiments for the different material conditions. In the process chain of a crash box, cold rolling and heat treatment are the relevant process steps to define the sheet thickness as well as the material condition and therefore, the mechanical properties. Additionally, all different material conditions have to be tested in tensile tests to make a first choice based on simplified criteria like ECO-index, the product of ultimate tensile strength and uniform elongation, or buckling stress [2]. Both criteria allow for a first approximation of the energy absorption of different materials, while the buckling stress also considers the simplified geometry of a crash box. However, both criteria cannot replace the analysis of crash behavior under all boundary conditions. Another criterion, for example, is the mean deceleration to ensure passenger safety. A maximum value of 15 g (gravitational acceleration) for deceleration was identified as a limit by biomechanics specialists to avoid long-term damages due to whiplash in [8]. In order to be able to evaluate the crash behavior of the different material conditions, it is necessary establish their performance in drop tower tests.

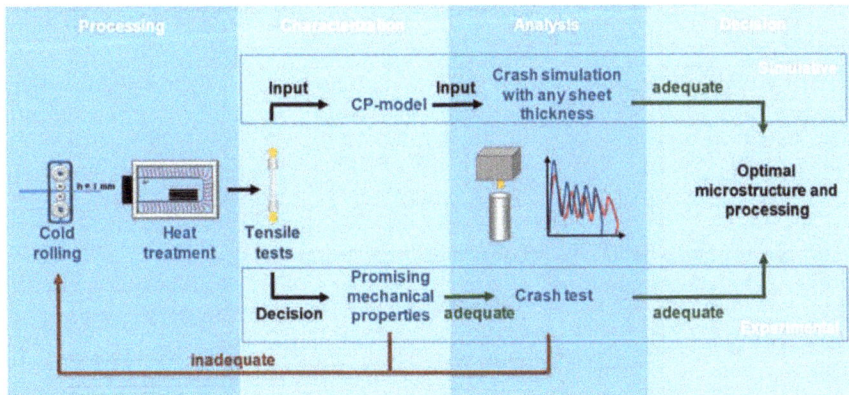

Figure 2. Schematic overview of the process chain and comparison of experimental and simulation-based design of material conditions for high manganese crash boxes.

The simulation-based design of a crash box should reduce the effort of drop tower tests and should give a direct correlation between crash behavior and microstructure. Based on that, a suggestion for the optimal processing strategy can be given. Also, for the simulation-based design, experimental tensile tests are necessary to validate the material model. However, the decision of which material conditions are suitable for the crash, can be made without further experimental crash tests. For the simulation-based design, only specific material conditions have to be tested in tensile tests to set up the material model, which can then simulate the plastic behavior of other material conditions. The optimal combination of material condition and box geometry can then be identified by simulation and, if necessary, validated for selected conditions with drop tower tests. In contrast, the experimental-based design needs an iterative adaption of the processing during the design process to identify an adequate material condition for the crash behavior.

In this work, the crash behavior of high manganese TWIP steel is optimized combining multi-scale simulation tools as well as knowledge from different material science and engineering disciplines to an ICME approach. Within this ICME approach, the crash behavior of the different material conditions was analyzed using flow curves of HMnS subjected to different degrees of cold rolling and additional RV annealing in FE simulation of drop tower tests. In addition to the previously mentioned criteria of crash distance, passenger safety in terms of deceleration was considered as a criterion for ideal crash performance. The aim of the work was to identify the ideal combination of material strength and sheet thickness of crash boxes to fulfill the criteria of passenger safety and lightweight construction at the same time.

2. Materials

2.1. Material and Characterization

The investigated material is a HMnS TWIP steel with 0.3 wt.% carbon, 23 wt.% manganese, 1 wt.% aluminum (X30MnAl23-1), and a stacking fault energy of ~25 mJm^{-2} [3]. The material was cast as 100 kg ingots at the Steel Institute (IEHK) of Rheinisch-Westfälische Technische Hochschule (RWTH) Aachen University (Aachen, Germany), homogenization-annealed at 1150 °C, and forged at 1150 °C at the Institute of Metal Forming (IBF). After an additional homogenization annealing, the 55 mm thick slabs were hot rolled at 1150 °C to 2.4 mm and then cold rolled in several passes to final thickness reductions of 30%, 40%, and 50%. Heat treatment parameters to achieve the recovered material condition for the different rolling degrees and the recrystallized material condition are based on the work of Haase et al. [3], given in Table 1.

Table 1. Cold rolling degree and heat treatment parameters for recovery annealing and recrystallization. CR: cold rolled; RV: recovery; RX: recrystallized.

Material Condition	Rolling DEGREE (%)	Annealing Temperature (°C)	Annealing Time (min)
30% CR + RV	30	630	10
40% CR + RV	40	550	60
50% CR + RV	50	550	30
50% CR + RX	50	700	10

Uniaxial tensile tests (IEHK, Aachen, Germany) were performed to validate the material model for the different conditions. The result of two experimentally tested material conditions 30% CR and 40% CR + RV, with similar yield strength in the tensile test but different hardening rates are shown in Figure 3a. The experimental stress–strain curves for the different material conditions after recovery annealing with different yield strength are illustrated in Figure 3b. Moreover, simulated stress–strain curves, which are described in detail in Section 2.2, are shown.

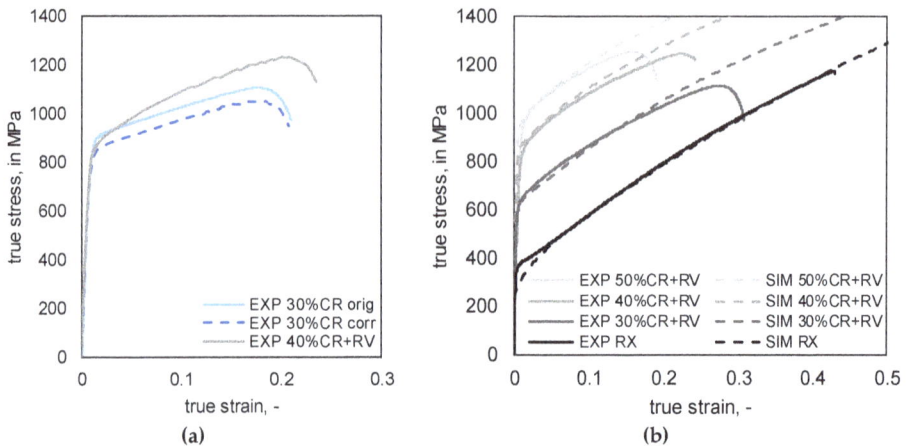

Figure 3. (**a**) True stress–strain curves of 30% CR and 40% CR + RV from experiments with similar yield strength but different hardening behavior with original data from tensile test (orig) and corrected data with offset (corr); (**b**) True stress–strain curves from tensile tests (EXP) and crystal plasticity (CP) simulation (SIM) of recovery annealed (RV) and recrystallized (RX) material conditions.

2.2. Simulation of Plastic Behavior

The plastic deformation behavior of the high manganese steel in FE simulation can be described by the true stress–strain curve from the tensile test. For the investigation of the influence of hardening rate on the crash behavior, the flow curve of 30% CR (orig) in Figure 3a has been reduced manually (corr) by an offset of 50 MPa to achieve the same yield strength as 40% CR + RV.

The main advantage of the usage of a universal material model in the simulation chain is that different material conditions can be derived from one set of parameters after model calibration. The stress–strain behavior of the different material conditions after recovery annealing, as shown in Figure 3b, was simulated with the help of a representative volume element (RVE) using the spectral solver of the crystal plasticity (CP) package DAMASK (Düsseldorf Advanced Material Simulation Kit) developed by Max-Planck-Institut für Eisenforschung GmbH (Düsseldorf, Germany) [9]. In addition to dislocation density, the CP model considers the evolution of the twin volume fraction. Due to the dependence of twinning on the SFE, this temperature dependent parameter was calculated by a combination of Calculation of Phase Diagrams (CALPHAD) method and ab initio calculations [6]. The CP model was used to simulate the deformation behavior under plane strain compression conditions and resulted in the determination of the dislocation densities and twin volume fractions after cold rolling degrees of 30%, 40%, and 50%. A similar procedure is described in [3,10,11]. Afterwards, the corresponding dislocation densities for each cold rolling degree were used as input parameters for full field RVE simulations (using 100 grains) of the deformation behavior under uniaxial tensile loading conditions.

3. Methods

3.1. Drop Tower Experiments

To validate the FE model used for simulation of the drop tower test, experimental crash tests were performed by the Forschungsgesellschaft Kraftfahrwesen mbH (FKA) (Aachen, Germany). For production of the crash boxes, the material was cold rolled to 1 mm sheet thickness with a thickness reduction of 50%. After heat treatment (RX and RV), 275 mm × 14 mm samples were bent to half hexagons, which were laser welded at the Fraunhofer Institute for Laser Technology (ILT) (Aachen, Germany) to hexagonal crash boxes with a diameter of 82 mm. The physical simulation of crash deformation of the box during uniaxial loading was performed using a drop tower. A mass of 253 kg and a height of 2.0 m resulting in a kinetic energy of ~5 kJ and an impact velocity close to 22 km/h were chosen for the tests. The crash box was fixed on the non-impact side. The top of the crash box was triggered by alternate bending of the edges 20 mm from the upper border, resulting in the opposite edges being bent in the same direction. The trigger is supposed to ease the initial bending, which leads to more regular folding.

3.2. FE Simulation of Drop Tower Test

The explicit FE simulation of the drop tower test was performed by the software ABAQUS 6.14-6 from Dassault Systemes Simulia Corp. (Johnston, IA, USA). Due to the high velocity of the crash tests, no mass scaling was needed for the performance of the simulations. The model consisted of a 3D deformable shell model of the hexagonal geometry meshed by 4-node general-purpose shell elements with reduced integration (S4R) of 2 mm edge length with hourglass control. The velocity of the 253 kg drop mass was set to 6.2 m/s for the moment of contact with the crash box. The contact between crash box and drop mass was defined as hard contact. The deceleration velocity of the drop mass was calculated in FE simulation by the converse effects of energy absorption of the crash box and the acceleration by gravity.

The geometry of the hexagonal crash box in Figure 4a,b was extended by a trigger following the trigger of the experimental crash boxes. In addition, a pre-strain, calculated by bending theory, at the vertical bending edges was used to consider the hardening due to bending during the production of the crash boxes. The plastic behavior of the simulated materials was assumed to be isotropic and was defined by the tabulated stress–strain curves from CP simulations given in Figure 3. The material behavior was measured under quasi-static conditions but used in a process with high strain rate, i.e., the influence of the strain rate on the yield strength of HMnS was neglected in this work.

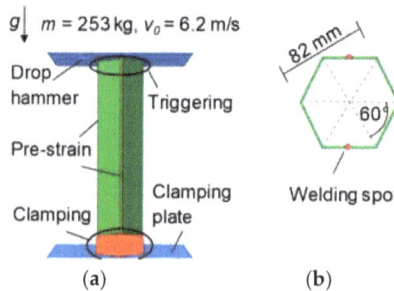

Figure 4. (**a**) Crash box geometry with boundary conditions of the FE simulation; (**b**) Hexagonal profile in mm with marked plane where the half hexagons were laser welded.

Simulations of drop tower tests were performed for the material conditions RX, 30% CR, 50% CR + RV, 40% CR + RV, and 30% CR + RV in a first step by applying the original crash box geometry with a sheet thickness of 1 mm. The crash behavior of the different material conditions was then compared under consideration of the criteria of final crash distance and mean deceleration during the simulated test. The mean value of the deceleration was considered neglecting the sinus shape of the deceleration curve assuming that, due to damping and short effective time, maxima do not affect the passengers' bodies [12]. In the next step, the sheet thickness of the crash boxes was individually reduced for each material condition except for the RX condition. The aim of the reduction of the sheet thickness was to utilize the full length of the crash box as a deformation path for energy absorption.

4. Results

4.1. Validation of FE Crash Model

The comparison of force–distance curves of the material conditions RX and 50% CR + RV from drop tower simulations and experiments is shown in Figure 5a. Those two material conditions, crushed with the original geometry of 1 mm sheet thickness, were used to validate the FE model for the parameter study. The good agreement between simulation and experiment with respect to the final crash distance revealed the applied pre-strain and geometrical trigger in the FE model to be valid for the different material conditions. Figure 5b shows the final geometries of the crash boxes from the simulation with the final crash distance. For the RX condition, the whole crash box length was used for folding, using the full energy absorption potential of this crash box geometry, while for the RV condition half of the crash box length remained unfolded.

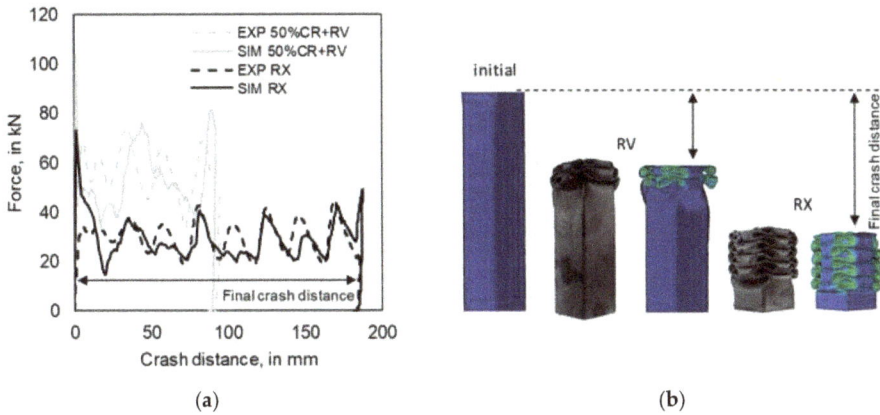

Figure 5. (**a**) Experimental and simulated force–distance curves from drop tower tests with sheet thickness t = 1 mm of the RX and 50% CR + RV conditions; (**b**) Geometries of the crash boxes before (initial) and after simulated and experimental drop tower tests of the RX and 50% CR + RV conditions.

4.2. Influence of Material Condition on Behavior for Constant Sheet Thickness

Using the experimental stress–strain curves from Figure 3a, the crash behavior of 30% CR and 40% CR + RV was simulated. The results are illustrated in Figure 6a,b. Due to recovery annealing of the 40% CR condition, the density of dislocations is reduced and strain hardening appears stronger than in the case of 30% CR. Using flow curves of both conditions in FE crash simulation of crash boxes with 1 mm sheet thickness shows that there is only a slight influence of hardening rate on the final crash distance and deceleration. The final crash distance is reduced and the mean deceleration is increased for the 40% CR + RV condition with the higher hardening rate.

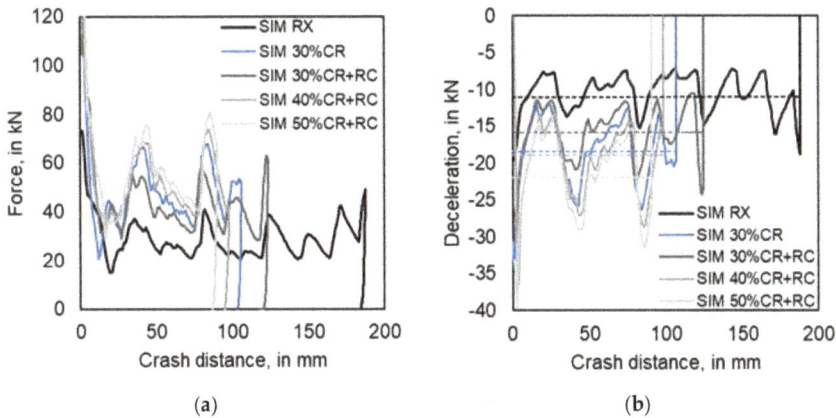

Figure 6. (**a**) Simulated force–distance curves of drop tower tests with sheet thickness t = 1 mm for different material conditions; (**b**) Simulated deceleration during crash for the different material conditions with mean deceleration (dashed lines).

In addition to the hardening rate, the influence of the yield strength on the simulated crash behavior was investigated with different combinations of cold rolling degrees and recovery annealing. With increasing rolling degree, the material's yield strength in Figure 3b increases, since the twin volume fraction increases with the increase of strain and remains stable during the recovery annealing [3]. In Figure 6a,b it is shown that with increasing strength of the material the final crash distance decreases and the mean deceleration increases. In Figure 7a both criteria are plotted for all material conditions including the limit of 15 g for deceleration. Only the RX condition satisfies the criterion of maximum deceleration. As a result, lightweight construction realized by reducing the necessary crash distance by increasing the strength of the material is not feasible due to the considered safety criterion, i.e., deceleration <15 g.

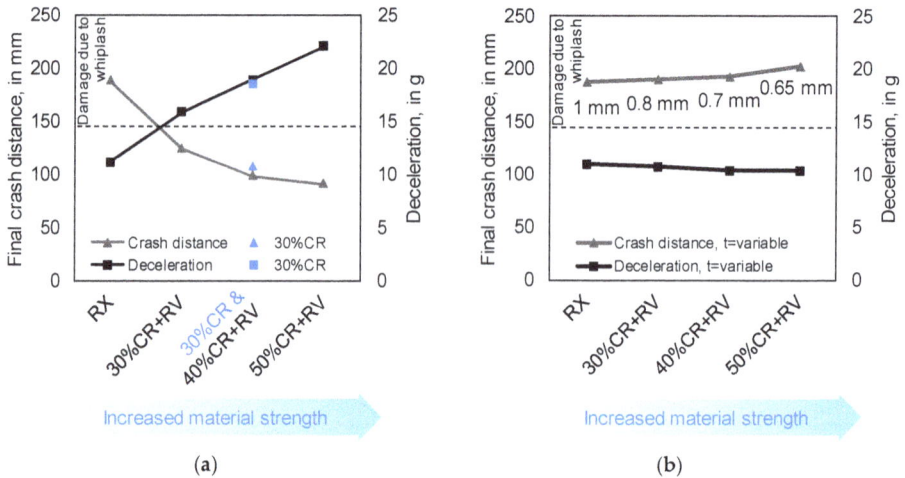

Figure 7. (**a**) Final crash distance and mean deceleration from FE parameter study for t = 1 mm for all material conditions; (**b**) Final crash distance and mean deceleration from FE parameter study for individual t for each material condition.

4.3. Optimization of Material and Sheet Thickness Combination for Crash Box

For the optimization of the combination of material condition and geometry for ideal crash behavior, the possible material conditions for crash boxes were reduced to the recovery-annealed conditions, as these conditions show a significant influence on final crash distance and enough strain reserves to guarantee a failure-free folding during the crash.

The sheet thickness of each material condition was reduced individually in order to enable the usage of the complete length of the crash box as well as reduced mean deceleration. In Figure 7b it is shown that with increasing material strength, the sheet thickness can be reduced to 0.8 mm for 30% CR + RV, 0.7 mm for 40% CR + RV, and 0.65 mm for 50% CR + RV, while using the complete length of the crash box as crash distance, similar to the RX condition, and while keeping the deceleration below 15 g.

The different material conditions with the according yield strength predicted by the CP simulations, the reduced sheet thickness, and the possible weight reduction are given in Table 2. A significant weight reduction of the crash boxes was achieved by reducing the sheet thickness of each material condition. For instance, the combination of 874 MPa yield strength of the 50% CR + RV condition and reduction of sheet thickness results in a weight reduction of 35%.

The final crash box geometries after drop tower simulation of the material conditions with reduced sheet thickness are shown in Figure 8. However, it was observed that with increasing material strength and decreasing sheet thickness, the pattern of the folds became more irregular.

Table 2. Yield strength of simulated flow curves of different material conditions, individual sheet thickness and weight reduction of different material conditions.

Material Condition	Yield Strength SIM (MPa)	Sheet Thickness (mm)	Weight Reduction (%)
50% CR + RX	256	1.00	Reference
30% CR + RV	640	0.80	20
40% CR + RV	819	0.70	30
50% CR + RV	874	0.65	35

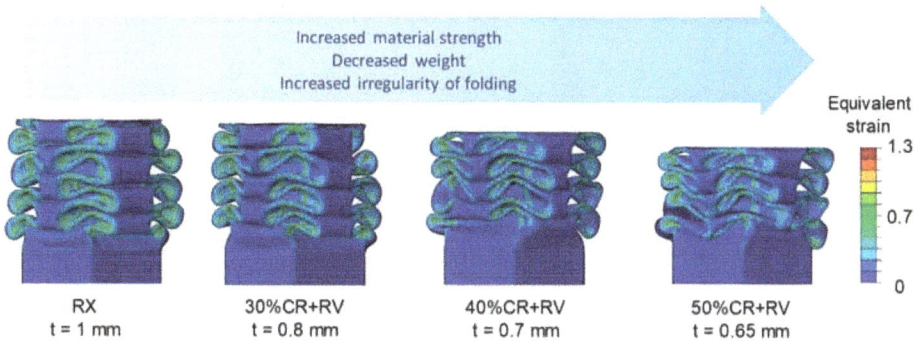

Figure 8. Final crash box geometry after crash simulation for individual t for each material condition.

5. Discussion

Different mechanical properties of the investigated X30MnAl23-1 HMnS could be designed by different procedures of cold rolling and heat treatment from Table 1. There is the possibility of adjusting the yield strength as well as the hardening rate. With increasing hardening rate, the final crash distance decreases but the influence of hardening rate on crash behavior is not very significant. Due to the dependence of the mechanical properties of the different material conditions on the process-specific microstructures, the range of different hardening rates combined with a similar yield strength are not independently adjustable. The yield strength of the different material conditions has a more pronounced influence on the crash behavior, as it significantly influences the final crash distance and therefore the deceleration. Moreover, the yield strength can be adjusted in a wider range by the degree of cold rolling.

Increasing the material strength by cold rolling, without consideration of crash box geometry, can successfully decrease the final crash distance and therefore contributes to the lightweight design by reducing the length of the crash box. However, as the FE simulation of a drop tower test showed, decreasing the final crash distance results in increased deceleration at the same time, which endangers the passenger safety due to increased risk of whiplash. The adaption of the sheet thickness as an additional degree of freedom in crash box geometry design can be used to optimize crash behavior for the different material conditions in order to fulfill all safety criteria. In this case, lightweight design is realized by reducing the sheet thickness, while using the original crash box length for folding. Another criterion, which has to be taken into account during optimization of material strength and sheet thickness, is the irregularity of the folding. To ensure regular folding for reproducible energy absorption, the sheet thickness can only be reduced within certain limits.

Based on tailored processing, the materials' properties can be adjusted with respect to strength and strain-hardening potential. The strain-hardening behavior, which influences the crash behavior, can be described by the CP model. By applying the FE simulation, it is possible to predict the crash behavior in terms of the safety limit of 15 g deceleration and lightweight criteria, taking into account the influence of microstructure and box geometry. As a result, the experimental effort for crash box design could be reduced, since not every combination of material condition and sheet thickness has to be processed and tested in drop tower tests. With the correlation between processing and microstructure, the CP simulation can be used in further works to simulate material conditions, which have not been tested in tensile tests, to investigate the hardening behavior and increase the reduction of the experimental effort for the design of material conditions.

6. Conclusions

In this study, strain-hardening engineering was used to design an HMnS crash box by the application of a combination of multiscale characteristics. Considering safety and lightweight criteria, the ideal combination of material condition and sheet thickness was identified.

With the help of the simulation-based crash box design, the experimental effort could be reduced significantly and a reduction of weight up to 35% could be achieved by combining ideal material condition and sheet thickness.

Due to the multiscale simulation of CP simulation and FE simulation, the macroscopic crash behavior can be correlated with the microstructure, represented by the relation of dislocation density and twin density, and therefore with the processing of the materials.

Author Contributions: Conceptualization, A.Q., C.H., F.R. and M.T.; software, F.R.; methodology, A.Q. and M.M.; investigation, A.Q., and M.M.; writing—original draft preparation, A.Q.; writing—review and editing, M.T., C.H., F.R., M.M., and G.H.; visualization, A.Q.; supervision, M.T., and G.H.; project administration, G.H. and M.T.; funding acquisition, G.H.

Funding: This research was funded by the "Deutsche Forschungsgemeinschaft" (DFG) within the Sonderforschungsbereich (Collaborative Research Center) 761 "Steel–ab initio" with project number 29898171 within a collaboration of part project B2, B4, A7, and C6.

Conflicts of Interest: The authors declare no conflict of interest. The funders had no role in the design of the study; in the collection, analyses, or interpretation of data; in the writing of the manuscript, or in the decision to publish the results.

References

1. De Cooman, B.C.; Kwon, O.; Chin, K.-G. State-of-the-knowledge on TWIP steel. *Mater. Sci. Technol.* **2012**, *28*, 513–527. [CrossRef]
2. Bambach, M.; Conrads, L.; Daamen, M.; Güvençb, O.; Hirt, G. Enhancing the crashworthiness of high-manganese steel by strain-hardening engineering and tai-lored folding by local heat-treatment. *Mater. Des.* **2015**, *110*, 157–168. [CrossRef]
3. Haase, C.; Barrales-Mora, L.A.; Molodov, D.A.; Roters, F.; Gottstein, G. Applying the texture analysis for optimizing thermomechanical treatment of high manganese twinning-induced plasticity steel. *Acta Mater.* **2014**, *80*, 327–340. [CrossRef]
4. Haase, C.; Barrales-Mora, L.A.; Molodov, D.A.; Gottstein, G. Tailoring the Mechanical Properties of a Twinning-Induced Plasticity Steel by Retention of Deformation Twins During Heat Treatment. *Metall. Mater. Trans. A* **2013**, *44*, 4445–4449. [CrossRef]
5. Haase, C.; Ingendahl, T.; Güvenc, O.; Bambach, M.; Bleck, W.; Molodova, D.A.; Barrales-Mora, L.A. On the applicability of recovery-annealed Twinning-Induced Plasticity steels: Potential and limitations. *Mater. Sci. Eng. A* **2015**, *649*, 74–84. [CrossRef]
6. Güvenç, O.; Roters, F.; Hickel, T.; Bambach, M. ICME for Crashworthiness of TWIP Steels: From Ab Initio to the Crash Performance. *JOM* **2015**, *67*, 120–128. [CrossRef]
7. Wong, S.L.; Madivala, M.; Prahl, U.; Roters, F.; Raabe, D. A crystal plasticity model for twinning- and transformation-induced plasticity. *Acta Mater.* **2016**, *118*, 140–151. [CrossRef]
8. Schmidt, G. Grundlegendes zum Unfallmechanismus. In *Die Beschleunigungsverletzung der Halswirbelsäule*; Moorahrend, U.: Stuttgart, Germany, 1993; pp. 25–37.
9. Roters, F.; Diehl, M.; Shanthraj, P.; Eisenlohr, P.; Reuber, C.; Wong, S.L.; Maitic, T.; Ebrahimid, A.; Hochrainere, T.; Fabritiusa, H.-O.; et al. DAMASK-The Düsseldorf Advanced Material Simulation Kit for modeling multi-physics crystal plasticity, thermal, and damage phenomena from the single crystal up to the component scale. *Comput. Mater. Sci.* **2019**, *158*, 420–478. [CrossRef]

10. Diehl, M.; Groeber, M.; Haase, C.; Molodov, D.A.; Roters, F.; Raabe, D. Identifying Structure–Property Relationships Through DREAM.3D Representative Volume Elements and DAMASK Crystal Plasticity Simulations: An Integrated Computational Materials Engineering Approach. *JOM* **2017**, *69*, 848–855. [CrossRef]

11. Haase, C.; Kühbach, M.; Barrales-Mora, L.A.; Wong, S.L.; Roters, F.; Molodova, D.A.; Gottstein, G. Recrystallization behavior of a high-manganese steel: Experiments and simulations. *Acta Mater.* **2015**, *100*, 155–168. [CrossRef]

12. Gauss, W. Unfallrekonstruktion und biomechanische Begutachtung bei HWS-Verletzungen durch Heckaufprall. In *Beschleunigungsverletzung der Halswirbelsäule*; Graf, M., Grill, C., Eds.; Wedig, H.-D.: Darmstadt, Germany, 2009; pp. 78–91.

![metals logo] *metals*

MDPI

Article

Comparison of the Dislocation Structure of a CrMnN and a CrNi Austenite after Cyclic Deformation

Rainer Fluch [1,*], Marianne Kapp [1], Krystina Spiradek-Hahn [2], Manfred Brabetz [2], Heinz Holzer [3] and Reinhard Pippan [4]

[1] Product and Process Development, voestalpine BÖHLER Edelstahl GmbH & Co KG, 8605 Kapfenberg, Austria
[2] AGD Seibersdorf, University of Leoben, 2444 Seibersdorf, Austria
[3] Österreichisches Gießerei-Institut, University of Leoben, 8700 Leoben, Austria
[4] Erich Schmid Institute of Materials Science, Austrian Academy of Sciences, 8700 Leoben, Austria
* Correspondence: rainer.fluch@bohler-edelstahl.at; Tel.: +43-3862-20-36062

Received: 14 May 2019; Accepted: 19 June 2019; Published: 13 July 2019

Abstract: In the literature, the effects of nitrogen on the strength of austenitic stainless steels as well as on cold deformation are well documented. However, the effect of N on fatigue behaviour is still an open issue, especially when comparing the two alloying concepts for austenitic stainless steels—CrNi and CrMnN—where the microstructures show a different evolution during cyclic deformation. In the present investigation, a representative sample of each alloying concept has been tested in a resonant testing machine at ambient temperature and under stress control single step tests with a stress ratio of 0.05. The following comparative analysis of the microstructures showed a preferred formation of cellular dislocation substructures in the case of the CrNi alloy and distinct planar dislocation glide in the CrMnN steel, also called high nitrogen steel (HNS). The discussion of these findings deals with potential explanations for the dislocation glide mechanism, the role of N on this phenomenon, and the consequences on fatigue behaviour.

Keywords: austenitic high nitrogen steel (HNS); cold deformation; fatigue

---

1. Introduction

In the last twenty five years, a lot of research has been conducted regarding interstitially alloyed high nitrogen steels (HNS), especially on high nitrogen austenitic steels, which still have not emerged from their status as highly specialised products for niche markets where a combination of strength, toughness, corrosion resistance, and non-magnetic behaviour is required. Those classical applications are medical implants, drill collars, and retaining rings. To achieve the high strength values required for these applications ($R_m > 1000$ MPa), a further characteristic of that alloying system is taken advantage of: its exceptional work hardening rate. As a consequence, the alloys are often deployed in a work hardened condition. Additionally, it is common requirement for all these applications that the material has to withstand cyclic loads. Therefore, fatigue resistance is an additional important characteristic of these alloys.

In the case of austenitic steels, the arrangement of dislocations during cyclic deformation is of crucial significance to their fatigue behaviour. The first attempts to define this behaviour in alloys with face centered cubic (FCC) microstructure through their dislocation structures independent of the stacking fault energy (SFE) were done by Feltner and Laird [1] and Lukáš and Klesnil [2]. They distinguished between wavy slip for high SFE and planar slip for low SFE materials. Wavy slip implies that cross slip of dislocations can occur easily; in the case of planar slip, cross slip is impeded. Figure 1a shows the "dislocation distribution map" proposed by Lukáš and Klesnil [2] (from Reference [3]).

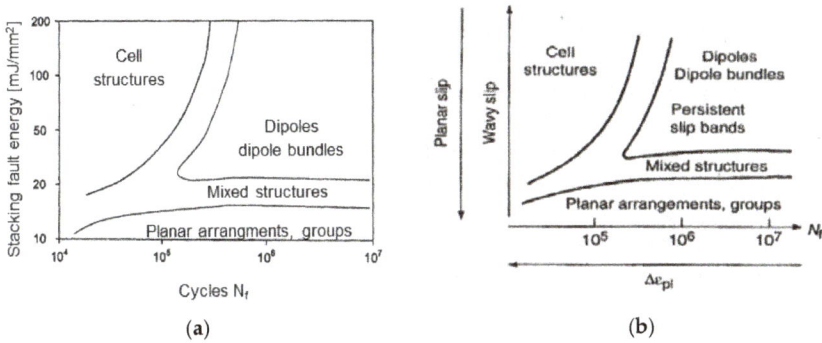

Figure 1. (a) Dependence of stacking fault energy on cyclic slip mode and fatigue life [3], with permission from DGM, 2019; (b) modified diagram proposed by Mughrabi [4], with permission from Elsevier, 2019.

Parallel to this observation, many authors have reported that nitrogen reduces the stacking fault energy in austenitic steels [5–9]. The majority of those investigations have been carried out on CrNi-austenitic steels [10–12]. However, there have been other studies which described an increase of stacking fault energy with N [13], or even a non-monotonous dependence [14]. Moreover, in recent publications it was stated that—in contrast to twinning induced plasticity (TWIP) and transformation induced plasticity (TRIP) steels—for the calculation of SFE in austenitic alloys containing N, the existing methods are is still not reliable [15–17].

Therefore, as far as N is concerned, no clear relationship with the stacking fault energy could be established. A different approach to explain the planar dislocation behaviour of austenitic steels is through short range ordering (SRO), as was already mentioned by Douglas et al. [18] in 1964. Gerold and Karnthaler stated that there might be an overestimation of the influence of SFE on the mobility of dislocations [19]. They mentioned that SRO leads to a glide plane softening effect in solid solution. SRO has also been discussed by Grujicic et al. [20] and Gavriljuk and Berns [21] in connection with high nitrogen austenitic steels, which would explain the planar slip behaviour of high nitrogen austenitic steel and, as a consequence, its high work hardening rate. Based on these findings, Mughrabi suggested a modified version of Figure 1a where he replaced the stacking fault energy on the y-axis, indicating that wavy slip is dominating higher in the diagram, and planar slip is preferred towards the bottom of the diagram (Figure 1b) [4].

Comparing the two alloying concepts for austenitic stainless steels, different mechanisms for dislocation slip can be identified:

- Wavy slip due to high SFE is expected for CrNi alloys.
- Planar slip appears in CrMnN alloys which show beside a lower SFE also SRO owing to the interstitially alloyed nitrogen

The goal of the investigation was to analyse the dislocation slip behaviours of a CrNi alloy and a CrMnN alloy in a cold worked (CW) condition under cyclic loading.

2. Materials and Methods

2.1. Test Materials

The investigated materials were a CrNi alloy, of the standard austenitic grade UNS S31673, and a CrMnN alloy. The exact chemical composition can be found in Table 1. Additionally, the SFE values were calculated with formula (1) from Reference [22], which Noh et al. [23] found to be the most adequate for different CrNiMnN alloys.

$$SFE = 5.53 + 1.4[Ni] - 0.16[Cr] + 17.1[N] + 0.72[Mn] \tag{1}$$

For CrMnN, the SFE = 35 mJ·m^{-2}, which is a high value and would assume a cellular gliding for dislocations (Figure 1a). This would be in contrast to the above mentioned literature findings where mainly planar glide was reported for CrMnN steels.

The rods of the test materials were hot rolled to a 26.5 mm diameter, solution annealed (SA), and machined. Cold deformation was performed by cold drawing with a reduction of area of 14%, which resulted in residual tensile stresses in the tangential and also axial directions. The residual stresses have been observed by spring back after cutting exposure samples but have not been measured. The grain size of the CrMnN steel was approximately 40 μm, and the grain size of the reference grade CrNi was approximately 80 μm [12]. No precipitations were detected. The mechanical properties of the tested materials in solution annealed and cold worked (CW) states at room temperature under atmospheric conditions are presented in Table 2.

Table 1. Chemical composition of the test grades in wt%.

Grade	%C	%Mn	%Cr	%Mo	%Ni	%N	SFE [mJ·m^{-2}]
CrNi S31673	<0.04	1.80	17.50	2.80	14.70	0.10	26
CrMnN	<0.04	23.00	21.10	0.30	1.50	0.85	35

Table 2. Mechanical properties of tested grades in solution annealed (SA) and cold worked (CW) condition at room temperature.

Grade	Condition	$R_{p0.2}$ [MPa]	R_m [MPa]	A_5 [%]
S31673	SA	410	637	54
	CW	653	718	32
CrMnN	SA	563	949	62
	CW	941	1165	38

2.2. Fatigue Experiments

A Rumul Testronic 8601/44 resonant testing machine with a nominal load of 100 kN was used. The sample geometry can be gathered from Figure 2.

As testing conditions, the following parameters were chosen:

Frequency: $f \approx 170$ Hz
Temperature: $T = 23\,°C$
Max. load cycles: $N = 10^7$
Stress ratio: $R = 0.05$

A resonant testing machine reacts to changes in the system, like cracks, by a shift in the natural resonance. Therefore, a frequency drop of 0.2 Hz was defined for the shutdown parameter. The fatigue limit was set as the maximum stress σ_{max} where three specimens did not show rupture after 10^7 cycles. In order to avoid a rise in temperature caused by the high frequency during cycling in the gauge length, the sample geometry was adapted to an hourglass-shaped gauge length with a radius of 40 mm (R40). Additionally, the samples were purged with air for cooling. The maximum temperature measured with an emission-adsorption pyrometer during cycling was 60 °C. No difference between the two alloys concerning heating behaviour could be detected.

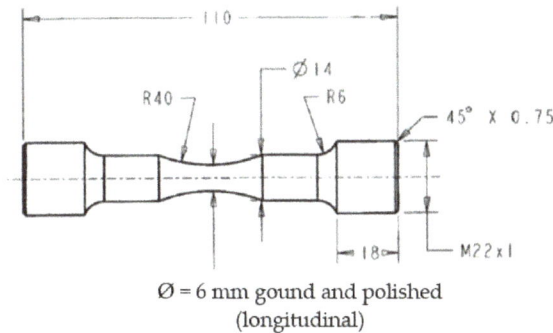

Ø = 6 mm gound and polished
(longitudinal)

Figure 2. Drawing of the sample geometry.

2.3. TEM Investigations

The electron microscopic investigations have been carried out on thinned foil of the cross section of cold worked fatigue specimens with a transmission electron microscope Philips CM20 STEM and Tecnai F20 (high resolution Philips, FEI Company, Eindhoven, the Netherlands). The acceleration voltage was 200 kV. Thinning was achieved by mechanical thinning followed by electrolytic polishing. The samples were taken transverse to the axis, one from the grip section as an unloaded reference and one from the gauge section after 10^7 cycles. The CrNi sample was loaded with a maximum stress of 446 MPa, and the CrMnN sample with 526 MPa. The dislocation structure was analysed by TEM with bright and dark field imaging. To determine the crystallographic parameters, electron diffraction was applied.

3. Results and Discussion

3.1. Fatigue Experiments

Figure 3 displays the fatigue results of S31673 in the solution annealed and cold worked condition. The specimen which lasted till 10^7 cycles did not show rupture as the loading set up shut down after that number of cycles. The fatigue limit for the 14% CW specimen was about 50 MPa higher than for the solution annealed specimen. This value represents well the curve of the CW material, which seems almost parallel to the solution annealed curve in the tested area. Considering the $R_{p0.2}$ value—often looked to as the reference strength for the fatigue limit, and which is 243 MPa higher for the CW than for the SA condition—it is obvious that a softening of the CW material occurs during cyclic loading. This softening of cold hardened austenitic steels has been described in the literature [24–26] and has been attributed to the reversibility of dislocation slip.

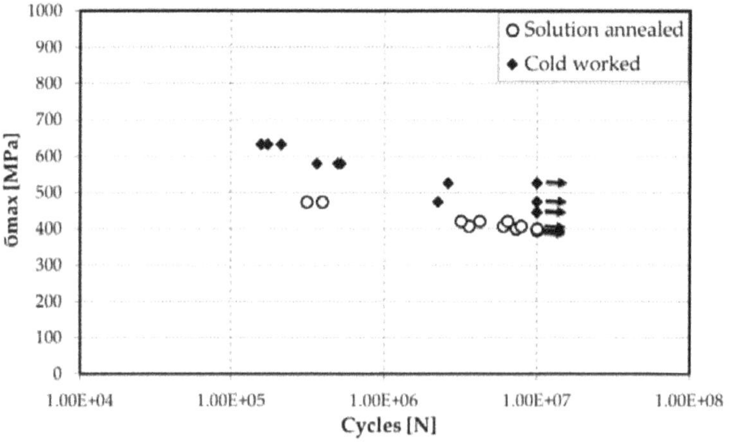

Figure 3. Plot of an S-N curve for S31673 at room temperature in the solution annealed and cold worked condition.

As expected, the CrMnN grade showed higher fatigue limits (>500 MPa, Figure 4) than S31673 since this grade possesses higher strength owned especially to the high interstitial alloyed N content. Although $R_{p0.2}$ of the CW material is 378 MPa higher, the fatigue limit at 10^7 cycles is not significantly higher than for the SA material. This implies that the reduction of the fatigue strength relating to $R_{p0.2}$ of cold worked material during cycling loading is more pronounced than for the CrNi grade. As mentioned earlier, this occurrence can be explained by the different dislocation slide behaviour of the two alloys, which leads to a higher reduction of dislocation density in the CrMnN steel. Nonetheless, at lower cycle numbers (e.g., 5×10^6) the cold worked CrMnN material shows higher fatigue stress values than the SA material.

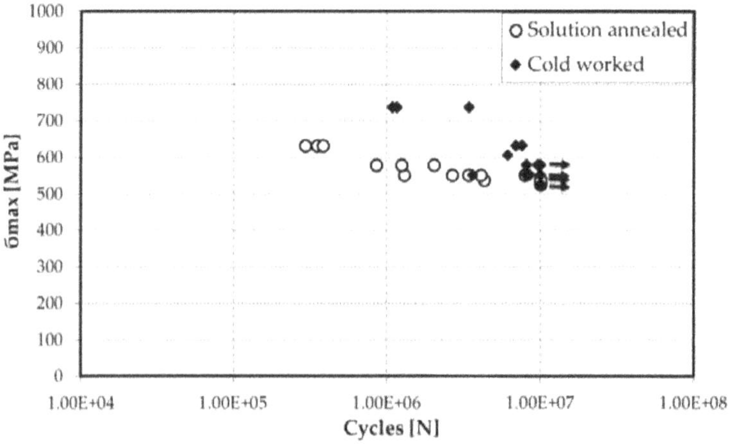

Figure 4. Plot of an S-N curve for CrMnN at room temperature in the solution annealed and cold worked condition.

3.2. TEM Investigations

To get a better understanding of the fatigue results, and to gain insight of the dislocation structures, TEM investigations have been carried out. The main points of interest were the cold deformed specimens and the different behaviour of the two alloying systems, CrNi and CrMnN, under cyclic loading. Thus, the TEM investigations were executed on a S31673 and a CrMnN fatigue sample, both previously 14% cold worked.

The images of the samples taken from the unloaded head of the cold worked fatigue samples show the following microstructural features:

S31673 exhibits a high dislocation density and, in some areas, twinning with tendencies of cell formation between the twins (see marked area in Figure 5a). The twins are relatively broad and can be found in two orientation systems. To some extent, planar slip is also recognizable.

The dislocation density in CrMnN specimen is significantly higher than in the CrNi grade, and the dislocations are arranged along planar dislocation glide systems which constitute the predominant structure Figure 5b. The glide systems are arranged in three directions parallel to the {111} planes. Additionally, a high twin density can be found where the twins are thin compared to S31673.

To investigate the influence of the cyclic load on the dislocation structure of the 14% CW material, samples from the gauge length were investigated.

In the S31673 sample, more distinct cell structures were observed (see Figure 5c). Only residuals of the planar slip system with a low dislocation density in the area between can be found compared to the images of the unloaded S31673 sample.

No obvious differences could be found in the fatigued CrMnN specimen compared to the unloaded CrMnN specimen (from the head) (Figure 5d). The planar glide bands are arranged along the {111} planes. To some extent, there is room for the interpretation that the planes are not as defined as in the unloaded specimen. In Figure 5e, no distinct glide bands are visible, but a dislocation accumulation in one direction (red arrow) can be seen. The red ellipses indicate the beginnings of cell formation.

To summarise the findings of the TEM investigations, it can be said that there is a clear difference in the dislocation structures between the two alloying systems in the cold worked condition. S31673 shows a pronounced cell structure of the cold deformed material, which is caused by wavy slip and is attributed to the high stacking fault energy of CrNi-alloys. Under cyclic load, this dislocation structure is further amplified. Whereas for the cold worked CrMnN grade, a dominant planar dislocation arrangement occurs, which can be found in both states and is attributed mainly to short range ordering. This is in contrast to the results of the calculation of the SFE in Table 1, where the CrMnN alloy has the higher SFE with 35 mJ·m^{-2} compared to CrNi with SFE = 26 mJ·m^{-2}. This proves that calculating the SFE is not a practicable way to estimate the gliding behaviour of N alloyed steels.

On the one hand, this planar slip is responsible for the high cold hardening rate [21], but in the case of cyclic loading, it has been suggested [27] that this planar dislocation arrangement promotes the reversibility of dislocation glide and, in consequence, to the annihilation of dislocations, which could justify the distinct cyclic softening of the tested CrMnN. Nevertheless, a softening occurs also in the case of S31673 considering that Rm rises 130 MPa from the solution annealed to the cold worked condition, whereas the fatigue limit σ_{max} increases only about 50 MPa.

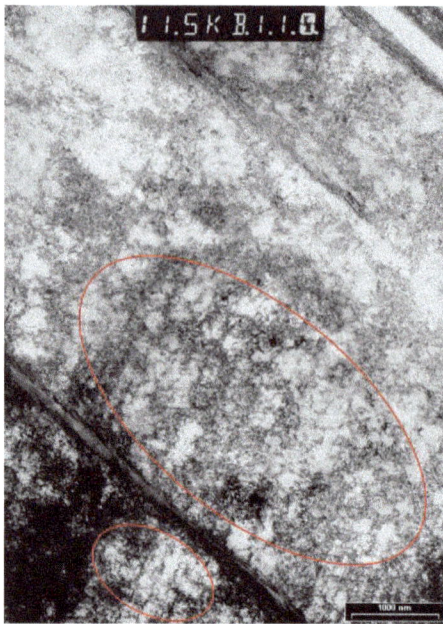

(a)

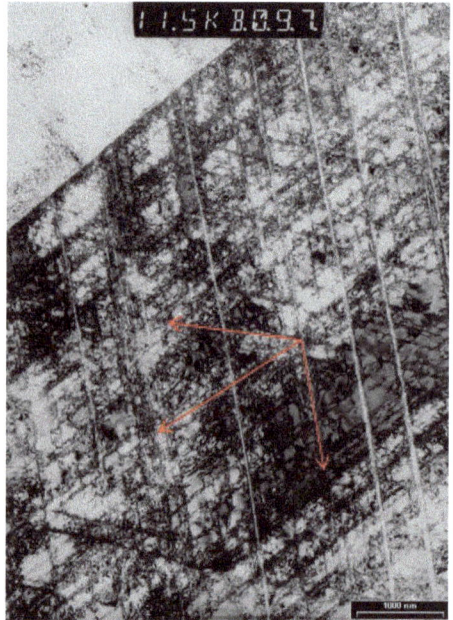

(b)

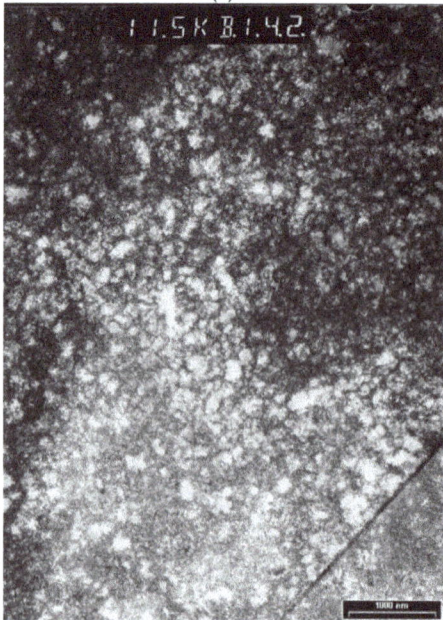

(c)

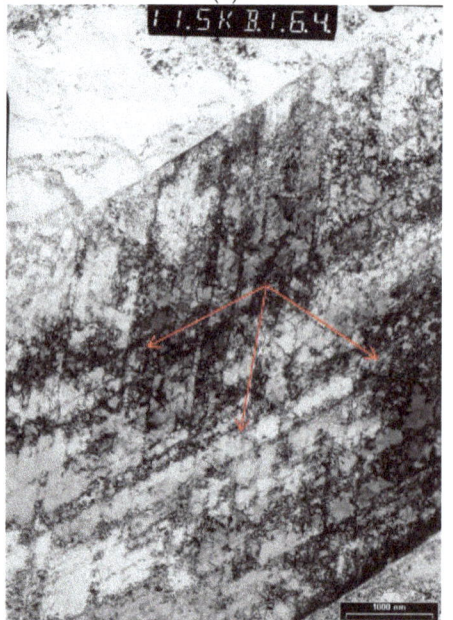

(d)

Figure 5. *Cont.*

(e)

Figure 5. (**a**) TEM image of S31673, 14% CW taken from unloaded head of fatigue sample; (**b**) TEM image of CrMnN, 14% CW taken from unloaded head of fatigue sample; (**c**) TEM image of S31673, 14% CW taken from the gauge length of fatigued sample. (**d**) TEM image of CrMnN, 14% CW taken from the gauge length of fatigued sample. (**e**) TEM image of CrMnN, 14% CW taken from the gauge length of fatigued sample. No distinct glide bands are visible, but there is dislocation accumulation in one direction (red arrow). Red ellipses indicate the beginning of cell formation.

4. Conclusions

The fatigue behavior and the microstructural evolution during cyclic loading of the two austenitic stainless steel alloying concepts CrNi and CrMnN have been compared in the cold worked condition. For each concept, a representative alloy has been chosen. The CrMnN steel shows superior fatigue behavior, which is attributed to the high strength of the material caused by the interstitially alloyed nitrogen. The CrMnN grade showed a more pronounced reduction of the fatigue strength relating to $R_{p0.2}$ of the cold worked material than the CrNi grade (S31673). A possible explanation for that behavior could be found in the TEM investigations. S31673 showed a cellular dislocation structure which intensified during cyclic loading (see also Figure 1b) and characterises wavy slip. Conversely, the CrMnN grade demonstrated distinct planar dislocation glide, which has been proposed as a cause for dislocation annihilation [27], thus explaining the stronger reduction of the fatigue strength related to $R_{p0.2}$ during cyclic loading. An alternative explanation for the reduction of fatigue strength could be the strain localization on the surface through the planar dislocation glide in the CrMnN grade which can facilitates crack initiation.

Author Contributions: Conceptualization, R.F.; investigation, K.S.-H., M.B., and H.H.; methodology, R.F.; project administration, R.F.; resources, R.F.; supervision, M.K. and R.P.; visualization, R.F.; writing—original draft, R.F.; writing—review and editing, M.K., K.S.-H., and R.P.

Funding: This research received no external funding.

Conflicts of Interest: The authors declare no conflict of interest.

References

1. Feltner, C.E.; Laird, C. The role of slip character in steady state cyclic stress-strain behavior. *Trans. TMS-AIME* **1969**, *245*, 1372–1373.
2. Lukáš, P.; Klesnil, M. Dislocation structures in fatigued single crystals of Cu-Zn system. *Phys. Stat. Sol.* **1971**, *5*, 247–258. [CrossRef]
3. Mughrabi, H. Mikrostrukturelle Ursachen der Ermüdungsrissbildung. In *Ermüdungsverhalten Metallischer Werkstoffe, Symposium der Deutschen Gesellschaft für Materialkunde*; Bad Nauheim, Germany, 1984.
4. Mughrabi, H. Fatigue, an everlasting materials problem—Still en vogue. *Procedia Eng.* **2010**, *2*, 3–26. [CrossRef]
5. Uggowitzer, P.J. Stickstofflegierte Stähle. In *Ergebnisse der Werkstofforschung*; Thubal-Kain: Zürich, Switzerland, 1991; pp. 87–101.
6. Simmons, J.W. Strain hardening and plastic flow properties of nitrogen-alloyed Fe-17Cr-(8–10) Mn-5Ni austenitic stainless steels. *Acta Mater.* **1997**, *45*, 2467–2475. [CrossRef]
7. Llewellyn, D.T. Work hardening effects in austenitic stainless steels. *Mater. Sci. Technol.* **1997**, *13*, 389–400. [CrossRef]
8. Kikuchi, M.; Mishima, Y. (Eds.) *High Nitrogen Steels HNS95*; ISIJ International: Kyoto, Japan, 1996.
9. Hänninen, H. Application and performance of high nitrogen steels. In *High Nitrogen Steels 2004*; GRIPS Media: Ostend, Belgium, 2004.
10. Fawley, R.; Quader, M.A.; Dodd, R.A. Compositional effects on the deformation modes, annealing twin frequencies, and stacking fault energies of austenitic stainless steels. *TMS AIME* **1968**, *242*, 771–778.
11. Schramm, R.E.; Reed, R.P. Stacking fault energies of seven commercial austenitic stainless steels. *Metall. Trans. A* **1975**, *6*, 1345–1351. [CrossRef]
12. Stoltz, R.E.; Sande, J.B.V. The effect of nitrogen on stacking fault energy of Fe-Ni-Cr-Mn steels. *Metall. Trans. A* **1980**, *11*, 1033–1037. [CrossRef]
13. Petrov, Y. *Defects and Diffusionless Transformation in Steel (Russian)*; Naukova Dumka: Kiew, Ukraine, 1978; p. 262.
14. Fujikura, M.; Takada, K.; Ishida, K. Effect of manganese and nitrogen on the mechanical properties of Fe-18%Cr-10%Ni stainless steels. *Trans. Iron Steel Inst. Jan.* **1975**, *15*, 464–469.
15. Mosecker, L.; Saeed-Akbari, A. Nitrogen in chromium–manganese stainless steels: A review on the evaluation of stacking fault energy by computational thermodynamics. *Sci. Technol. Adv. Mater.* **2013**, *14*, 033001. [CrossRef] [PubMed]
16. Razumovskiy, V.I.; Hahn, C.; Lukas, M.; Romaner, L. Ab Initio Study of Elastic and Mechanical Properties in FeCrMn Alloys. *Materials* **2019**, *12*, 1129. [CrossRef] [PubMed]
17. Lee, S.-J.; Fujii, H.; Ushioda, K. Thermodynamic calculation of the stacking fault energy in Fe-Cr-Mn-C-N steels. *J. Alloys Compd.* **2018**, *749*, 776–782. [CrossRef]
18. Douglas, D.L.; Thomas, G.; Roser, W.R. Ordering, Stacking Faults and Stress Corrosion Cracking In Austenitic Alloys. *Corrosion* **1964**, *20*, 15–28. [CrossRef]
19. Gerold, V.; Karnthaler, H.P. On the origin of planar slip in F.C.C. alloys. *Acta Metall.* **1989**, *37*, 2177–2183. [CrossRef]
20. Grujicic, M.; Nilsson, J.-O.; Owen, W.S.; Thorvaldsson, T. Basic deformation mechanisms in nitrogen strengthened stable austenitic stainless steels. In *High Nitrogen Steels, HNS 88*; The Institute of Metals: London, UK, 1988.
21. Gavriljuk, V.G.; Berns, H. *High Nitrogen Steels: Structure, Properties, Manufacture, Applications*; Springer-Verlag: Berlin/Heidelberg, Germany, 1999.
22. Ojima, M.; Adachi, Y.; Tomota, Y.; Katada, Y.; Kaneko, Y.; Kuroda, K.; Saka, H. Weak Beam TEM Study on Stacking Fault Energy of High Nitrogen Steels. *Steel Res. Int.* **2009**, *80*, 477–481.
23. Noh, H.; Kang, J.-H.; Kim, K.-M.; Kim, S.-J. Different Effects of Ni and Mn on Thermodynamic and Mechanical Stabilities in Cr-Ni-Mn Austenitic Steels. *Metall. Mater. Trans. A* **2019**, *50*, 616–624. [CrossRef]
24. Degallaix, S.; Foct, J.; Hendry, A. Mechanical behaviour of high-nitrogen stainless steels. *Mater. Sci. Technol.* **1986**, *2*, 946–950. [CrossRef]
25. Taillard, R.; Foct, J. Mechanisms of the action of nitrogen interstitials upon low cicle fatigue behaviour of 316 stainless steels. In *High Nitrogen Steels, HNS88*; The Institute of Metals: London, UK, 1988.

26. Panzenböck, M.H. Ermüdungsverhalten Stickstofflegierter Cr-Mn-Austenite. Ph.D. Thesis, University of Leoben, Leoben, Austria, 1995.
27. Degallaix, S. Role of nitrogen on the monotonic and cyclic plasticity of Type 316L SS at Room temperature. In *Basic Mechanisms in Fatigue of Metals*; Lukas, P., Polak, J., Eds.; Academia: Prague, Czech Republic, 1988; pp. 65–72.

![metals logo] *metals*

MDPI

Article

Strain Hardening, Damage and Fracture Behavior of Al-Added High Mn TWIP Steels

Manjunatha Madivala [1],* [ORCID], Alexander Schwedt [2], Ulrich Prahl [3] [ORCID] and Wolfgang Bleck [1]

[1] Steel Institute (IEHK), RWTH Aachen University, Intzestraße 1, 52072 Aachen, Germany;
bleck@iehk.rwth-aachen.de

[2] Central Facility for Electron Microscopy (GFE), RWTH Aachen University, Ahornstraße 55, 52074 Aachen,
Germany; schwedt@gfe.rwth-aachen.de

[3] Institute of Metal Forming (imf), TU Bergakademie Freiberg, Bernhard-von-Cotta-Straße 4, 09599 Freiberg,
Germany; Ulrich.Prahl@imf.tu-freiberg.de

* Correspondence: manjunatha.madivala@iehk.rwth-aachen.de; Tel.: +49-241-80-90137

Received: 15 February 2019; Accepted: 19 March 2019; Published: 21 March 2019

Abstract: The strain hardening and damage behavior of Al-added twinning induced plasticity (TWIP) steels were investigated. The study was focused on comparing two different alloying concepts by varying C and Mn contents with stacking fault energy (SFE) values of 24 mJ/m^2 and 29 mJ/m^2. The evolution of microstructure, deformation mechanisms and micro-cracks development with increasing deformation was analyzed. Al-addition has led to the decrease of C diffusivity and reduction in tendency for Mn-C short-range ordering resulting in the suppression of serrated flow caused due to dynamic strain aging (DSA) in an alloy with 0.3 wt.% C at room temperature and quasi-static testing, while DSA was delayed in an alloy with 0.6 wt.% C. However, an alloy with 0.6 wt.% C showing DSA effect exhibited enhanced strain hardening and ductility compared to an alloy with 0.3 wt.% C without DSA effect. Twinning was identified as the most predominant deformation mode in both the alloys, which occurred along with dislocation glide. Al-addition has increased SFE thereby delaying the nucleation of deformation twins and prolonged saturation of twinning, which resulted in micro-cracks initiation only just prior to necking or failure. The increased stress concentration caused by the interception of deformation twins or slip bands at grain boundaries (GB) has led to the development of micro-cracks mainly at GB and triple junctions. Deformation twins and slip bands played a vital role in assisting inter-granular crack initiation and propagation. Micro-cracks that developed at manganese sulfide and aluminum nitride inclusions showed no tendency for growth even after large deformation indicating the minimal detrimental effect on the tensile properties.

Keywords: TWIP steel; deformation twinning; serrated flow; dynamic strain aging; damage; fracture

---

1. Introduction

Through many years of development and application, advanced high strength steels (AHSS), aluminum, magnesium and titanium alloys have proved themselves to be versatile and effective materials for automotive parts [1–6]. However, the demand for safety and cost-effectiveness of the components have raised enormously [7]. Thus it is necessary to develop new materials focusing on the special requirements. Through careful control of chemistry and processing, materials can be tailored to provide optimum performance, keeping an aim on specific applications. By choosing different alloying elements and careful processing, the required microstructural constituents with unique mechanical behavior can be obtained resulting in many types of materials for automotive applications [8]. Thus variations in chemistry, microstructure, and deformation mechanisms are crucial aspects in the design and performance of metallic materials. The current work reveals the mechanical behavior of the Al-added high Mn twinning induced plasticity (TWIP) steels designed for safety-relevant automotive parts and other applications.

TWIP steels exhibit extraordinary mechanical properties such as exceptional strength (>1000 MPa), high ductility (>50%) and excellent energy absorption capacity (>55 kJ/kg) making them highly suitable for automotive applications [9–15]. The sustained and high strain hardening rate (SHR) of TWIP steels can be ascribed to their deformation mechanisms such as dislocation glide, deformation twinning, and ε-martensite transformation. The materials behavior strongly dependent on their chemical composition, microstructure and deformation conditions [15–19]. The unique characteristic behavior of these steels is the nucleation and propagation of deformation bands during deformation, which results in serrations on the σ–ε curves, known as Portevin-Le Chatelier (PLC) effect [20–22]. The serrations in TWIP steels are classified as Type A, depending on their morphology. They are known as locking serrations because of repeated locking and unlocking of dislocations by solute atoms commonly known as dynamic strain aging (DSA) [16,22–24]. It was stated that DSA effectively modified the SHR but its influence on the SHR is limited due to the lower activation energy attained in these steels [21]. Many authors stated that mechanical twinning and its associated dynamic Hall-Petch effect remain the only mechanisms that can be responsible for sustained and high SHR of TWIP steels [10,13,21,25]. However, the influence of DSA on the deformation behavior and its role in manifesting the SHR cannot be ignored. Thus in this study, an alloy showing DSA effect is compared with an alloy without DSA effect. The SHRs of the alloys were correlated to the evolution of twin volume fraction (TVF) with increasing deformation.

The appearance of serrations on the σ–ε curves requires the material to be plastically deformed to a certain deformation known as critical strain (ε_c). A detailed study on the X60Mn22 alloy revealed repeated serrations on the σ–ε curves at different strain rates and temperatures. With an increase in temperature or grain size, the ε_c for the initiation of serrations decreased and with the increase of strain rate ($\dot{\varepsilon}$), the ε_c increased [14,19]. In Fe-Mn-C alloys, it was shown that with the increase in C content from 0.0 wt.% to 1.1 wt.% C, the ε_c for triggering serrations has decreased [26]. A comparative study on Fe-Mn-C-x alloys with and without Al showed that with Al-addition the ε_c has increased [27]. It was also shown that with increasing Al content from 0.0 to 3.0 wt.%, the initiation of serrated flow shifted to higher strains or completely suppressed [28]. A similar study also revealed severe serrated flow in an X60Mn22 alloy at different $\dot{\varepsilon}$, whereas in an X60MnAl22-1.5 alloy serrations appeared only at quasi-static $\dot{\varepsilon}$ and the initiation of serrations shifted to high strains [29]. It was shown in X60MnAl18-x alloys with 0.0, 1.5 and 2.5 wt.% Al contents, that severe serrated flow also occurred in Al-containing alloys at elevated temperatures [16]. Thus, it can be clearly understood that ε_c for the initiation of the serrations in Fe-Mn-C-Al alloys is mainly influenced by stacking fault energy (SFE), the alloying elements such as C and Al, deformation temperature, $\dot{\varepsilon}$ and microstructure of the material. From the experimental observations in the literature, it can be stated that in the alloys of Fe-Mn-C and Fe-Mn-C-Al with different chemical compositions the serrated σ–ε behavior can be observed but the range of temperature, $\dot{\varepsilon}$ regimes, microstructure states and the strain ε_c at which serrations appears could be different. In the current study, an X60MnAl17-1 and X30MnAl23-1 alloys were investigated in detail to study the influence of alloying elements such as C and Al on the serrated flow behavior.

Macroscopic observations revealed an abrupt failure in TWIP steels without any strain localization or necking, which occurred at the intersection of two deformation bands close to the specimen edges. It was claimed that the increased stress concentration caused by the intercepting deformation twins and the slip band extrusions at grain boundary (GB) has led to micro-crack formation. Inter-granular micro-crack initiation and propagation events were observed in the microstructure at GBs and triple junctions along with the rapid nucleation of minute voids [14]. A similar study also showed micro-cracks at the intersection of deformation bands mainly at the edge and side surfaces of the tensile specimen. The cracks were observed at GB junctions and the mechanical twin boundaries [30]. It was claimed that the exponential increase of the macroscopic void volume fraction led to the sudden failure or decrease of SHR in TWIP steels [31]. A detailed study on the evolution of damage in TWIP steel by 3D X-ray tomography experiments has shown that the average void diameter and the stress triaxiality remained constant throughout the deformation until failure. It was claimed that the damage development process involved rapid nucleation of minute voids combined with substantial growth of the largest voids [32]. The distribution of tiny voids with few elongated cavities close to the fracture surface in shear tests clearly indicated localized failure [33]. The above-mentioned studies were on TWIP steels showing DSA effect, where an abrupt failure occurred in the material. However, the addition of Al not only increases ductility but also changes the local deformation and fracture behavior. Thus in this study damage and failure behavior of Al-added TWIP steels were investigated both at the micro- and macro-scale. The current work focuses on the relationship between the deformation mechanisms and the mechanism of micro-crack formation.

In the current study, the strain hardening, damage and fracture behavior of Al-added TWIP steels with different chemical compositions were investigated. To ascertain the role of the interstitial C atoms on the interaction between Mn-C short-range ordering (SRO) and dislocations in causing DSA phenomena, two alloys with different C contents were studied. By analyzing the local plastic strain (ε_{local}) evolution, the serrated flow behavior caused due to the nucleation and propagation of deformation bands and their subsequent influence on the failure initiation during the deformation was investigated. The measurement of a rise in temperature during deformation aided in the accurate estimation of SFE, which enabled the prediction of change in deformation mechanisms. The evaluated TVF and the predicted SFE was used to explain the SHR, the mechanical behavior of alloys. The evolution of microstructure, the mechanism of micro-crack initiation and their development with increasing strain were studied in detail. The influence of non-metallic inclusions on the mechanical properties and micro-crack formation was also investigated.

2. Materials and Experimental Methods

The materials investigated in this study are Al-added high Mn TWIP steels produced in the laboratory, designated as X60MnAl17-1 and X30MnAl23-1. The electron backscatter diffraction (EBSD) technique was used to characterize the initial microstructure such as the grain size and the evolution of twinning with increasing strain. The optical and field-emission scanning electron microscope (SEM) with energy dispersive X-ray spectroscopy (EDS/EDX) was used to qualitatively and quantitatively analyze the inclusions type, size, and distribution. Interrupted micro-tensile tests were carried out in SEM to study the evolution of microstructure and micro-cracks development during deformation. Tensile tests were carried out in conjunction with the digital image correlation (DIC) to investigate the macroscopic material behavior. An infrared thermography camera along with video extensometer was used to measure the temperature rise due to adiabatic heating (AH) during deformation. The elastic properties of the material were determined by the ultrasonic method. The fracture behavior of the macroscopic tensile specimens was analyzed in the SEM.

2.1. Materials Processing

The materials were produced by ingot-casting using a vacuum induction furnace. The cast ingots (each ~30 kg) were homogenized in a muffle furnace at 1150 °C for 5 h, in order to reduce the segregation of alloying elements, especially Mn. The homogenized ingots (140 mm in height) were then forged at 1150 °C to a height of 55 mm, followed by another homogenization and then hot rolled at 1150 °C to a sheet thickness of 2.8 mm. The hot rolled sheets were cold-rolled to 50% reduction to get a final sheet thickness of 1.4 mm. Finally, annealing heat treatment was carried out in a salt bath furnace at 900 °C for 20 min, followed by quenching in water to obtain a completely recrystallized microstructure. The major differences in alloy compositions are C and Mn contents. The alloy X60MnAl17-1 has 0.6 wt.% C, 17.17 wt.% Mn, whereas alloy X30MnAl23-1 has 0.3 wt.% C, 22.43 wt.% Mn. The chemical compositions and SFE values of the alloys are presented in Table 1. The chemical composition was determined using optical emission spectroscopy. The Mn content from 0.001 to 20 wt.% with an accuracy of ±0.05, C from 0.001 to 1.2 wt.% with an accuracy of ±0.01 and Al from 0.001 to 2 wt.% with an accuracy ±0.01 can be determined. The Mn content above

20 wt.% is verified by melt extraction analysis. The SFE was calculated using the sub-regular solution thermodynamic model of Saeed-Akbari et al. [11]. The interface energy ($\sigma^{\gamma/\varepsilon}$) value of 10 mJ/m^2 as recommended in [11,34] was used for calculating the SFE. The model is implemented as a MATLAB$^\circledR$ program to evaluate the SFE as explained in [19]. Based on their chemical composition, the evaluated SFE value of X60MnAl17-1 and X30MnAl23-1 are 29 and 24 mJ/m^2 respectively.

Table 1. Chemical compositions (in wt.%) and the stacking fault energy (SFE) values (in mJ/m^2) of the investigated alloys.

Alloy	C	Si	Mn	P	S	Cr	Ni	Cu	Al	V	N	Fe	SFE
X60MnAl17-1	0.60	0.06	17.17	<0.009	<0.006	0.05	0.04	0.03	1.50	0.07	0.015	Bal.	29
X30MnAl23-1	0.30	0.04	22.43	<0.009	<0.005	0.05	0.04	0.02	1.39	0.10	0.013	Bal.	24

2.2. Microstructure Characterization

2.2.1. Micro-Tensile Tests

The un-deformed and deformed microstructures were analyzed by using EBSD on the RD-TD plane (RD/TD: rolling/transverse direction). The micro-tensile test specimens with a gauge length of 2 mm, a width of 1 mm and having a thickness of 1.4 mm as shown in Figure 1 were obtained by wire erosion cutting from the cold-rolled and annealed sheet. The samples were initially mechanically ground and then polished up to 1 µm using diamond suspension. They were further electro-polished by applying a voltage potential of 35 V for 15 s using an electrolyte consisting of 95% acetic acid and 5% perchloric acid. A field-emission JEOL JSM-7000F SEM equipped with an EDAX-TSL Hikari detector operating at a voltage of 20 kV and a probe current of about 20–30 nA was used for the measurements. The data was analyzed using TSL OIM$^\circledR$ Analysis 7 software. A step size of 100 nm was chosen for the measurements and all the scanned points with a confidence index (CI) value of below 0.1 were not considered for the analysis. The measurements were carried out in the middle of the sample as shown in Figure 1c and at a true strain of 0.0, 0.1, 0.2, 0.3, 0.4 and $\varepsilon_{necking}$. The procedure used for processing EBSD data to identify the grains and twins is described in [14,19]. The TVF was evaluated based on the relative fraction of the pixel areas of Σ-3 twins to the measurement areas at different macroscopic strains [19].

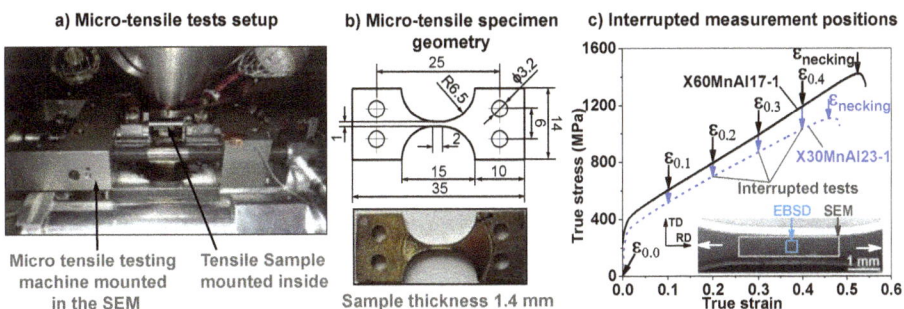

a) Micro-tensile tests setup   b) Micro-tensile specimen geometry   c) Interrupted measurement positions

Micro tensile testing machine mounted in the SEM   Tensile Sample mounted inside   Sample thickness 1.4 mm

Figure 1. In-situ micro-tensile tests in conjunction with scanning electron microscope (SEM) and electron backscatter diffraction (EBSD): (**a**) test setup, (**b**) specimen geometry, (**c**) true stress-strain curves obtained for X60MnAl17-1 and X30MnAl23-1; tests were carried out with a crosshead displacement rate of 0.2 mm/min, which corresponds to a quasi-static $\dot{\varepsilon}$ of about 0.001 s^{-1} and interrupted at different strains as pointed to observe the microstructure and micro-crack development; EBSD measurement area and the region of SEM investigations on the micro-tensile specimen is shown; white arrows indicate tensile loading direction.

The micro-tensile tests were carried out in a field emission SEM to investigate the mechanisms of micro-crack formation. The tests were interrupted at different strains to observe the micro-crack initiation and their development as shown in Figure 1. The tests were performed at room temperature (RT) and by applying a crosshead displacement rate of 0.2 mm/min which corresponds to a quasi-static $\dot{\varepsilon}$ of 0.001 s^{-1} using an in-situ micro-tensile testing device mounted in the SEM. The macroscopic fractured samples were also analyzed in SEM to study the failure behavior of the alloys. To observe the changes in dimple size and morphology at the inclusions several SEM images were taken at different locations. The average dimple size was determined by image processing as explained in [19].

2.2.2. Inclusions and Fracture Surface Analysis

The specimens of size 20×20 mm^2 (RD-TD plane) were embedded in an epoxy resin, ground and mechanically polished with a diamond suspension up to 1 µm for inclusions analysis. They were characterized quantitatively to determine the size and area fraction according to DIN EN 10247 standard by optical microscopy in conjunction with an image-analysis software. Due to the inherent high-contrast ratio between non-metallic inclusions and the matrix, inclusions were accurately determined by image-analysis and quantitative data was extracted. The inclusions were analyzed qualitatively based on their chemical composition by SEM/EDS technique.

2.3. Mechanical Tests

2.3.1. Macro-Tensile Tests with Digital Image Correlation (DIC)

Macroscopic tensile tests were carried out according to DIN EN ISO 6892 standard. A sample geometry of gauge length 75 mm, a width of 12.5 mm and having a thickness of 1.4 mm as shown in Figure A1 was used. The samples were obtained by water-jet cutting along the RD and the machined edges were polished up to a surface roughness of 0.125 µm. The tests were performed at RT in conjunction with DIC or video extensometer using a Zwick/Roell Z100 machine with a crosshead displacement rate of 1.8 mm/min, which corresponds to a quasi-static $\dot{\varepsilon}$ of about 0.001 s^{-1}. A force transducer was used to measure force and a video extensometer was used to measure the elongation. The GOM Aramis 12M 2D DIC system (see Figure A1) was used for evaluating the ε_{local} distribution and to investigate the initiation and propagation of deformation bands during straining. A high contrast black and white stochastic pattern was prepared on the samples for the measurements. The images were captured at a rate of 1 Hz during the test and a facet size of 100 µm was set for the ε_{local} evaluation. The force data was imported into the ARAMIS DIC software for calculating the σ–ε curves. The mechanical properties were evaluated following the ASTM E517, ASTM E646, and DIN ISO 10275 standards. The properties evaluated using video extensometer and DIC were almost similar. The SHR curves were evaluated from the smoothened σ–ε curves.

2.3.2. Macro-Tensile Tests with High-Speed Thermocamera

The rise in temperature due to adiabatic heating during deformation was measured by high-speed infrared thermography camera of type varioCAM® hr along with video extensometer as shown in Figure A2. The temperature measurement of the camera is from -40 °C to 1200 °C. The overall accuracy of temperature measurement within the range of calibration is ± 1 °C. A tensile sample with a gauge length of 30 mm and a width of 6 mm as shown in Figure A2 was tested using a Zwick/Roell Z100 machine at a $\dot{\varepsilon}$ of 0.001 s^{-1}. The emissivity was also calibrated before the measurements and the specimen was coated with black lacquer to minimize the reflections from the surroundings. The images were acquired at a frequency of 1 Hz in full-frame-mode with a maximum of 384×288 pixels resolution using the IRBIS® online software. The data was analyzed to extract the temperature variation within the gauge length and co-related with the strain measured from the video extensometer.

2.3.3. Elastic Constants Measurement

Elastic properties of the alloys were determined by using the GE USM35 ultrasonic testing machine. Longitudinal and transversal pulses were measured by using normal incident 20 MHz (CLF4) longitudinal and 10 MHz (K7KY) shear transducers respectively. The presence of air can disrupt the propagation of the acoustic waves into the material, therefore a viscous liquid such as oil (longitudinal)/honey (transversal) was applied as contact material between the transducer and the sample surface. With the accurate measurement of the sample thickness (l) and the transit time (t) between the peaks of consecutive echoes, the longitudinal (v_l) and transversal (v_t) sound velocities of the material was calculated by using $v_{l/t} = l/(\frac{t}{2})$. The density ($\rho$) of the alloys was determined by using AccuPyc 1330 pycnometer. The elastic properties of the alloys such as Young's modulus (E), shear modulus (G) and Poisson's ratio (ν) were calculated using the relations given by [35]

$$E = 3\rho v_t^2 (v_l^2 - \frac{4}{3}v_t^2)/(v_l^2 - v_t^2), \qquad G = \rho v_t^2, \qquad \nu = \frac{1}{2}(v_l^2 - 2v_t^2)/(v_l^2 - v_t^2) \qquad (1)$$

3. Results

3.1. Microstructure Analysis

The initial microstructure of both alloys X60MnAl17-1 and X30MnAl23-1 consisted of fully austenitic (γ) grains with an average grain size of 16.0 µm and 12.0 µm respectively. The inverse pole figure (IPF) maps with respect to the RD depicting the grain sizes and the crystal orientation distribution are shown in Figure 2a,b. It can be observed from Figure 2d that for an alloy X30MnAl23-1, the grain size varies within a range of about 1–40 µm, with small fractions of grains larger than 25 µm, whereas for an alloy X60MnAl17-1 shown in Figure 2c, the grain size varies with a large range of about 1–50 µm, with small fractions of grains larger than 35 µm. The initial texture is rather weak for both alloys with a texture index value of 1.33 for X60MnAl17-1 and 1.18 for X30MnAl23-1, as shown in Figure 2a,b.

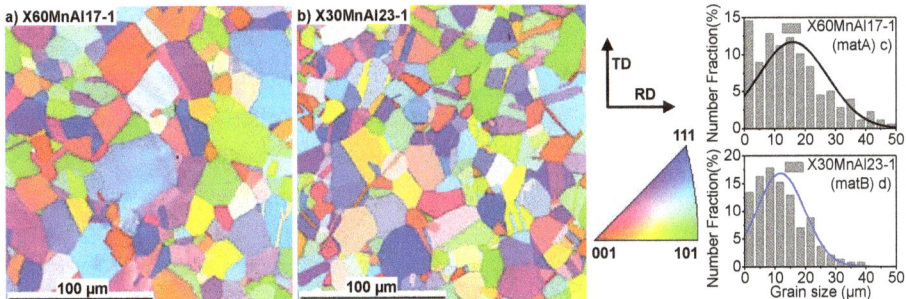

Figure 2. The EBSD inverse pole figure (IPF) maps of the undeformed samples showing grain size and orientations: (**a**) X60MnAl17-1, (**b**) X30MnAl23-1. Grain size fitted using normal distribution function shown in: (**c**) X60MnAl17-1, (**d**) X30MnAl23-1; microstructures of both alloys is completely austenite (γ); the average grain size of X60MnAl17-1 is 16.0 µm and X30MnAl23-1 is 12.0 µm.

The estimated area fraction and the size of inclusions present in the materials are presented in Table 2. The measured area fraction of inclusions in both alloys is ~0.1%. The average size of the inclusions in X60MnAl17-1, X30MnAl23-1 is about 1.2 µm and 1.5 µm respectively. The distribution of different type of inclusions present in the materials is shown in Figure 3. EDS analysis performed on the inclusions at different locations indicate the presence of manganese sulfides (MnS) and aluminum nitrides (AlN), both in globular and elongated shapes as shown in Figure 2. The globular MnS inclusions have slightly high fractions of Mn and S contents compared to the base material, whereas the elongated MnS contains a higher fraction of Mn and S.

Table 2. Cleanliness analysis: area fraction, average and largest size of the inclusions. Size of the smallest inclusions found in both alloys is ∼0.5 μm. The scatter was estimated by analyzing four to five images taken at different locations.

Alloy	Area Fraction (%)	Average Size (μm)	Largest Size (μm)
X60MnAl17-1	0.10 ± 0.01	1.23 ± 0.05	8.0 ± 1.5
X30MnAl23-1	0.11 ± 0.02	1.50 ± 0.10	12.0 ± 2.0

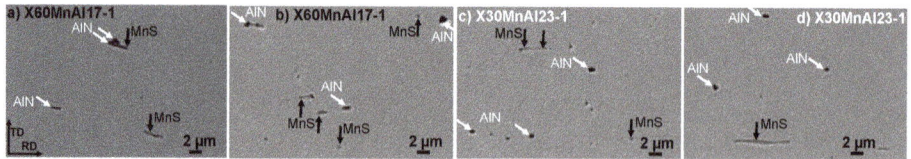

Figure 3. SEM micrographs depicting non-metallic inclusions present in the investigated alloys: X60MnAl17-1 (**a,b**); X30MnAl23-1 (**c,d**); manganese sulphide (MnS) (black arrows) and aluminium nitride (AlN) (white arrows).

3.2. Mechanical Properties

The mechanical properties of the TWIP steels vary with chemical composition due to change in SFE resulting in different deformation behavior. The stress-strain curves and the SHR for X60MnAl17-1 and X30MnAl23-1 obtained by uni-axial tensile tests at RT are shown in Figure 4. The σ–ε curve of X30MnAl23-1 exhibit homogenous flow behavior throughout the deformation, whereas periodic serrations of type A can be observed in the flow curve of X60MnAl17-1. Both X60MnAl17-1 and X30MnAl23-1 show linear flow behavior and slight necking just before the final failure. Initially, strain hardening for both alloys is characterized by a sharp drop of the SHR and then it increases to reach a constant value with small increase in deformation. Thereafter, both alloys show a general drop with a pronounced multistage character at different strains. The first marked inflection in SHR after the sudden drop occurs at ∼0.04 true strain for X60MnAl17-1 and at ∼0.03 true strain for X30MnAl23-1 to attain the highest point of strain hardening. The peak value of SHR for X60MnAl17-1 is about 2460 MPa and X30MnAl23-1 is about 2360 MPa and after that SHR started to decrease steadily for both alloys. The second marked inflection in SHR occurs at ∼0.15 true strain for X60MnAl17-1 and at ∼0.18 true strain for X30MnAl23-1. The SHR increases at second inflection for X60MnAl17-1 compared to a steady decrease in SHR for X30MnAl23-1 until failure. The SHR of X60MnAl17-1 attains another maximum between 0.3 and ∼0.4 true strain, and then it decreases continuously until failure. The SHR for both alloys at the beginning of the deformation is above 2350 MPa and it decreases to about 1400 MPa at failure. Since the SHR of both alloys shows multistage hardening behavior with increasing macroscopic strain indicating a change in active deformation mechanisms at different strains. The serrated flow caused due to DSA in X60MnAl17-1 alloy after a strain of ∼0.15 could have played a crucial role in enhancing the strain hardening and ductility of the material.

The variation in the mechanical properties of X60MnAl17-1 and X30MnAl23-1 are presented in Table 3. The X60MnAl17-1 with 0.6 wt.% C has a higher yield strength of about 294 MPa compared to 246 MPa of X30MnAl23-1 with 0.3 wt.% C. Similarly, the tensile strength of X60MnAl17-1 and X30MnAl23-1 are very different, which are about 844 MPa and 693 MPa respectively. The TE of X60MnAl17-1 is 70%, which is larger compared to 63% of X30MnAl23-1. The density of X60MnAl17-1 is slightly lower and E, G are much larger compared to X30MnAl23-1. The mechanical properties of X60MnAl17-1 and X30MnAl23-1 presented in Table 3 are very different even though both the designed alloys are very similar in terms of SFE values, heat treatment, and microstructure.

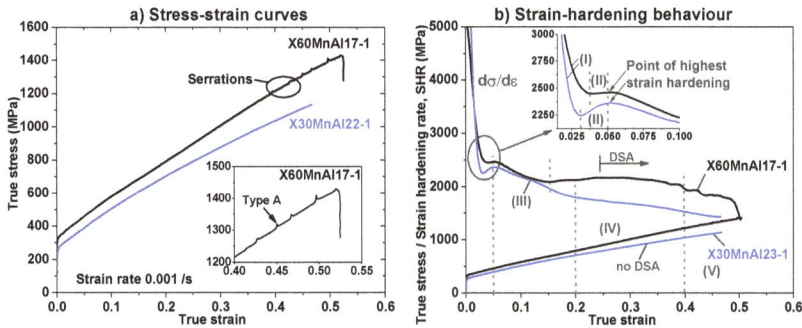

Figure 4. Macroscopic material response: (**a**) True stress-strain curves (**b**) Strain-hardening behavior.

Table 3. Mechanical properties of Al-added TWIP steels: Yield/ultimate tensile strength (YS/UTS), uniform/total elongation (UE/TE), Lankford coefficient (r-value), strain hardening exponent (n-value), density (ρ), Young's modulus (E), shear modulus (G) and Poisson's ratio (ν). All data are average values determined from at least three parallel experiments.

Alloy	YS	UTS	UE	TE	r-Value	n-Value	ρ	E	G	ν
Unit	MPa	MPa	%	%	-	-	kg/m^3	GPa	GPa	-
X60MnAl17-1	294 ± 10	844 ± 15	65 ± 5	70 ± 5	0.90 ± 0.01	0.35 ± 0.01	7700 ± 10	188 ± 2	75 ± 1	0.267 ± 0.01
X30MnAl23-1	246 ± 10	693 ± 15	62 ± 2	63 ± 2	0.83 ± 0.01	0.37 ± 0.01	7715 ± 5	161 ± 1	63 ± 1	0.274 ± 0.01

3.3. Deformation Mechanisms

The chemical composition of an alloy has a significant influence on the SFE, which controls the activation of different deformation mechanisms during deformation. Hence SFE of an alloy plays a crucial role in the evolution of microstructure and its phases. The SFE value of 29 mJ/m^2 for X60MnAl17-1 and 24 mJ/m^2 for X30MnAl23-1, indicate that both alloys are designed to exhibit twinning induced plasticity in combination with dislocation glide as a major deformation mechanism. Activation of DSA in an X60MnAl17-1 alloy can be observed from Figure 4, whereas in an X30MnAl23-1 alloy DSA is not activated.

The evolution of microstructure and deformation mechanisms is shown in Figure 5. Annealing twins can be observed in the initial stages of deformation as shown in Figure 5a,e (black arrows). Mechanical twins or twin bundles could not be observed in EBSD image quality maps in both alloys at a strain of $\sim$0.1. However, an advanced technique such as transmission electron microscopy is essential for observing the initiation of twins. With the small increase in macroscopic strain to $\sim$0.2, mechanical twins can be observed in the grains oriented along <111> (see Figures 5b,f and 13). It could be observed that the fraction of twinned grains or twin area fraction in an X60MnAl17-1 alloy is much high compared to an X30MnAl23-1 alloy. Increasing the deformation further to $\sim$0.3 lead to further nucleation and growth of mechanical twins (white arrows). More intense twinning can be observed in grains of an X60MnAl17-1 alloy compared to an X30MnAl23-1 alloy. The grains which are oriented along the <101> direction reorient toward the <111> direction during deformation and twins initiate in these grains as shown in Figure 13. At a macroscopic strain of $\sim$0.4, both primary and secondary twins can be observed in both alloys. In Figure 5d,h, mechanical twins can be observed in most of the grains indicating large deformation at micro-scale. The twin bundles can be very well identified and the increase in density of mechanical twins can be seen with increasing macroscopic strain. At a strain of $\sim$0.4, the mechanical twins are saturated in the grains in both alloys with no scope for further twinning. The area fraction of detected twins or twin bundles with increasing macroscopic strain is shown in Figure 13. The intense twinning in the grains can be observed before the failure initiation. High TVF in an X60MnAl17-1 alloy has led to better SHR and increased ductility compared to an X30MnAl23-1 alloy. There are many instances in the microstructure where twinning occurs in both the neighboring grains and share a common GB (light blue arrows). The heavily deformed grains can also have a common junction (yellow arrow). The stress level in such GBs and triple junctions is very high compared to other grains. Such high stressed regions of the microstructure are prone to initiation of micro-cracks.

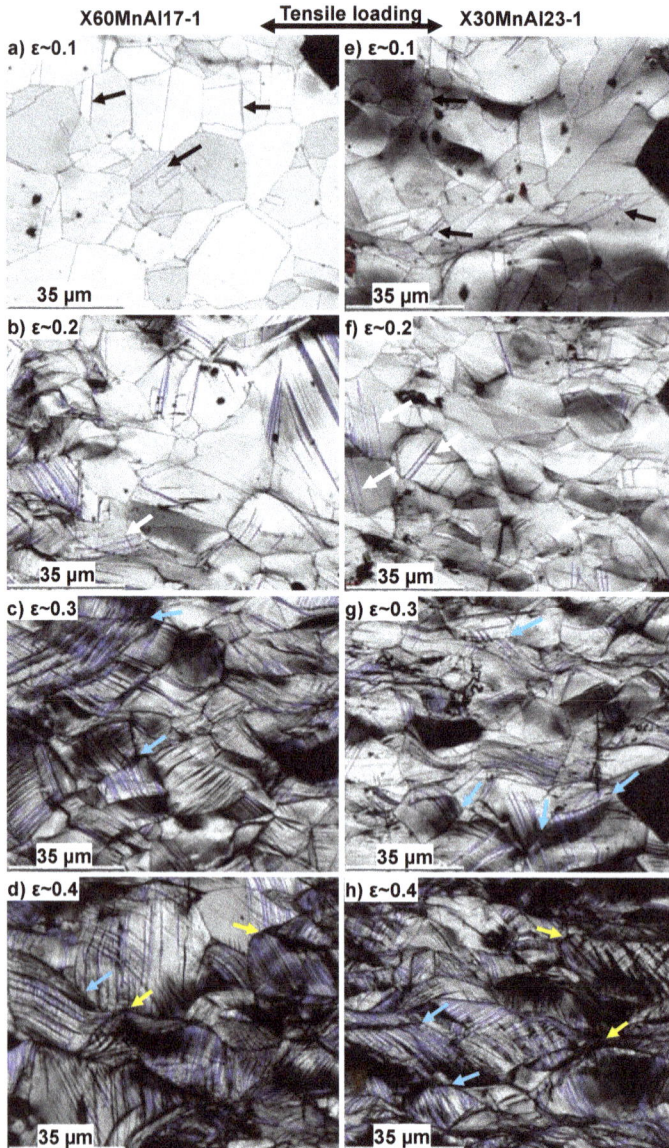

Figure 5. EBSD image quality (IQ) maps depicting the evolution of twinning with increasing macroscopic strain: X60MnAl17-1 (**a**) $\varepsilon_{\sim 0.1}$ (**b**) $\varepsilon_{\sim 0.2}$ (**c**) $\varepsilon_{\sim 0.3}$ (**d**) $\varepsilon_{\sim 0.4}$ (left column), X30MnAl23-1 (**e**) $\varepsilon_{\sim 0.1}$ (**f**) $\varepsilon_{\sim 0.2}$ (**g**) $\varepsilon_{\sim 0.3}$ (**h**) $\varepsilon_{\sim 0.4}$ (right column); blue color in IQ maps in (**a**,**e**) indicates annealing twin boundaries and detected Σ-3 deformation twin boundaries in (**b**–**d**,**f**–**h**); The arrows indicate annealing twin boundaries (black), nucleation of deformation twins (white), intercepting twins at grain boundary (light blue) and triple junctions (yellow); pixels in black are unindexed points.

3.4. Local Deformation Behavior

The chemical composition of the alloy mainly C content, has a significant influence on the local deformation behavior. The σ–ε behavior extracted from the quasi-static tensile tests at RT carried out in conjunction with DIC is shown in Figure 6a. The serrations can be observed on the σ–ε curve caused due to DSA in an X60MnAl17-1 alloy, whereas smooth σ–ε curve without any serrations in an X30MnAl23-1 alloy. The DSA in X60MnAl17-1 alloy has led to the plastic instability in the form of initiation and propagation of the deformation bands. The deformation bands initiated at a macroscopic strain of $\sim$24% during deformation propagate until failure. The strain at which deformation bands initiated is shown in Figure 7I and the region at which deformation bands were active is indicated by a rectangle in Figure 6a. After a strain of $\sim$24%, deformation bands nucleated steadily and continuously. The plateau between the two consecutive serrations peaks corresponds to the nucleation and propagation of deformation band within the gage length. The serration peak corresponds to the initiation/disappearance of the deformation band outside the gage length. The initiation and propagation of deformation bands during deformation for X60MnAl17-1 is shown in Figure 7i–k. The enlarged view of σ–ε curve from Figure 6a, where deformation bands were active is shown in Figure 7I. It can be observed that stress jumps at serrations at the beginning are smaller compared to the final stages of deformation. The deformation bands nucleated at one shoulder end or in the middle of the specimen propagate to the other end. For up to $\sim$28% strain, only one deformation band nucleated and propagated, whereas after 28% strain, two intersecting deformation bands nucleated and propagated during the deformation (see Figure 7j,k). It is clear from the local strain rate ($\dot{\varepsilon}_{local}$) distribution in Figure 7i–k that, the deformation is localized in the deformation bands, whereas it is uniform in other regions. During initiation of deformation bands, the ε_{local} accumulation in deformation bands is quite low, but in the subsequent bands, large strain localization can be observed (see Figure 7l–n). The ε_{local} is larger in the regions where deformation band has already passed compared to the region where deformation band is moving towards or yet to pass through. However, the difference in ε_{local} between the regions where the deformation band already passed by compared to the regions where the deformation band moving towards is $\leq$5% strain. Thus nucleation and propagation of deformation bands during deformation has resulted in inhomogeneous deformation behavior in X60MnAl17-1 alloy. The velocity at which deformation bands propagate decrease exponentially with increasing strain as shown in Figure 7II. The band velocity at the beginning is $\sim$2 mm/s, which is drastically reduced to $\sim$0.5 mm/s in the end. The failure in X60MnAl17-1 alloy occurred at the intersection of deformation bands when their motion is hindered significantly as shown in Figure 7k,n. The localization of strain within the deformation bands is larger compared to the ε_{local} distribution in other regions. Hence strain localization caused at the deformation bands influenced the failure initiation in X60MnAl17-1 alloy. For X30MnAl23-1 alloy, the $\dot{\varepsilon}_{local}$ distribution is shown in Figure 7u–w and ε_{local} distribution is shown in Figure 7x–z. It could be observed that the ε_{local} and $\dot{\varepsilon}_{local}$ is quite uniform until the beginning of necking. The necking begins at a macroscopic strain of $\sim$60% and with further straining failure occurs due to strain localization. Even though both alloys deform by twinning in combination with dislocation glide, the local deformation behavior in X30MnAl23-1 alloy is homogenous compared to inhomogeneous behavior in X60MnAl17-1 alloy due to the activation of DSA.

The rise in temperature due to adiabatic heating could influence the local deformation behavior. The temperature distribution over the entire gage length of the specimen at different elongations is shown in Figure 8b,d and the temperature variation in the middle of the specimen is shown in Figure 8a,c. From Figure 8, it can be observed that the temperature of the specimen rises due to adiabatic heating (AH) with increasing strain. The temperature distribution is inhomogeneous in X60MnAl17-1 alloy due to the propagation of deformation bands compared to the homogenous temperature distribution in X30MnAl23-1 alloy. The temperature increased above RT by 14 °C in an X60MnAl17-1 alloy and by 12 °C in an X30MnAl23-1 alloy. Thus AH has resulted in an increase of SFE from 24 to 25 mJ/m^2 in an X30MnAl23-1 alloy and from 29 to 31.5 mJ/m^2 in an X60MnAl17-1 alloy. However, the SFE increase due to AH is not significant for the change in deformation mechanisms.

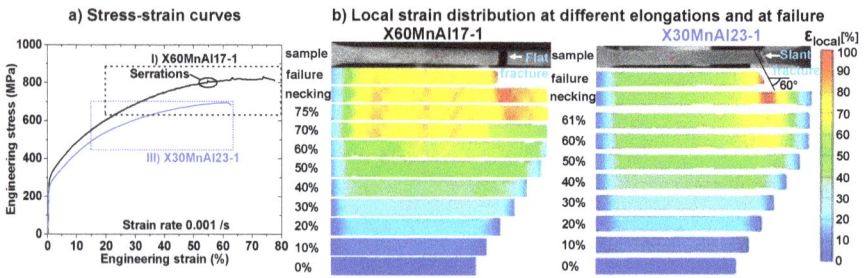

Figure 6. (a) Stress-strain behavior extracted from the macroscopic tensile tests carried out in conjunction with digital image correlation (DIC). (b) Local strain distribution: X60MnAl17-1—Failure at the intersection of deformation bands with large strain localization (see also Figure 7k), X30MnAl23-1—Failure due to diffuse necking and large strain localization; rectangles indicate regions chosen to investigate deformation bands; left: macroscopic strain; right: legend for local von Mises effective plastic strain (ε_{local}) distribution.

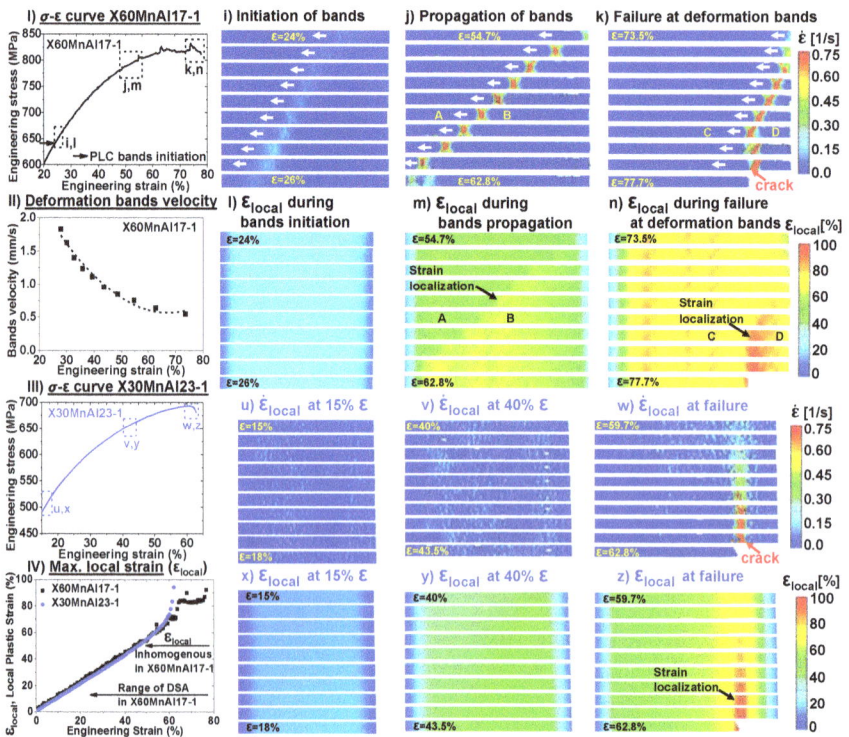

Figure 7. Deformation bands initiation, propagation and their location at the point of failure obtained from the tensile tests in conjunction with DIC: (**I**) X60MnAl17-1: enlarged view of $\sigma-\varepsilon$ curve depicting the initiation of deformation bands (**i,l**), propagation of deformation bands (**j,m**) and failure at deformation bands (**k,n**), (**II**) plot of the deformation bands velocity versus strain, (**III**) X30MnAl23-1: enlarged view of $\sigma-\varepsilon$ curve depicting no deformation bands initiation and propagation (**u,v,x,y**) and failure due to diffuse necking and strain localization (**w,z**), (**IV**) plot of the maximum local plastic strain versus strain; (**i–k,u–w**) shows local strain rate ($\dot{\varepsilon}_{local}$) distribution; (**l–n,x–z**) shows local strain (ε_{local}) distribution; the legend for $\dot{\varepsilon}_{local}$ and ε_{local} distribution (**right**); white arrows indicate the direction of propagation of the deformation bands; the macroscopic strain at the start and end of deformation band is given; images were shown in steps of ~0.4–0.6% macroscopic strain.

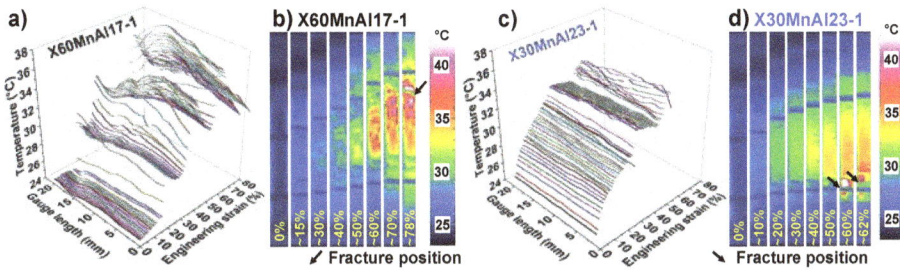

Figure 8. Temperature increase due to adiabatic heating in the macroscopic tensile tests carried out in conjunction with infrared thermography camera and video extensometer: (**a**,**c**) Rise in surface temperature during deformation, (**b**,**d**) Distribution of local temperature at different elongations and at failure.

The ε_{local} distribution along the tensile direction at different elongations is shown in Figure 6b. It can be observed that over the entire gage length of the tensile specimen, the ε_{local} distribution is uniform until 50% elongation. In an X60MnAl17-1 alloy, inhomogeneous distribution of ε_{local} can be observed due to deformation bands propagation. The failure initiated at the point of intersection of deformation bands close to the edges of the tensile specimen. In an X30MnAl23-1 alloy for the elongations above 50%, the strain starts to localize within the gage length of the tensile specimen. The region of strain localization has increased further with the increase in deformation and then finally leading to abrupt failure in the localized region.

Slant fracture with pronounced necking can be observed on the broken tensile sample. The macroscopic failure strain (ε_f) in an X60MnAl17-1 alloy is ~77%, whereas the local failure strain (ε_{f_local}) in the deformation band is ~98%. Similarly in an X30MnAl23-1 alloy ε_f is ~64% and the ε_{f_local} is ~98%. This large difference between the ε_{f_local} and the ε_f indicates that a large amount of strain localization occurred before the final failure. The major differences in failure initiation can be observed in the alloys with and without DSA effect. In an X60MnAl17-1 alloy with DSA effect, failure occurred at the intersection of deformation bands close to edges, whereas in an X30MnAl23-1 alloy without any DSA effect, failure occurred by classical necking and strain localization.

3.5. Mechanisms of Damage and Failure

The damage initiation is mainly influenced by the active deformation mechanisms and the presence of non-metallic inclusions in the material. The evolution of microstructure with increasing strain is shown in Figure 9. The micro-crack formation events in the microstructure are shown in Figure 10 and at inclusions is shown in Figure 11. The damage development in both the alloys could be explained in three stages namely damage incubation stage ($\varepsilon_{0.10-0.3}$), damage nucleation stage ($\varepsilon_{0.3-necking}$) and damage growth stage ($\varepsilon_{necking-failure}$). In an X60MnAl17-1 alloy, the major active deformation mechanisms were twinning and slip, activation of DSA and occurrence of intense twinning can be observed at lower strains, hence the damage nucleation begins at earlier stages of deformation. Whereas in an X30MnAl23-1 alloy, only twinning and slip were the active deformation mechanisms and the occurrence of intense twinning can be observed at later stages of deformation close to necking, which could have delayed the beginning of damage nucleation.

During the damage incubation stage, the interaction of the various active deformation mechanisms with the microstructure constituents and non-metallic inclusions play a significant role. Twinning in combination with dislocation slip within the grains has resulted in large local deformation as shown in Figure 9a,d. With further increase in deformation (see Figure 9b,e), nucleation and growth of multiple deformation twins within a single grain result in a large pileup of dislocations at GBs or twin boundaries. It can be observed from Figure 9 that, there are many instances in microstructure, where two neighboring grains which deform by twinning share a common GB or a grain which deform by twinning share a GB with another grain which deforms by slip or three differently deformed grains share a common junction (triple points) can be observed. A large pileup of dislocations in such instances of microstructure could be observed at $\varepsilon_{>0.2}$, which could result in high stress at GBs. The stress level will be even higher if the GB lay $\perp$ to the loading direction. Such instances of microstructure could be prone to damage nucleation. The presence of inclusions such as MnS and AlN as shown in Figure 3 could also play a role at this stage. Micro-crack formation events at the MnS inclusions can be observed at $\varepsilon_{0.3}$ in Figure 9d. Many such instances of micro-cracks were also observed in AlN and MnS inclusions at very early stages of deformation at $\varepsilon_{0.10-0.3}$.

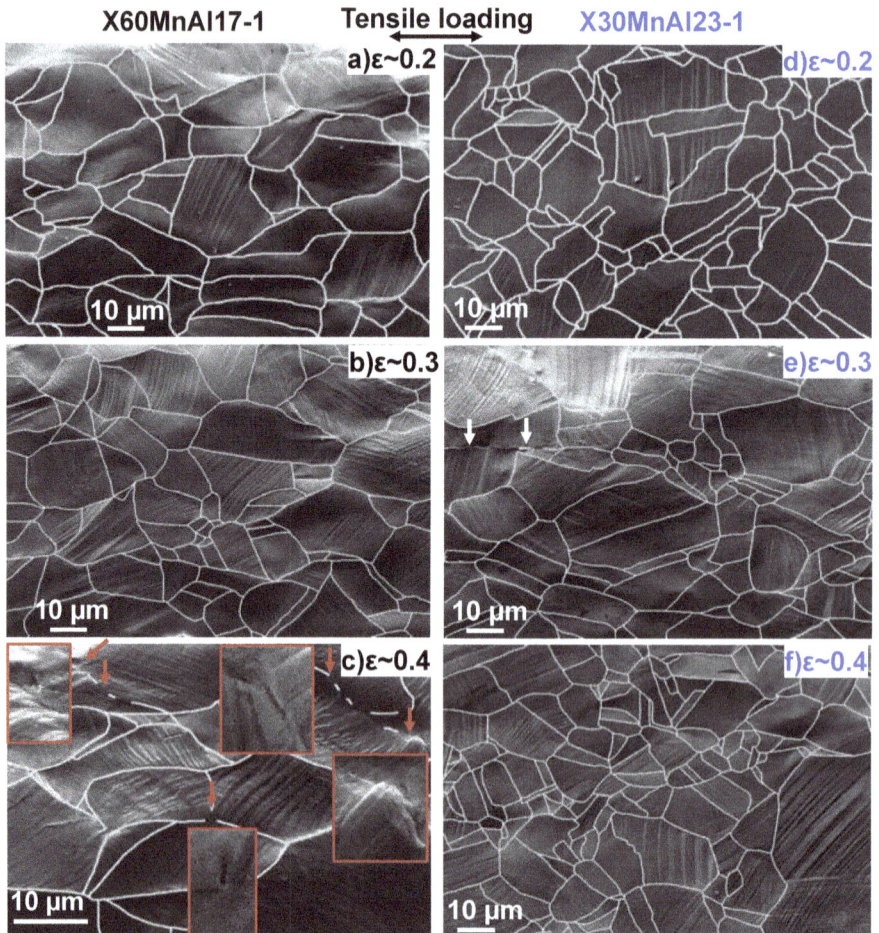

Figure 9. Evolution of microstructure and micro-crack formation with deformation. X60MnAl17-1: (**a**) $\varepsilon_{\sim0.2}$ (**b**) $\varepsilon_{\sim0.3}$ (**c**) $\varepsilon_{\sim0.4}$ (left column), X30MnAl23-1: (**d**) $\varepsilon_{\sim0.2}$ (**e**) $\varepsilon_{\sim0.3}$ (**f**) $\varepsilon_{\sim0.4}$ (right column); red arrows indicate micro-cracks; white arrows indicate cracks in MnS inclusions at grain boundary (GB); macroscopic true strain is indicated at the top right corner.

The saturation of the twinning within the microstructure is the precursor to the initiation of damage nucleation stage. Both the alloys deform by dislocation slip and twinning as shown in Figures 5 and 9. However, more pronounced twinning can be observed at earlier strains in X60MnAl17-1 alloy compared to an X30MnAl23-1 alloy. From Figure 5, it is very clear that at $\varepsilon_{0.3}$ large deformation due to slip and twinning within the grains can be observed in X60MnAl17-1 alloy, which increases stress at the GB. In such high stressed GBs shown in Figure 9c micro-voids nucleate and coalesce together leading to micro-crack formation. The micro-cracks are mainly observed at GBs, triple junctions and especially in boundaries which are ⊥ to loading direction as shown in Figure 10a,b. Thus it is very clear that stress concentration caused by the intersection of slip bands at GB lead to intergranular micro-crack formation. Many instances of micro-cracks at triple points can also be observed. However, in X30MnAl23-1 alloy the nucleation of micro-cracks did not start at $\varepsilon_{\sim 0.4}$, but at $\varepsilon_{necking}$. Many instances of inter-granular micro-crack formation events can be observed as shown in Figure 10c,d. The majority of cracks observed in the microstructure are at GB and triple points. The enlarged regions of microstructure in Figure 10 clearly indicates that micro-cracks initiated at the interception of slip bands and deformation twins at GB. From Figure 10a–d, it can be observed that the mechanism of micro-crack nucleation in both the alloys is the same. The micro-cracks which nucleated at the MnS and AlN inclusions did not grow much even after deforming to $\varepsilon_{>0.4}$ and does not propagate to the matrix (see Figure 11). Such micro-cracks can be observed in the AlN inclusions present at GB in an X60MnAl17-1 alloy as shown in Figure 11a. Similarly, in an X30MnAl23-1 alloy, micro-cracks were also observed at the MnS inclusion present at GB shown in Figure 11b and within the grain in Figure 11c. These micro-cracks were within the inclusion even after large deformation. Thus it can be stated that, micro-cracks nucleated in all inclusions present in the material, much before the beginning of the necking in both the alloys. The coalescence of micro-cracks at the inclusions present at different locations in the material could play a vital role in macroscopic failure initiation. The micro-cracks in X60MnAl17-1 alloy initiated earlier compared to X30MnAl23-1 alloy. The activation of DSA in an X60MnAl17-1 alloy could have led to an early start of damage nucleation.

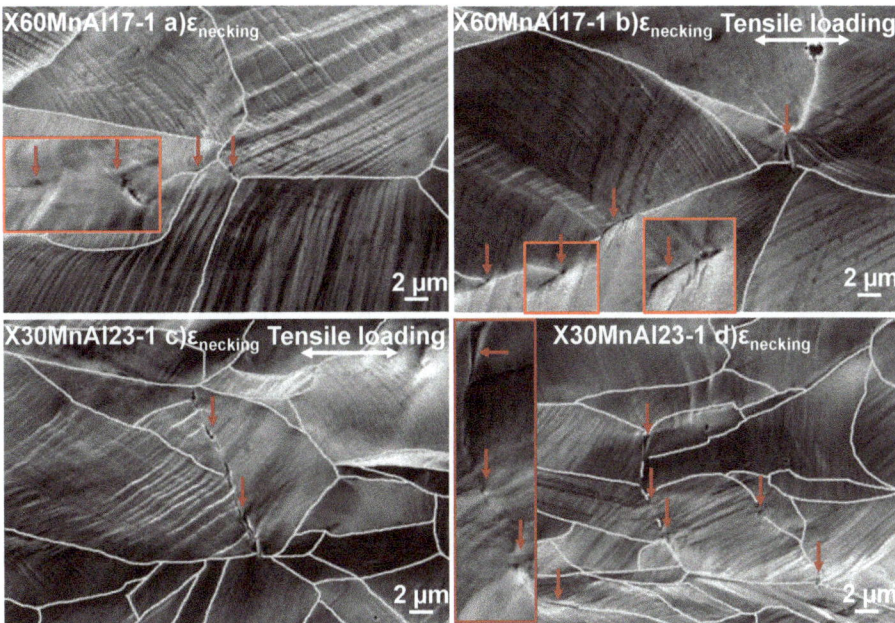

Figure 10. Mechanism of micro-cracks development at a strain of $\varepsilon_{necking}$. X60MnAl17-1 (**a**) Micro-crack initiation at triple junction and crack propagation into the neighboring grains assisted by the intercepting deformation twins, (**b**) Intergranular cracks nucleating at GB, X30MnAl23-1: (**c**) Micro-cracks initiation at GB and their propagation, (**d**) Intergranular cracks nucleation at grain boundaries and triple junctions; red arrows indicate micro-cracks.

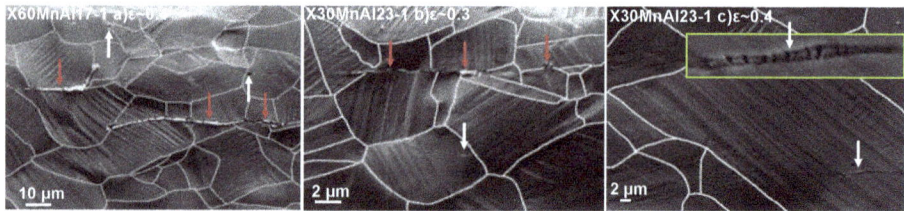

Figure 11. Micro-cracks at inclusions. X60MnAl17-1: (**a**) Micro-cracks in AlN inclusions at GB and in MnS inclusions close to GB and also inside the grains, X30MnAl23-1: (**b**) Micro-cracks in MnS inclusion at GB, (**c**) Micro-cracks in MnS inclusion inside a grain; red arrows indicates cracks in inclusions at GB; white arrows indicate cracks in inclusions inside the grains.

The damage growth stage starts with the coalescence of micro-cracks at various constituents of the microstructure in the material leading to macroscopic failure initiation. From the observations shown in Figures 10 and 11, it is very clear that there is a great interplay between the deformation twins and slip bands at micro-scale until the saturation of twinning. The saturation of twinning within the microstructure has led to the nucleation of micro-cracks at the high stressed GB and triple junctions. Since deformation by twinning and slip is quite uniform in almost all grains in the entire microstructure, the cracks were initiated all over the cross-section. The micro-cracks formed at MnS and AlN inclusions in the material also play a significant role in the damage growth. The variety of micro-cracks formed at different microstructure constituents coalescence together rapidly to develop macro-cracks leading to localized necking. It can be understood from the σ–ε curve in Figures 4 and 6 that in both alloys large strain localization in the material occurred resulting in necking and failure. The appearance of localized necking before the failure shows that there is a plastic localization before the final failure as shown in Figure 12. This indicates that nucleation of micro-voids in the material is followed by the coalescence and growth of many micro-voids leading to micro-crack formation. The role of MnS and AlN inclusions on the failure initiation can be visualized on the fracture surfaces. The presence of large elliptical or round voids (white arrows) shows that coalescence of many smaller voids has occurred at the MnS or AlN inclusions. The large number of minute voids of size $\leq$2.5 µm can be seen on the fracture surfaces for both alloys. Thus it is clear that the predominant fracture mode in both alloys is a ductile failure with the formation of fine dimples.

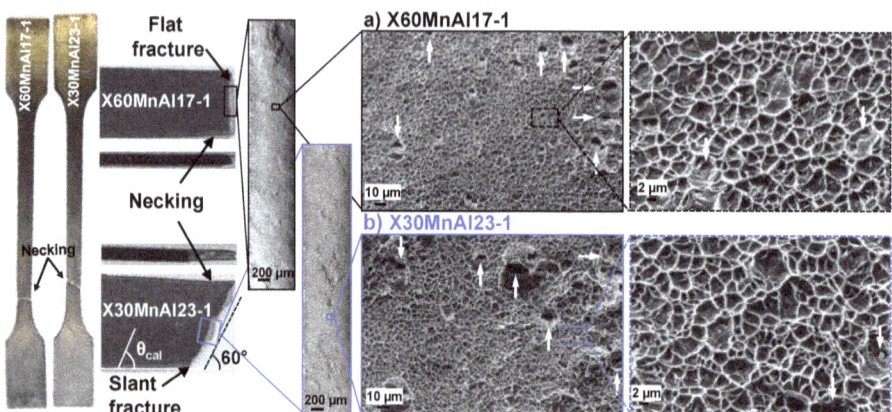

Figure 12. Fracture behavior of tensile specimens. Diffuse necking and fracture surface appearance (**left**) and SEM images of the fracture surfaces at different magnifications (**right**). Average dimple size of both X60MnAl17-1 and X30MnAl23-1 is approximately $\leq$2.5 µm. White arrows indicate large elliptical voids at MnS and AlN inclusions.

4. Discussion

4.1. Serrated Flow and PLC Effect

In an X60MnAl17-1 alloy repeated serrations can be observed on the $\sigma-\varepsilon$ curves during the plastic deformation, whereas in an X30MnAl23-1 alloy, no serrations can be observed at RT and quasi-static $\dot{\varepsilon}$ (see Figure 4). The manifestations in $\sigma-\varepsilon$ behavior in the form of serrations such as a change in mechanical properties and negative SRS is mainly due to DSA. The mechanism of DSA in C containing high Mn TWIP steels can be attributed to the formation of Mn-C short-range order or short-range cluster (SRO/SRC) [11,22,24]. It is proposed that DSA occurs when the C atoms in the Mn-C octahedral clusters reorient themselves in the core of the dislocations, thereby locking the dislocations leading to high dislocation density [36]. During this dynamic interaction between the diffusing atoms and the mobile dislocations during plastic flow, C atoms pin strongly to the dislocations increasing their pile up. When the applied stress is sufficiently high, the dislocations overcome the obstacles in the form of SRO/SRC at once by a single diffusive jump of C atoms and move to next obstacle, where they are stopped again and the process is repeated [22]. The formation of Mn-C octahedral clusters result in increased lattice resistance for dislocation glide as the passage of partial dislocation will change the local position of both substitutional and interstitial atoms [20].

The serrations were initiated at a ε_c of ~12.5% in an X60Mn22 alloy without Al [14], whereas in alloys with 1.5 wt.% Al-addition, the serrations appeared at a ε_c of ~24% in X60MnAl17-1 and serrations are completely suppressed in X30MnAl23-1 at RT and quasi-static $\dot{\varepsilon}$ (see Figures 6 and 7). In the Fe–Mn–Al–C alloys, the serrations are reported to occur mainly in the steels with C content above 0.6 wt.% at RT tensile testing [14,24,27–29,37]. It was stated that with the addition of Al, the diffusivity of the carbon will be reduced and the serrated flow will be shifted to higher temperatures [38]. The addition of Al has suppressed the DSA phenomena by increasing the activation energy for C diffusion and reducing the interaction time between stacking faults and the Mn-C SRO [16]. However, even in Al-added alloys, the DSA could occur at elevated temperatures because of the increased defect mobility resulting in serrated $\sigma-\varepsilon$ curves.

Twinning behavior in both Al-added alloys are quite similar, except for the initiation and saturation of twinning and TVF (see Figures 5 and 13). Even though twinning occurs in both alloys, the serrated $\sigma-\varepsilon$ flow is observed only in X60MnAl17-1 alloy and not in X30MnAl23-1 alloy. The local deformation behavior is inhomogeneous in an X60MnAl17-1 alloy due to the initiation and propagation of deformation bands during deformation commonly known as Portevin-LeChatelier (PLC) effect, whereas in an X30MnAl23-1 alloy homogenous $\sigma-\varepsilon$ behavior can be observed. The nucleation and propagation of deformation bands result in serrations on the $\sigma-\varepsilon$ curves which can be ascribed to the DSA. Kang et al. showed that the number of Mn-C SRO clusters increases with increasing macroscopic strain in X60Mn18 alloy [23]. Song and Houston reported that with an increase in ε, the volume fraction of Mn-C SRO/SRC increases and the mean radius of clusters decreases [24]. Madivala et al. investigated an X60Mn22 alloy and showed that serrations disappear at elevated temperatures and at high strain rates. It was also stated that twinning occurs at these temperatures and at high strain rate tests [14,19]. Koyama et al. also indicated that the deformation twinning and martensitic transformation did not cause the serrations [39]. Lebedkina et al. suggested that twinning starting at some point in the specimen will initiate twins in the neighboring grains and advance into the unoccupied regions of the specimen. However, they claimed that twinning cannot alone responsible for the entire strain accumulation during strain jumps due to low TVF observed in TWIP Steel [40]. Recently, Sevsek et al. proposed a multi-scale-bridging model for the formation and propagation of the deformation bands and the resulting occurrence of the serrated flow in high Mn TWIP steels. However, they also stated that neither the SRO-based theory nor the deformation twinning based theory proposed could explain the experimental observations completely [37]. Based on the experimental observations from this study and the from the literature, it can be clearly stated that deformation twinning and ε-martensite transformation cannot be responsible for causing serrations in Fe-Mn-C-Al alloys. Thus, DSA is the dominant factor for causing the serrations in an X60MnAl17-1 alloy.

The strain localization within the deformation bands in X60MnAl17-1 alloy is slightly different over the gauge length during the transition regime (see Figure 7l–n). After their complete development, the strain localization in them increases drastically compared to other areas (see Figure 7n). The ε_{local} is much larger in the region where deformation band has already passed by (see in Figure 7m,n marked by 'B', 'D') compared to another region where the deformation bands has to pass through (see in Figure 7m,n marked by 'A', 'C'). The velocity of propagation of the deformation bands decreases with increase in macroscopic strain. Song and Houston showed by SANS experiments that the size and number density of SRO is much larger in the regions where deformation band has already passed by compared to the region where it has to pass through [24]. Also due to the fact that SRO/SRC offers higher resistance to dislocation glide thereby leading to (a) aging of dislocations by C-atoms during their interaction time at obstacles (b) enhancement of diffusion coefficient by vacancies generated during plastic flow and (c) increase in mobile dislocation density with strain. This results in increased resistance offered by the Mn-C clusters and also by the deformation twins for the bands propagation leading to decrease in velocity and difference local deformation behavior.

4.2. Strain Hardening and Twinning Evolution

Deformation twinning during plastic deformation plays a significant role in enhancing the strain hardening behavior of materials through the creation of new boundaries thereby subdividing the original grains commonly known as dynamic Hall-Petch effect [10,13,39]. Twin boundaries act as the barriers to dislocation motion thereby progressively reducing the effective mean free path (MFP) of dislocations resulting in enhanced strain hardening. Bouaziz et al., De Cooman et al., Koyama et al. claim that mechanical twinning and the related dynamic Hall-Petch effect associated with TWIP effect remain the only mechanisms that can explain the high strain hardening of TWIP steel [10,13,39]. However, the role of DSA in manifesting the SHR of TWIP steels as explained in Section 4.1 cannot be neglected.

An X60MnAl17-1 alloy, where DSA can be observed shows enhanced strain hardening and ductility compared to an X30MnAl23-1 alloy, where DSA does not occur (see Figure 13). Such sustained and high work hardening rates can also be seen in austenitic steels with high C content because of DSA [41]. The ab-initio based study combined with SANS experiments by Song et al. [42] and SANS experimental study by Kang et al. proved the presence of Mn-C octahedral clusters and their evolution, indicating the occurrence of DSA during deformation in high Mn steels. The occurrence of static strain aging due to the presence of Mn-C SRO was stated to be responsible for increased yield strength and pronounced yielding of annealed TWIP steel [43]. It was stated that deformation twinning in TWIP steels is promoted by thermally activated nature of dislocation slip due to dynamic interactions between solute C atoms and dislocations that inhibit dislocation slip due to lattice friction effects [44]. It was proposed that the trapping of slowly moving partial dislocations by C atoms results in increased inter-granular stress and reduction of MFP [37,39]. The increased localized stress near the GB facilitate grain boundary nucleation of deformation twins [45]. DSA is also showed to be crystallographic orientation dependent, could occur at elevated temperatures and is enhanced by increasing C content [39]. Thus in an X60MnAl17-1 alloy, the occurrence of DSA has led to increased stresses at GBs. This led to triggering of the nucleation of multiple deformation twins in the preferably oriented grains. Whereas in X30MnAl23-1 alloy, conventional twinning mechanisms occurs in combination with dislocation glide at RT (see Figure 5). Thus the yield strength increase by static strain aging and the indirect promotion of the mechanical twinning during deformation in an X60MnAl17-1 alloy, could have led to high SHR and enhanced ductility.

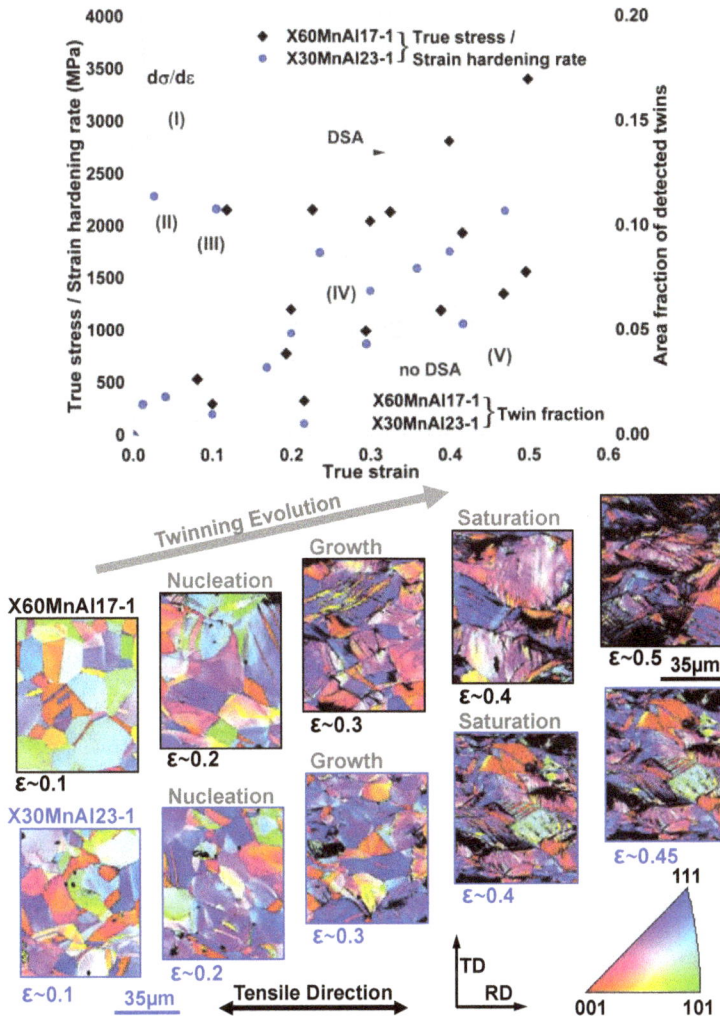

Figure 13. Strain hardening in relation to the evolution of twinning; strain hardening behavior (**top**) and EBSD IPF maps with increasing macroscopic true strain (**bottom**).

The evolution of twinning with increasing macroscopic strain is correlated with the SHR as shown in Figure 13. The RD-IPF maps at different strains are also shown. Based on the detailed work by De Cooman et al., Gutierrez-Urrutia and Raabe, the SHR in the current study is divided into five stages [13,46]. Stage (I) is the initial strain hardening stage controlled by the dislocation density evolution without the initiation of twins. The initial downfall of SHR is characterized by a continuously decreasing rate of dislocation storage and an increasing rate of dynamic recovery. The dislocation reactions in this stage are responsible for the nucleation of twins. Stage (II) is characterized by an increase of SHR in both alloys and is referred to as the primary twinning stage. It begins slightly above yielding and at a strain of ∼0.03 in an X30MnAl23-1 alloy and at a strain of ∼0.04 in an X60MnAl17-1 alloy (see Figure 4). The stress at the beginning of stage II is referred to as the twin initiation stress. It is about 345 MPa in an X30MnAl23-1 alloy and is about 435 MPa in X60MnAl17-1 alloy. The evolution of dislocation structures and the initiation of primary twins are supposed to be responsible for the rise of SHR in this stage. The maximum SHR for both the alloys occurs at ∼0.05 strain at the end of stage II.

Stage (III) is characterized by a decrease in SHR with increasing macroscopic strain. It can be attributed to the reduced rate of primary twins initiation. It was also stated that the initial twins subdivide the original grains thereby reducing the MFP of dislocations, which results in increased stress for twin nucleation. At the end of this stage, twins can be clearly observed in the preferably oriented grains as shown in Figure 13 at a strain of 0.2. Stage (IV) and (V) are quite different in both the alloys. Stage (IV) and (V) in an X30MnA123-1 alloy marked by a continuous decrease in SHR, ascribed to the reduced additional refinement of the dislocation and twin substructures, together with the increasing strengthening effect of the individual twins as obstacles to dislocation glide, reduce the capacity for trapping more dislocations. Stage (IV) in an X60MnAl17-1 alloy can be distinguished by again increase in SHR and reaching maximum between a strain of 0.3 and 0.4. This stage is characterized by the activation of the secondary twin systems and the formation of the multiple twin-twin interactions. This results in a further subdivision of grains and thus reduces the MFP of the dislocations considerably leading to increasing or constant SHR. During stage (V) in X60MnAl17-1 alloy, the SHR decreases continuously. In this stage TVF increases significantly and the twin bundles get thicker and denser. This can be observed in the microstructure at a strain of 0.4 and 0.5. The major difference in SHR between two alloys can be observed at stage (III), at which DSA gets activated in X60MnAl17-1 alloy which has indirectly led to significant increase in TVF, whereas in X30MnAl23-1 alloy, the TVF increase was marginal.

4.3. Damage and Fracture

Heterogeneous deformation twinning and dislocation slip could be observed in the microstructure in the early stages of deformation (see Figures 5 and 13). Micro-cracks could be observed in the MnS and AlN inclusions at strain below 0.2 (see Figure 11). The nucleation and propagation of deformation bands due to DSA has resulted in localization of strain within the bands leading to failure initiation (see Figure 7). Thus twins, slip bands, deformation bands, and non-metallic inclusions play a significant role in the fracture phenomena in TWIP steels.

In both alloys, dislocation glide and deformation twinning are the major active deformation mechanisms. Even though both these deformation mechanisms did not induce damage directly, but they acted as a source for increasing the stress concentration within the microstructure, which is a precursor step for damage initiation. A schematic diagram illustrating the mechanism of micro-crack formation in TWIP steels based on the experimental observations is depicted in Figure 14. The development of large inter-granular regions with high dislocation density within the microstructure due to the subdivision of grains due to twinning and also deformation by slip resulted in pronounced inter-granular stress. Deformation by twinning and dislocation slip can be observed in the microstructure in the preferably oriented grains leading to the localization of the plastic strain. It could be observed that <111> tensile orientation is the preferential orientation for the deformation twinning and <100> is the non-preferential tensile orientation for twinning. Many bundles of primary and secondary deformation twins can be observed in <111>-oriented grains and slip bands in <100>-oriented grains (see Figure 13). Largest plastic strain concentration can occur in the <111>-oriented grains, which are preferred for the deformation twinning and the second largest strain concentration could occur in the <100>-oriented grains, which are preferred for the dislocation slip. The large plastic localization associated to accommodate the plasticity within the grains results in high-stress concentration at the GB. This results in the initiation of voids which combine together leading to initiation of inter-granular micro-cracks (see Figures 10 and 14). Even though micro-cracks initiated at inclusions at an early stage, their growth during deformation was limited (see Figure 11). Thus at a micro-scale, stress concentration caused by the interception of deformation twins and slip bands at GB lead to damage initiation. This results in the intergranular crack propagation with a fracture surface characterized by shallow dimples (see Figure 12).

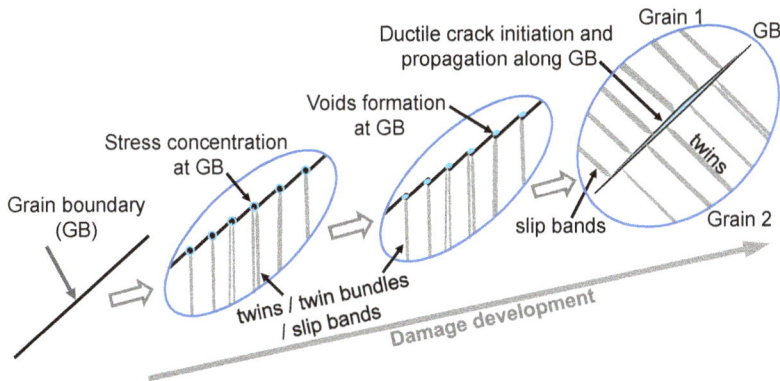

Figure 14. A schematic illustration of micro-cracking mechanisms in TWIP steels based on the experimental observations from interrupted micro-tensile tests carried out in the SEM. Stress concentration caused by the intercepting deformation twins at GB. Void formation and their growth leading to inter-granular micro-crack initiation at GBs.

5. Conclusions

The Al-added TWIP steels X60MnAl17-1 and X30MnAl23-1 were investigated by carrying out uni-axial tensile tests. The local plastic strain (ε_{local}) and temperature evolution during deformation were monitored in conjunction with digital image correlation and synchronous temperature measurements using thermocamera. Interrupted tensile tests were carried out in SEM along with EBSD measurements to study the microstructure evolution and micro-cracks development. The main conclusions can be drawn as follows:

- Strain hardening rate of an X60MnAl17-1 alloy is extraordinarily high compared to an X30MnAl23-1 alloy. An X60MnAl17-1 alloy showed higher yield strength, tensile strength and elongation compared to an X30MnAl23-1 alloy. The enhanced mechanical properties of the X60MnAl17-1 alloy is mainly due to the enhanced deformation twinning in addition to dislocation glide and also activation of dynamic strain aging (DSA). DSA is completely suppressed in an X30MnAl23-1 alloy at room temperature and quasi-static strain rate due to lower carbon content.
- Twining is the most predominant deformation mechanism occurred along with dislocation slip in both the alloys. The addition of Al has led to increased stacking fault energy thereby delaying nucleation of deformation twins and prolonged the saturation of twinning.
- Micro-cracks are observed at elongated MnS inclusions or at AlN inclusions at a relatively small strain of ~2/3 of total strain. However, these micro-cracks showed no tendency to grow.
- Large heterogeneous deformation within the grains by twinning or dislocation slip has led to a high-stress concentration at grain boundaries (GBs) due to the interception of deformation twins and slip band extrusions at GBs. Hence micro-cracks in Al-added TWIP steels originated mainly at grain boundaries and triple junctions.
- In an X60MnAl17-1 alloy, the occurrence of DSA has led to inhomogeneous flow behavior due to the nucleation and propagation of deformation bands during deformation. This resulted in large strain localization within the deformation bands and the velocity of band motion decreased with increasing strain due to the intersection of two bands. The ε_{local} accumulation within the intersecting bands resulted in a macroscopic crack initiation close to the edges of the tensile specimen. In an X30MnAl23-1 alloy, homogenous ε_{local} distribution throughout the gauge length could be observed until the beginning of necking. Thereafter failure in the material occurred by classical necking and strain localization.
- The ductile failure mode is the most predominant mode of failure in Al-added TWIP steels, mainly characterized by the formation of very fine dimples with a crack propagation along GBs.

Author Contributions: M.M. designed, performed and analyzed the experimental data such as SEM, tensile tests and mechanical properties. A.S. performed the EBSD measurements and analyzed the data. W.B. and U.P. contributed with ideas and intensive discussions. M.M. wrote the initial draft version. All authors contributed equally to the interpretation of results and writing the final version of the manuscript.

Funding: This research was funded by the German Research Foundation (DFG) within the framework of the SFB 761 "Steel-ab initio".

Acknowledgments: This work was carried out within the framework of the SFB 761 "Steel-ab initio" under the sub-project C6: Damage and failure. We would like to thank all the members of the SFB 761 "Steel-ab initio" project for their valuable cooperation and tremendous support.

Conflicts of Interest: The authors declare no conflict of interest.

Appendix A. Macro-Tensile Tests Setup for the Local Strain Analysis Using Digital Image Correlation

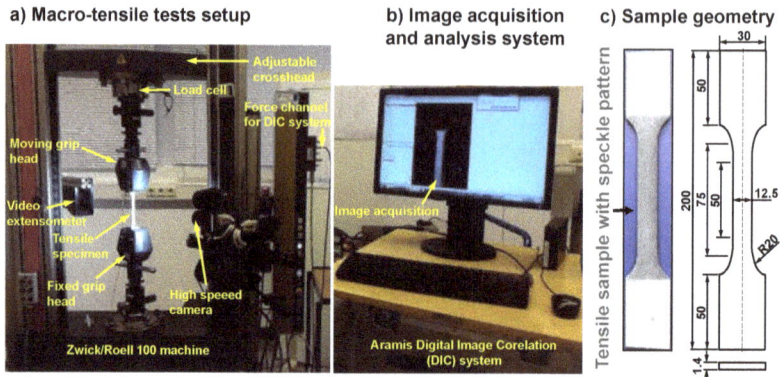

Figure A1. Macro-Tensile tests carried out in conjunction with DIC for investigating the local plastic strain (ε_{local}) evolution, nucleation, and propagation of deformation bands with increasing strain. (**a**) Tensile testing machine depicting the adjustable crosshead, grips for clamping, load cell, video extensometer and high-speed camera for image acquisition, (**b**) Aramis DIC system used for analysis, (**c**) Tensile sample with black and white speckle pattern prepared for ε_{local} measurement.

Appendix B. Macro-Tensile Tests Setup for Temperature Measurement using High Resolution Thermocamera

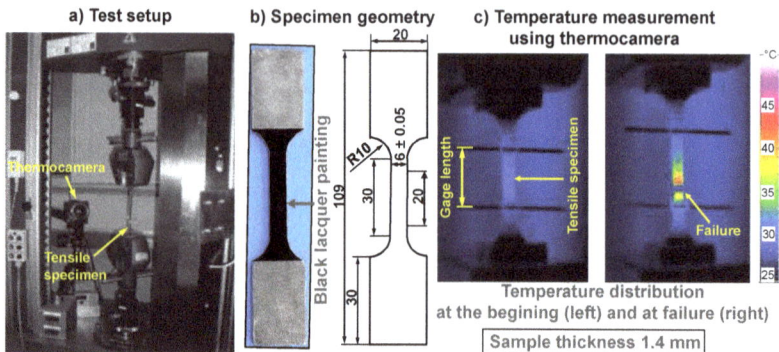

Figure A2. Temperature measurement setup using Macro-Tensile tests carried out in conjunction with high-resolution thermocamera for investigating the rise in temperature due to adiabatic heating. (**a**) Tensile testing machine depicting the thermocamera, tensile sample, (**b**) sample painted with black lacquer painting to minimize the reflections from the surroundings, (**c**) Image acquisition and analysis using IRBIS® online software.

References

1. Horvath, C.D. Advanced steels for lightweight automotive structures. In *Materials, Design and Manufacturing for Lightweight Vehicles*; Elsevier: Amsterdam, The Netherlands, 2010; pp. 35–78. [CrossRef]
2. Hirsch, J. Aluminium in Innovative Light-Weight Car Design. *Mater. Trans.* **2011**, *52*, 818–824. [CrossRef]
3. Hirsch, J.; Al-Samman, T. Superior light metals by texture engineering: Optimized aluminum and magnesium alloys for automotive applications. *Acta Mater.* **2013**, *61*, 818–843. [CrossRef]
4. Sherman, A.M.; Sommer, C.J.; Froes, F.H. The use of titanium in production automobiles: Potential and challenges. *JOM* **1997**, *49*, 38–41. [CrossRef]
5. Luo, A.; Sachdev, A. 12—Applications of magnesium alloys in automotive engineering. In *Advances in Wrought Magnesium Alloys*; Bettles, C., Barnett, M., Eds.; Woodhead Publishing Series in Metals and Surface Engineering; Woodhead Publishing: Cambridge, UK, 2012; pp. 393–426. [CrossRef]
6. Kantanen, P.; Somani, M.; Kaijalainen, A.; Haiko, O.; Porter, D.; Kömi, J. Microstructural Characterization and Mechanical Properties of Direct Quenched and Partitioned High-Aluminum and High-Silicon Steels. *Metals* **2019**, *9*, 256. [CrossRef]
7. Hall, J.N.; Fekete, J.R. Steels for auto bodies. In *Automotive Steels: Design, Metallurgy, Processing and Applications*; Woodhead Publishing: Cambridge, UK, 2017; pp. 19–45.
8. Krauss, G. Physical metallurgy of steels. In *Automotive Steels: Design, Metallurgy, Processing and Applications*; Woodhead Publishing: Cambridge, UK, 2017; pp. 95–111.
9. Grässel, O.; Krüger, L.; Frommeyer, G.; Meyer, L. High strength Fe–Mn–(Al, Si) TRIP/TWIP steels development–properties–application. *Int. J. Plast.* **2000**, *16*, 1391–1409. [CrossRef]
10. Bouaziz, O.; Allain, S.; Scott, C.; Cugy, P.; Barbier, D. High manganese austenitic twinning induced plasticity steels: A review of the microstructure properties relationships. *Curr. Opin. Solid State Mater. Sci.* **2011**, *15*, 141–168. [CrossRef]
11. Saeed-Akbari, A.; Mosecker, L.; Schwedt, A.; Bleck, W. Characterization and Prediction of Flow Behavior in High-Manganese Twinning Induced Plasticity Steels: Part I. Mechanism Maps and Work-Hardening Behavior. *Metall. Mater. Trans. A* **2012**, *43*, 1688–1704. [CrossRef]
12. Lee, S.J.; Han, J.; Lee, S.; Kang, S.H.; Lee, S.M.; Lee, Y.K. Design for Fe-high Mn alloy with an improved combination of strength and ductility. *Sci. Rep.* **2017**, *7*, 3573. [CrossRef] [PubMed]
13. De Cooman, B.C.; Estrin, Y.; Kim, S.K. Twinning-induced plasticity (TWIP) steels. *Acta Mater.* **2018**, *142*, 283–362. [CrossRef]
14. Madivala, M.; Schwedt, A.; Prahl, U.; Bleck, W. Anisotropy and strain rate effects on the failure behavior of TWIP steel: A multiscale experimental study. *Int. J. Plast.* **2019**, *115*, 178–199. [CrossRef]
15. Madivala, M.; Bleck, W. Strain Rate Dependent Mechanical Properties of TWIP Steel. *JOM* **2019**, *71*, 1291–1302. [CrossRef]
16. Jung, I.C.; De Cooman, B.C. Temperature dependence of the flow stress of Fe–18Mn–0.6C–xAl twinning-induced plasticity steel. *Acta Mater.* **2013**, *61*, 6724–6735. [CrossRef]
17. Pierce, D.T.; Jiménez, J.A.; Bentley, J.; Raabe, D.; Wittig, J.E. The influence of stacking fault energy on the microstructural and strain-hardening evolution of Fe–Mn–Al–Si steels during tensile deformation. *Acta Mater.* **2015**, *100*, 178–190. [CrossRef]
18. Kim, J.K.; Kwon, M.H.; De Cooman, B.C. On the deformation twinning mechanisms in twinning-induced plasticity steel. *Acta Mater.* **2017**, *141*, 444–455. [CrossRef]
19. Madivala, M.; Schwedt, A.; Wong, S.L.; Roters, F.; Prahl, U.; Bleck, W. Temperature dependent strain hardening and fracture behavior of TWIP steel. *Int. J. Plast.* **2018**, *104*, 80–103. [CrossRef]
20. Chen, L.; Kim, H.S.; Kim, S.K.; De Cooman, B.C. Localized Deformation due to Portevin–LeChatelier Effect in 18Mn–0.6C TWIP Austenitic Steel. *ISIJ Int.* **2007**, *47*, 1804–1812. [CrossRef]
21. Allain, S.; Cugy, P.; Scott, C.; Chateau, J.P.; Rusinek, A.; Deschamps, A. The influence of plastic instabilities on the mechanical properties of a high-manganese austenitic FeMnC steel. *Int. J. Mater. Res.* **2008**, *99*, 734–738. [CrossRef]
22. Lee, S.J.; Kim, J.; Kane, S.N.; De Cooman, B.C. On the origin of dynamic strain aging in twinning-induced plasticity steels. *Acta Mater.* **2011**, *59*, 6809–6819. [CrossRef]

23. Kang, M.; Shin, E.; Woo, W.; Lee, Y.K. Small-angle neutron scattering analysis of Mn–C clusters in high-manganese 18Mn–0.6C steel. *Mater. Charact.* **2014**, *96*, 40–45. [CrossRef]
24. Song, W.; Houston, J. Local Deformation and Mn-C Short-Range Ordering in a High-Mn Fe-18Mn-0.6C Steel. *Metals* **2018**, *8*, 292. [CrossRef]
25. Koyama, M.; Sawaguchi, T.; Tsuzaki, K. TWIP Effect and Plastic Instability Condition in an Fe-Mn-C Austenitic Steel. *ISIJ Int.* **2013**, *53*, 323–329. [CrossRef]
26. Koyama, M.; Sawaguchi, T.; Tsuzaki, K. Overview of Dynamic Strain Aging and Associated Phenomena in Fe–Mn–C Austenitic Steels. *ISIJ Int.* **2018**, *58*, 1383–1395. [CrossRef]
27. Kim, J.; Estrin, Y.; De Cooman, B.C. Application of a Dislocation Density-Based Constitutive Model to Al-Alloyed TWIP Steel. *Metall. Mater. Trans. A* **2013**, *44*, 4168–4182. [CrossRef]
28. Kim, J.K.; De Cooman, B.C. Stacking fault energy and deformation mechanisms in Fe-xMn-0.6C-yAl TWIP steel. *Mater. Sci. Eng. A* **2016**, *676*, 216–231. [CrossRef]
29. Yang, H.K.; Zhang, Z.J.; Dong, F.Y.; Duan, Q.Q.; Zhang, Z.F. Strain rate effects on tensile deformation behaviors for Fe–22Mn–0.6C–(1.5Al) twinning-induced plasticity steel. *Mater. Sci. Eng. A* **2014**, *607*, 551–558. [CrossRef]
30. Yu, H.Y.; Lee, S.M.; Nam, J.H.; Lee, S.J.; Fabrègue, D.; Park, M.h.; Tsuji, N.; Lee, Y.K. Post-uniform elongation and tensile fracture mechanisms of Fe-18Mn-0.6C-xAl twinning-induced plasticity steels. *Acta Mater.* **2017**, *131*, 435–444. [CrossRef]
31. Yang, C.L.; Zhang, Z.J.; Zhang, P.; Zhang, Z.F. The premature necking of twinning-induced plasticity steels. *Acta Mater.* **2017**, *136*, 1–10. [CrossRef]
32. Fabrègue, D.; Landron, C.; Bouaziz, O.; Maire, E. Damage evolution in TWIP and standard austenitic steel by means of 3D X ray tomography. *Mater. Sci. Eng. A* **2013**, *579*, 92–98. [CrossRef]
33. Lorthios, J.; Nguyen, F.; Gourgues, A.F.; Morgeneyer, T.F.; Cugy, P. Damage observation in a high-manganese austenitic TWIP steel by synchrotron radiation computed tomography. *Scr. Mater.* **2010**, *63*, 1220–1223. [CrossRef]
34. Allain, S.; Chateau, J.P.; Bouaziz, O.; Migot, S.; Guelton, N. Correlations between the calculated stacking fault energy and the plasticity mechanisms in Fe–Mn–C alloys. *Mater. Sci. Eng. A* **2004**, *387–389*, 158–162. [CrossRef]
35. Ledbetter, H.M.; Frederick, N.V.; Austin, M.W. Elastic–constant variability in stainless–steel 304. *J. Appl. Phys.* **1980**, *51*, 305–309. [CrossRef]
36. Dastur, Y.N.; Leslie, W.C. Mechanism of work hardening in Hadfield manganese steel. *Metall. Trans. A* **1981**, *12*, 749–759. [CrossRef]
37. Sevsek, S.; Brasche, F.; Haase, C.; Bleck, W. Combined deformation twinning and short-range ordering causes serrated flow in high-manganese steels. *Mater. Sci. Eng. A* **2019**, *746*, 434–442. [CrossRef]
38. Shun, T.; Wan, C.M.; Byrne, J.G. A study of work hardening in austenitic Fe-Mn-C and Fe-Mn-Al-C alloys. *Acta Metall. Mater.* **1992**, *40*, 3407–3412. [CrossRef]
39. Koyama, M.; Sawaguchi, T.; Tsuzaki, K. Deformation Twinning Behavior of Twinning-Induced Plasticity Steels with Different Carbon Concentrations. *Tetsu-to-Hagane* **2014**, *100*, 1253–1260. [CrossRef]
40. Lebedkina, T.A.; Lebyodkin, M.A.; Chateau, J.P.; Jacques, A.; Allain, S. On the mechanism of unstable plastic flow in an austenitic FeMnC TWIP steel. *Mater. Sci. Eng. A* **2009**, *519*, 147–154. [CrossRef]
41. Ogawa, T.; Koyama, M.; Tasan, C.C.; Tsuzaki, K.; Noguchi, H. Effects of martensitic transformability and dynamic strain age hardenability on plasticity in metastable austenitic steels containing carbon. *J. Mater. Sci.* **2017**, *52*, 7868–7882. [CrossRef]
42. Song, W.; Bogdanovski, D.; Yildiz, A.; Houston, J.; Dronskowski, R.; Bleck, W. On the Mn–C Short-Range Ordering in a High-Strength High-Ductility Steel: Small Angle Neutron Scattering and Ab Initio Investigation. *Metals* **2018**, *8*, 44. [CrossRef]
43. Wesselmecking, S.; Song, W.; Ma, Y.; Roesler, T.; Hofmann, H.; Bleck, W. Strain Aging Behavior of an Austenitic High-Mn Steel. *Steel Res. Int.* **2018**, *88*, 1700515. [CrossRef]
44. Allain, S.; Bouaziz, O.; Lebedkina, T.; Lebyodkin, M. Relationship between relaxation mechanisms and strain aging in an austenitic FeMnC steel. *Scr. Mater.* **2011**, *64*, 741–744. [CrossRef]

45. Gwon, H.; Kim, J.K.; Shin, S.; Cho, L.; De Cooman, B.C. The effect of vanadium micro-alloying on the microstructure and the tensile behavior of TWIP steel. *Mater. Sci. Eng. A* **2017**, *696*, 416–428. [CrossRef]

46. Gutierrez-Urrutia, I.; Raabe, D. Grain size effect on strain hardening in twinning-induced plasticity steels. *Scr. Mater.* **2012**, *66*, 992–996. [CrossRef]

metals

MDPI

Article

From High-Manganese Steels to Advanced High-Entropy Alloys

Christian Haase [1],* and Luis Antonio Barrales-Mora [2]

[1] Steel Institute, RWTH Aachen University, Intzestraße 1, 52072 Aachen, Germany
[2] George W. Woodruff School of Mechanical Engineering, Georgia Institute of Technology, 2 Rue Marconi, 57070 Metz, France
*   Correspondence: christian.haase@iehk.rwth-aachen.de; Tel.: +49-241-80-95821

Received: 7 June 2019; Accepted: 25 June 2019; Published: 27 June 2019

Abstract: Arguably, steels are the most important structural material, even to this day. Numerous design concepts have been developed to create and/or tailor new steels suited to the most varied applications. High-manganese steels (HMnS) stand out for their excellent mechanical properties and their capacity to make use of a variety of physical mechanisms to tailor their microstructure, and thus their properties. With this in mind, in this contribution, we explore the possibility of extending the alloy design concepts that haven been used successfully in HMnS to the recently introduced high-entropy alloys (HEA). To this aim, one HMnS steel and the classical HEA Cantor alloy were subjected to cold rolling and heat treatment. The evolution of the microstructure and texture during the processing of the alloys and the resulting properties were characterized and studied. Based on these results, the physical mechanisms active in the investigated HMnS and HEA were identified and discussed. The results evidenced a substantial transferability of the design concepts and more importantly, they hint at a larger potential for microstructure and property tailoring in the HEA.

Keywords: high-manganese steels; high-entropy alloys; alloy design; plastic deformation; annealing; microstructure; texture; mechanical properties

---

1. Introduction

One of the most important objectives of materials scientists is the improvement of the mechanical properties of materials. For this purpose, it is possible to make use of diverse physical phenomena that essentially modify the microstructure to either facilitate or complicate the motion and generation of dislocations depending on the requirements of the final component. In the same way as alloys aim to be more than the sum of their parts, modern strengthening concepts aim to combine and—perhaps more important—trigger strengthening mechanisms at the right times. This is the idea behind the high-manganese steels (HMnS), whose mechanical properties are improved by activating either martensitic transformations (transformation-induced plasticity, TRIP) or mechanical twinning (twinning-induced plasticity, TWIP) to achieve higher strength and larger elongation [1]. The combination of these mechanisms with strong planar dislocation glide makes these materials possess outstanding strain-hardening potential attributable to an observed dynamic Hall–Petch effect [2].

Another recently developed class of alloys are the so-called high-entropy alloys (HEA). HEAs are alloys with more than four elements in usually equiatomic or near-equiatomic compositions. The idea behind these alloys is that their high entropy stabilizes the solid solution against the formation of intermetallic phases. HEAs are less well studied than HMnS, but since their discovery by Cantor et al. [3] and Yeh et al. [4], HEAs have attracted much fascination within the research community due to the vast space of possible alloys and thus, potential microstructures and property combinations. Although several core effects (high-entropy effect, sever lattice distortion effect, sluggish diffusion effect, cocktail effect) of HEAs have been proposed [5], a clear design strategy is still missing.

The underlying physical mechanisms active in HMnS have already been studied for several decades [1,2,6], and therefore are comparatively well understood. One of the most important parameters when designing HMnS is the tailoring of their stacking fault energy (SFE). This property controls the dissociation distance of partial dislocations and determines whether TRIP and/or TWIP is activated/suppressed during plastic deformation. The role of the SFE during plastic deformation and subsequent heat treatment in metals has been studied for decades, and its effect is relatively well-known for face-centered cubic (fcc) metals [7]. In fcc HEAs, the mechanisms of microstructure formation have been found to occur in a similar way to the same class of alloys with comparable SFE [8]. For instance, the Cantor alloy (CoCrFeMnNi) with an estimated SFE between 18.3–27.3 mJ/m² [9,10] has been shown to develop the TWIP effect, depending on the processing conditions [11]. Evidently, the determination of the SFE in HEAs is fundamental, because this property indicates the possible acting mechanisms for microstructure development. However, as with steels, the complex chemistry of HEAs leads to almost unlimited combinations that make a systematic determination of this property difficult. Nevertheless, to accelerate the development of these alloys, several research groups have made use of ab initio calculations, e.g., [12]. The advantage of this method is that it is possible to calculate several alloy compositions with less effort than the one required for an experimental determination of the same alloys. So far, the systems CoCrFeMnNi [9,10,13,14], AlCoCrCuFeNi, and AlCoCrFeNi [15] have been investigated regarding their SFE. A comprehensive list can be found in Reference [12]. First-principle methods have even allowed the constructions of property maps, where the SFE can be read as a function of the chemical composition [14], which can be used to predict the expected physical acting mechanisms. The experimental determination of the SFE, although it is arguably more laborious, is as important as its computational calculation because, as accurate as they are, ab initio calculations still rely on a model, which may not be accurate for some conditions. Thus, experimentation provides the necessary validation of the models. For the experimental determination of the SFE, two methods have been used before: namely, the measurement of the dissociation width of partial dislocations [16] and a combination of X-ray diffraction analysis and first-principle calculations of elastic constants [9]. The strategy of obtaining similar microstructures as those in advanced steels has been successfully utilized in some HEAs. For instance, dual-phase fcc-hcp (hexagonal closest packed) HEAs have substantiated increased ductility and strength [17], owing to the activation of the TWIP and TRIP effects [18]. With this in mind, in the present contribution, the mechanisms active during deformation and annealing in low-SFE single-phase HMnS and HEAs will be compared and discussed in order to put forward an SFE-oriented design of fcc-based HEAs.

2. Materials and Methods

Two alloys were investigated: one HMnS and one HEA. The chemical composition and their SFEs are shown in Table 1. The chemical compositions were determined by inductively coupled plasma-optical emission spectrometry (ICP-OES) and combustion analysis. The SFE values of the HMnS were calculated using a subregular thermodynamic solution model [19], whereas the SFE of the HEA was taken from the literature [9,10].

Table 1. Chemical composition (in wt%) and stacking fault energy (SFE) values of the investigated alloys. The weight fractions of Al, Si, N, and P were not measured in the high-entropy alloys (HEA).

Alloy	Fe	Mn	Al	Co	Cr	Ni	Si	C	N	P	SFE (mJ/m²)
HMnS	Bal.	22.46	1.21	-	-	-	0.04	0.325	0.0150	0.01	25
HEA	20.3	18.8	-	21.8	17.9	21.0	-	0.038	-	-	18–27

Both alloys were melted in an air conduction furnace followed by ingot casting, and were subsequently homogenized at 1150 °C for 5 h under protective Ar atmosphere. Then, the ingots were forged and subsequently homogenized at 1150 °C. After hot rolling, the alloys were cold rolled at room temperature to achieve up to 50% thickness reduction. To investigate the influence of additional heat treatment on the material behavior, the HMnS was subjected to annealing treatments at 550 °C for different times. In turn, the HEA was produced using induction melting in Ar atmosphere followed by hot rolling at 1000 °C and homogenization at 1000 °C for 1 h. The hot-rolled sheet was further cold rolled up to 50% thickness reduction. Finally, the cold-rolled sheets were annealed in the range between 500–900 °C for 1 h in an air furnace. The specific parameters used for processing the HMnS and HEA in different conditions are given in Table A1.

For characterization of the HMnS and HEA, specimens with the dimensions 12 mm × 10 mm in the rolling direction (RD) and transverse direction (TD) were fabricated using electrical discharge machining. The samples were mechanically ground up to 4000 SiC grit paper followed by mechanical polishing using diamond suspensions of 3 μm and 1 μm. For X-ray diffraction (XRD) pole figure measurements, the middle layer of the RD–TD section was polished electrolytically at room temperature for 2 min at 22 V, whereas the RD–ND (ND—normal direction) section was electropolished for scanning electron backscatter diffraction (EBSD) using the same parameters as before. The electrolyte used for XRD and EBSD sample preparation consisted of 700 mL of ethanol (C_2H_5OH), 100 mL of butyl glycol ($C_6H_{14}O_2$), and 78 mL of perchloric acid (60%) (ClO_4). Transmission electron microscopy (TEM) samples (~100 μm thick, 3 mm in diameter) were prepared using the same electrolyte as for XRD and EBSD samples in a double jet Tenupol-5 electrolytic polisher with a voltage of 22–29 V and a flow rate of 10 at 15 °C (HMnS) and 4–6 °C (HEA).

EBSD analyses were performed in an LEO 1530 field emission gun scanning electron microscope (FEG-SEM) (Carl Zeiss AG, Oberkochen, Germany) operated at 20-kV accelerating voltage and a working distance of 10 mm. The HKL Channel 5 software and the MATLAB®-based toolbox MTEX [20–22] were utilized to post-process and visualize the ESBD data. The subdivision of EBSD mappings into subsets containing only non-recrystallized (non-RX) or recrystallized (RX) grains was realized, as described in [23]. TEM analyses were performed in a FEI Tecnai F20 TEM (FEI Company, Hillsboro, OR, USA) operated at 200 kV.

X-ray pole figures were acquired utilizing a Bruker D8 Advance diffractometer (Bruker Corporation, Billerica, MA, USA), equipped with a HI-STAR area detector, operating at 30 kV and 25 mA, using filtered iron radiation and polycapillary focusing optics. In order to characterize the crystallographic texture, three incomplete (0–85°) pole figures {111}, {200}, and {220} were measured. The macrotexture orientation distribution functions (ODFs) were also calculated and visualized using MTEX. The volume fractions of the corresponding texture components were calculated using a spread of 15° from their ideal orientation.

Mechanical properties were evaluated by uniaxial tensile tests at room temperature and a constant strain rate of 2.5×10^{-3} s^{-1} along the previous rolling direction on a screw-driven Zwick Z100 mechanical testing device (Zwick/Roell, Ulm, Germany). Flat bar tension specimens were used with a gauge length of 13 mm, gauge width of 2 mm, fillet radius of 1 mm, and a total length of 33 mm.

3. Results and Discussion

3.1. Behavior during Plastic Deformation

An excellent tool for identifying the mechanisms acting during plastic deformation is the analysis of the crystallographic texture. To exemplify, Figure 1 shows the development of the texture of the HMnS and HEA during cold rolling with thickness reductions from 10% to 50%. Both alloys evinced the evolution of the main rolling texture components of fcc alloys, namely the Cu, S, Goss, and Brass texture components, whereas the Cube component disappeared. The position and definition of the ideal texture components is given in Figure A1 and Table A2 in Appendix A. With increasing deformation, the orientations tended to cluster at positions between the Brass and Goss texture components. This effect intensified at larger strains. In addition, the Goss component spread toward the CuT texture component. Nevertheless, one major difference between the texture of the HMnS and HEA after 50% cold rolling can be recognized. The α-fiber texture components, Brass and Goss, were stronger than the Cu one in the HMnS, whereas, in the HEA, the Brass component was less pronounced, while the Cu component was stronger (Figure 1). Therefore, after 50% cold rolling, the HMnS had already developed a Brass-type texture, while the HEA was characterized by a mixture of Brass-type and Cu-type textures.

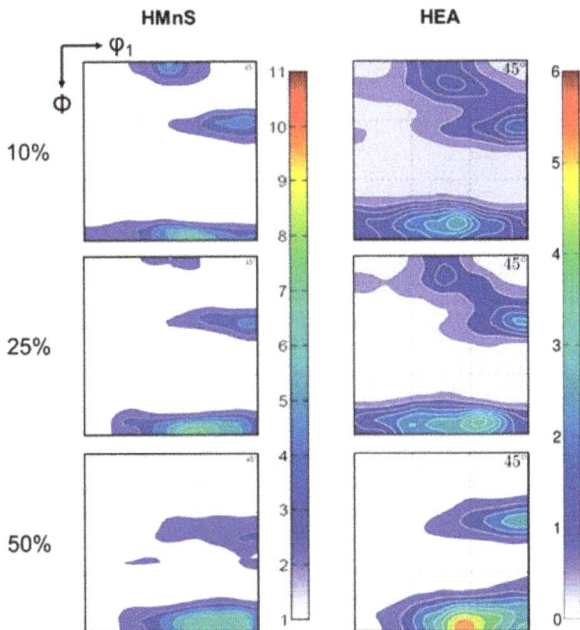

Figure 1. Texture evolution during cold rolling at room temperature of the investigated high-manganese steels (HMnS) and high-entropy alloy (HEA) illustrated by $\varphi_2 = 45°$ section of the ODF.

As shown in Figure 2, the microstructure of both alloys after 50% cold rolling consisted of grains elongated in RD and contained longitudinal features, such as slip bands. Deformation twins and microshear bands developed equally. Nevertheless, it has been proven that at low rolling degrees, deformation twinning in HEAs is less pronounced, and thus contributes to plastic deformation to a lesser extent, as compared to HMnS [8,24]. At high rolling degrees (80–90%), pronounced Brass-type textures have been observed in HMnS and HEAs [24,25]. Therefore, it can be concluded that the transition from Cu-type to Brass-type texture evolves in low-SFE HEAs in a comparable manner as to that in low-SFE HMnS, but it is shifted to higher degrees of plastic deformation. In terms of the activation of deformation twinning, the presence of C in the HMnS, and correspondingly the absence of C in the HEA, is presumably the main influential factor. On the one hand, C causes an increase of the flow stress, and therefore facilitates achieving the critical resolved shear stress for the activation of twinning. On the other hand, C-Mn short range ordered clusters contribute to the splitting of partial dislocations and ease the onset of deformation twinning [26]. In turn, in the C-free HEA, a more prolonged stage of deformation by dislocation slip retards the initiation of deformation twinning and delays the formation sequence of twin-matrix lamellae, their rotation into the rolling plane, and the subsequent onset of shear banding [27–33], which are thus shifted to higher strains. Hence, the related formation of a pronounced Brass-type texture with the typical texture components—Brass, Goss, CuT and γ-fiber—is shifted to higher degrees of thickness reduction during rolling of the HEA.

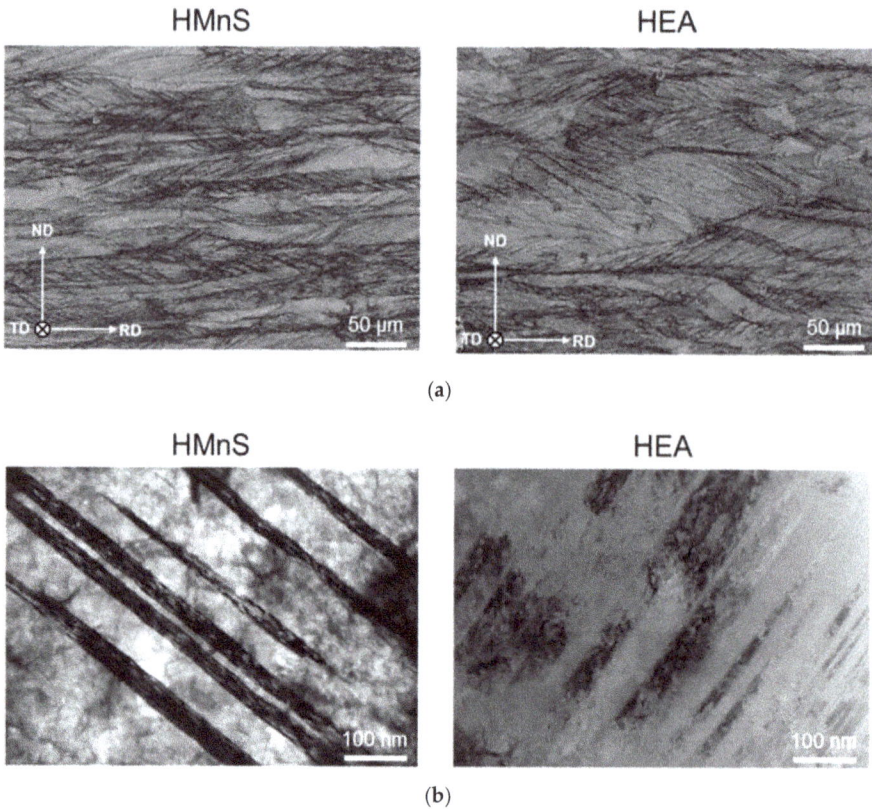

Figure 2. (**a**) Electron backscatter diffraction (EBSD) band contrast maps (top) and (**b**) TEM micrographs (bottom) of the HMnS and HEA after 50% cold rolling.

3.2. Behavior during Heat Treatment

The texture evolution during the heat treatment of the 50% cold-rolled HMnS and HEA is shown in Figures 3 and 4. Both alloy systems revealed a slight texture strengthening during static recovery, as substantiated by the increased volume fraction of the main texture components and a decreased fraction of randomly oriented grains (Figure 4). It has been shown before that the texture strengthening is caused by a reduced dislocation density due to recovery processes, which results in less scattering of radiation during XRD measurements [23]. During the recovery stage, the twins induced by cold rolling are thermally stable, and will not collapse upon annealing, but they can be consumed by newly formed grains during primary recrystallization (Figure 5). With the onset of recrystallization, the volume fraction of randomly oriented grains increased, whereas the fraction of the main rolling texture components decreased (Figure 4). This trend was further intensified with increasing recrystallized volume fraction. Nevertheless, the main rolling texture components were retained during recrystallization, although with lower intensity. In addition, a weak complete α-fiber was formed.

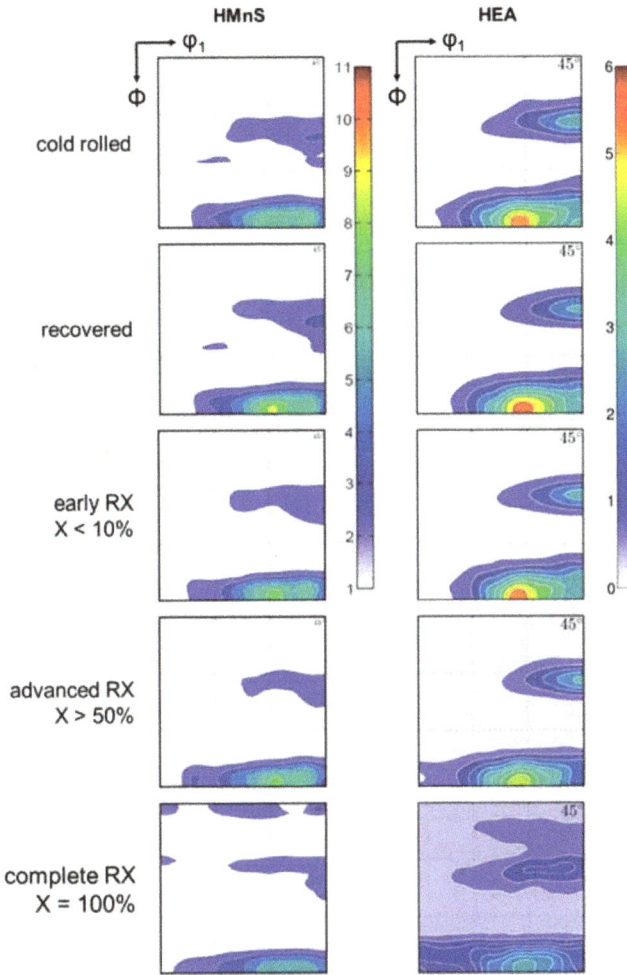

Figure 3. Texture evolution during annealing of the 50% cold-rolled HMnS and HEA illustrated by φ_2 = 45° section of the orientation distribution functions (ODF).

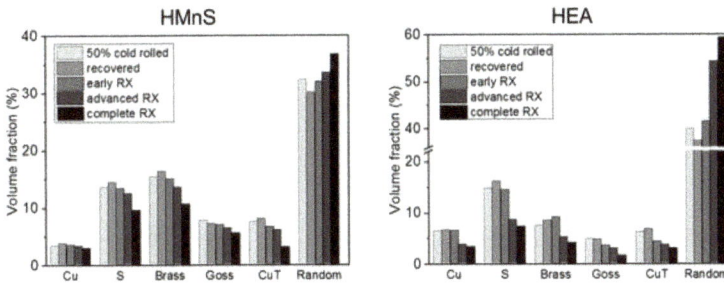

Figure 4. Evolution of the volume fractions of the main texture components during annealing of the 50% cold-rolled HMnS and HEA.

HMnS HEA

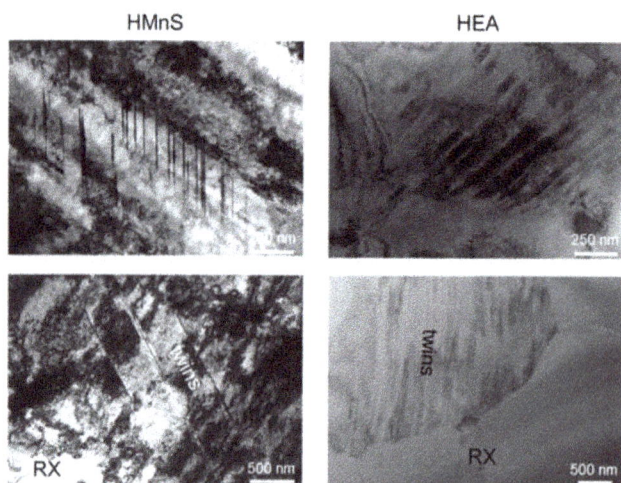

Figure 5. TEM bright-field images of the HMnS and HEA after 50% cold rolling and subsequent recovery annealing (**top**) and partial recrystallization (**bottom**).

In both alloys, the nucleation sites were found to be distributed heterogeneously at grain boundaries and triple junctions (Figure 6). New grains formed and grew primarily by strain-induced grain boundary migration. In this case, the orientations of nuclei were defined by the orientation of subgrains/dislocation cells in the deformed matrix. Additionally, extensive annealing twinning contributed to the formation of new texture components, and thus texture randomization [34]. As has been shown before, first-order twinning within Goss-oriented, Goss/Brass-oriented, and Brass-oriented nuclei resulted in the formation of a complete α-fiber due to the development of the A and rotated Goss (RtG) texture components [8,35,36]. Therefore, the evolution of the recyrstallization texture of the HMnS and HEA is a result of texture retention due to grain boundary nucleation and texture randomization due to annealing twinning [8,37].

HMnS HEA

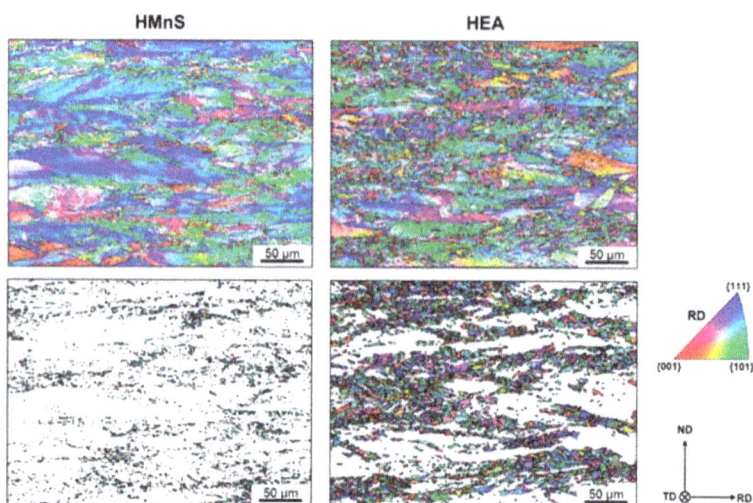

Figure 6. EBSD maps with color-coding according to the inverse pole figure of the HMnS and HEA after 50% cold rolling and partial recrystallization showing all grains (**top row**) and only recrystallized grains (**bottom row**).

3.3. Mechanical Properties

The results of uniaxial tensile tests of the HMnS and HEA after cold rolling and recrystallization are shown in Figure 7. In both cases, the cold-rolled conditions (CR) were characterized by a very high yield and ultimate tensile strength, but a strongly reduced total elongation. During subsequent recovery annealing (RC), the high yield strength was retained due to the thermal stability of deformation twins, whereas the total elongation was enhanced due to dislocations' annihilation [38,39]. After partial recrystallization (PRX) and complete recrystallization (RX), the soft recrystallized grains promoted increased total elongation at the expense of strength. In contrast to the deformed grains subjected to cold rolling, the soft RX grains facilitated regained formability due to the capability of accumulating new dislocations and undergoing deformation twinning [40,41].

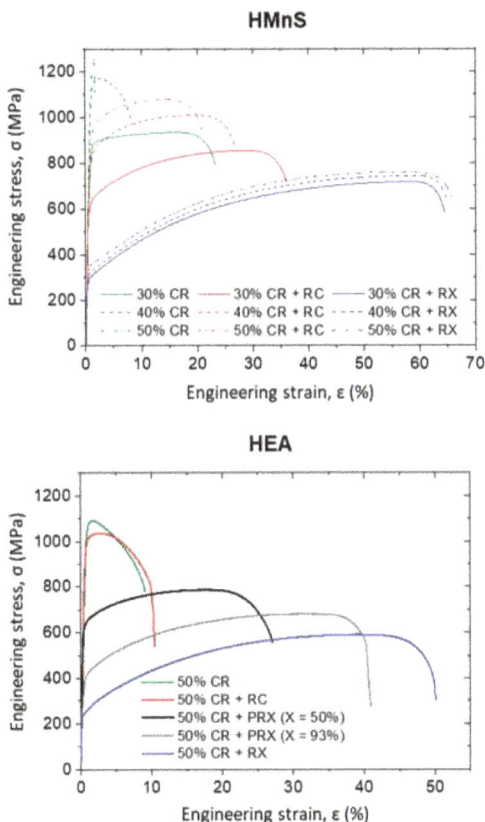

Figure 7. Engineering stress–strain curves of the HMnS and HEA in the states cold rolled (CR), recovered (RC), partially recrystallized (PRX with recrystallized fraction X), and recrystallized (RX).

4. Mechanism-Oriented Alloy Design

It has been shown and discussed in Sections 3.1–3.3 that the investigated HMnS and HEA behave very similarly. Based on the correlation between microstructure and texture evolution during deformation and heat treatment, it was demonstrated that the influence of the SFE on the material behavior of HMnS also holds true for fcc C-free HEA. The low SFE of both alloys facilitated the activation of twinning during deformation, albeit at different strains, as well as softening by static recovery and recrystallization during annealing. As a consequence, thermomechanical treatment resulted in comparable property profiles of the HMnS and HEA with minor differences due to the presence of C in the HMnS, which enabled higher strength and more pronounced twinning. Therefore, it can be asserted that SFE-based, mechanism-oriented alloy design, as has been used for tailoring the properties of HMnS for more than two decades, is a promising methodology to develop new advanced HEAs.

As shown in Figure 8, the activation/suppression of specific deformation mechanisms can be used to tailor the deformation behavior and mechanical properties of HMnS in a wide range. With an SFE ≤ 20 mJ/m^2, the TRIP effect promotes strong work hardening, whereas the TWIP (20 mJ/m$^2 \leq$ SFE ≤ 50 mJ/m^2) and slip-band refinement-induced plasticity (SRIP) (SFE ≥ 50 mJ/m^2) effects result in a lower work-hardening potential. Strength and work-hardening capability can be further modified by combining the aforementioned mechanisms with multi-phase (medium-manganese steels (MMnS)) and precipitation (κ-carbides) strengthening (Figure 8). Furthermore, the activation/suppression of specific mechanisms can also be adjusted by precise process design. These approaches include using the higher thermal stability of deformation twins over dislocations during recovery annealing, suppression and activation of twinning during, respectively, warm deformation, e.g., ECAP, and further room temperature deformation or tailoring the work-hardening behavior by using the strong orientation dependence of twinning in anisotropic additively manufactured specimens (Figure 8).

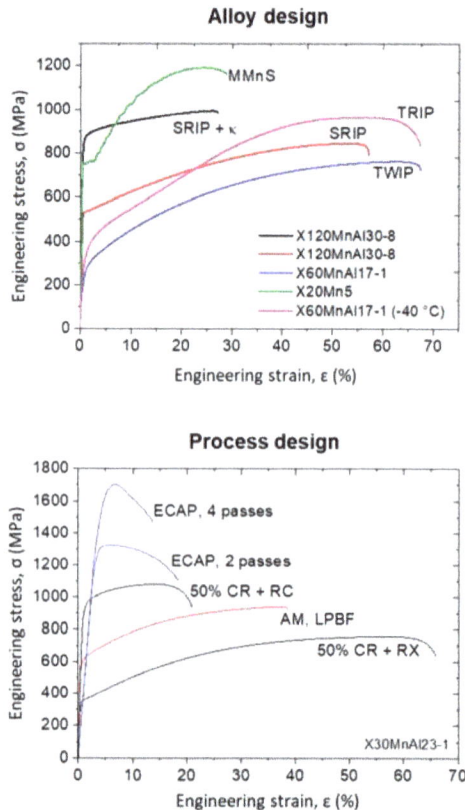

Figure 8. Engineering stress–strain curves of HMnS with various SFE/activated deformation mechanisms and the X30MnAl23-1 HMnS subjected to different processing techniques. MMnS—medium-manganese steel [42], SRIP—slip-band refinement-induced plasticity [43], TRIP—transformation-induced plasticity, TWIP—twinning-induced plasticity [44], ECAP—equal-channel angular pressing [45], SLM—selective laser melting [46].

Indeed, several of the approaches presented in Figure 8 have already been transferred to HEAs, such as TRIP and TWIP effects [17,47], precipitation hardening in compositionally complex alloys (CCAs) [48], recovery annealing [8], severe plastic deformation [49], additive manufacturing [50], etc. However, the further maturity of thermodynamics-based [51,52] and ab initio methods are required to explore the wide composition space of HEAs and CCAs. The mechanism-oriented design will certainly enable unveiling the full potential of these alloys.

5. Conclusions

A high-manganese steel and a high-entropy alloy with similar SFE were investigated regarding their microstructure, texture, and mechanical properties after cold rolling and heat treatment. The mechanisms for microstructure modification in both alloys were found to be similar due to the activation of the same physical mechanisms. During plastic deformation at low temperatures, the alloys substantiated initially deformation by dislocation slip and subsequently twinning. However, the onset of twinning in the HEA occurs at higher strains. This effect was attributed to the presence of carbon atoms in solid solution in the steel, whose hardening effect allows reaching the critical stress for twin formation earlier. Consequently, both alloys developed similar textures, but equivalent texture components developed at higher strains in the HEA. Upon annealing, the alloys substantiated the thermal stability of the twins formed during deformation in the recovery range. This effect permits the utilization of recovery annealing treatments in HEA for pronounced ductility and strength. In turn, during recrystallization, a retention of the texture due to grain boundary nucleation and texture randomization as a result of the formation and growth of annealing twins was exhibited. Finally, the mechanical characterization of the alloys evidenced the substantial range of variability of the properties that can be obtained from these alloys.

This suggests that fcc-HEAs can be almost arbitrarily tailored by a combination of the diverse physical mechanisms of microstructure modification that can be triggered in these alloys. The applicability and the potential of mechanisms-oriented alloys design have been already proven by the Collaborative Research Center (SFB) 761 "Steel—ab initio" [6], whose strategy can be used to design new HEAs and CCAs.

Author Contributions: Conceptualization, C.H. and L.A.B.-M.; formal analysis, C.H. and L.A.B.-M.; investigation, C.H. and L.A.B.-M.; visualization, C.H. and L.A.B.-M.; writing–original draft, C.H. and L.A.B.-M.; writing–review & editing, C.H. and L.A.B.-M.

Funding: This research was funded by the "Deutsche Forschungsgemeinschaft" (DFG) within the Sonderforschungsbereich (Collaborative Research Center) 761 "Steel–ab initio".

Acknowledgments: The authors are thankful for the help of Arndt Ziemons, Marie Luise Köhler, Mehran Afshar, Christian Schnatterer, Johannes Lohmar, and Stefan Senge with carrying out the experiments.

Conflicts of Interest: The authors declare no conflict of interest.

Appendix A

Table A1. Processing parameters used for the HMnS and HEA in different conditions. CR, RC, PRX, and RX denote cold rolled, recovered, partially recrystallized with recrystallized fraction X and recrystallized, respectively.

Alloy	Condition	Rolling Degree (%)	Annealing Temperature (°C)	Annealing Time (h)
HMnS	CR10	10	-	-
HMnS	CR25	25	-	-
HMnS	CR50	50	-	-
HMnS	CR50 + RC	50	500	1
HMnS	CR50 + PRX (X < 10%)	50	600	1
HMnS	CR50 + PRX (X > 50%)	50	650	1

Table A1. *Cont.*

Alloy	Condition	Rolling Degree (%)	Annealing Temperature (°C)	Annealing Time (h)
HMnS	CR50 + RX	50	700	1
HEA	20.3	10	-	-
HEA	CR25	25	-	-
HEA	CR50	50	-	-
HEA	CR50 + RC	50	550	0.5
HEA	CR50 + PRX (X < 10%)	50	550	1
HEA	CR50 + PRX (X > 50%)	50	550	2
HEA	CR50 + RX	50	700	8

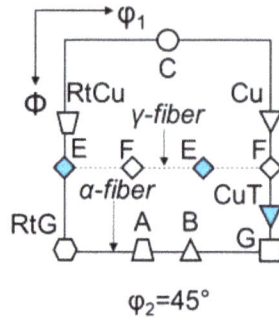

Figure A1. Schematic illustration of the ideal texture components and fibers, ODF section at $\varphi_2 = 45°$. The texture components are defined in Table A2.

Table A2. Definition of texture components illustrated in Figure A1.

Component	Symbol	Miller Indices	Euler Angles (φ_1, Φ, φ_2)	Fiber
Brass (B)	△	{110}<112>	(55, 90, 45)	α, β
Goss (G)	☐	{110}<100>	(90, 90, 45)	α, τ
Rotated Goss (RtG)	⬡	{110}<110>	(0, 90, 45)	α
A	⬠	{110}<111>	(35, 90, 45)	α
Cube (C)	◯	{001}<100>	(45, 0, 45)	/
E	◆	{111}<110>	(0/60, 55, 45)	γ
F	◇	{111}<112>	(30/90, 55, 45)	γ
Copper (Cu)	▽	{112}<111>	(90, 35, 45)	β, τ
Copper Twin (CuT)	▼	{552}<115>	(90, 74, 45)	τ
S	⊗	{123}<634>	(59, 37, 63)	β
α-fiber		<110> parallel to ND		
β-fiber		<110> tilted 60° from ND towards RD		
γ-fiber		<111> parallel ND		

References

1. Bouaziz, O.; Allain, S.; Scott, C.P.; Cugy, P.; Barbier, D. High manganese austenitic twinning induced plasticity steels: A review of the microstructure properties relationships. *Curr. Opin. Solid State Mater. Sci.* **2011**, *15*, 141–168. [CrossRef]
2. De Cooman, B.C.; Chen, L.; Kim, H.S.; Estrin, Y.; Kim, S.K.; Voswinckel, H. State-of-the-Science of High Manganese TWIP Steels for Automotive Applications. In Proceedings of the International Conference on Microstructure and Texture in Steels and Other Materials, Jamshedpur, India, 5–7 February 2008; Haldar, A., Suwas, S., Bhattacharjee, D., Eds.; Springer: London, UK, 2008; pp. 165–182.
3. Cantor, B.; Chang, I.T.H.; Knight, P.; Vincent, A.J.B. Microstructural development in equiatomic multicomponent alloys. *Mater. Sci. Eng. A* **2004**, *375*, 213–218. [CrossRef]
4. Yeh, J.W.; Chen, S.K.; Lin, S.J.; Gan, J.Y.; Chin, T.S.; Shun, T.T.; Tsau, C.H.; Chang, S.Y. Nanostructured high-entropy alloys with multiple principal elements: Novel alloy design concepts and outcomes. *Adv. Eng. Mater.* **2004**, *6*, 299–303. [CrossRef]
5. Murty, B.S.; Yeh, J.W.; Ranganathan, S. High-entropy alloys: Basic concepts. In *High Entropy Alloys*; Butterworth-Heinemann: Boston, MA, USA, 2014; pp. 13–35.
6. Sonderforschungsbereich 761. Available online: http://abinitio.iehk.rwth-aachen.de/ (accessed on 7 June 2019).
7. Gottstein, G. *Physical Foundations of Materials Science*; Springer: Berlin/Heidelberg, Germany, 2004.
8. Haase, C.; Barrales-Mora, L.A. Influence of deformation and annealing twinning on the microstructure and texture evolution of face-centered cubic high-entropy alloys. *Acta Mater.* **2018**, *150*, 88–103. [CrossRef]
9. Zaddach, A.J.; Niu, C.; Koch, C.C.; Irving, D.L. Mechanical properties and stacking fault energies of NiFeCrCoMn high-entropy alloy. *JOM* **2013**, *65*, 1780–1789. [CrossRef]
10. Huang, S.; Li, W.; Lu, S.; Tian, F.; Shen, J.; Holmström, E.; Vitos, L. Temperature dependent stacking fault energy of FeCrCoNiMn high entropy alloy. *Scripta Mater.* **2015**, *108*, 44–47. [CrossRef]
11. Otto, F.; Dlouhý, A.; Somsen, C.; Bei, H.; Eggeler, G.; George, E.P. The influences of temperature and microstructure on the tensile properties of a CoCrFeMnNi high-entropy alloy. *Acta Mater.* **2013**, *61*, 5743–5755. [CrossRef]
12. Ikeda, Y.; Grabowski, B.; Körmann, F. Ab initio phase stabilities and mechanical properties of multicomponent alloys: A comprehensive review for high entropy alloys and compositionally complex alloys. *Mater. Charact.* **2019**, *147*, 464–511. [CrossRef]
13. Zhao, S.; Stocks, G.M.; Zhang, Y. Stacking fault energies of face-centered cubic concentrated solid solution alloys. *Acta Mater.* **2017**, *134*, 334–345. [CrossRef]
14. Ikeda, Y.; Körmann, F.; Tanaka, I.; Neugebauer, J. Impact of chemical fluctuations on stacking fault energies of CrCoNi and CrMnFeCoNi high entropy alloys from first principles. *Entropy* **2018**, *20*, 655. [CrossRef]
15. Beyramali Kivy, M.; Asle Zaeem, M. Generalized stacking fault energies, ductilities, and twinnabilities of CoCrFeNi-based face-centered cubic high entropy alloys. *Scripta Mater.* **2017**, *139*, 83–86. [CrossRef]
16. Liu, S.F.; Wu, Y.; Wang, H.T.; He, J.Y.; Liu, J.B.; Chen, C.X.; Liu, X.J.; Wang, H.; Lu, Z.P. Stacking fault energy of face-centered-cubic high entropy alloys. *Intermetallics* **2018**, *93*, 269–273. [CrossRef]
17. Li, Z.; Pradeep, K.G.; Deng, Y.; Raabe, D.; Tasan, C.C. Metastable high-entropy dual-phase alloys overcome the strength-ductility trade-off. *Nature* **2016**, *534*, 227–230. [CrossRef] [PubMed]
18. Li, Z.; Tasan, C.C.; Springer, H.; Gault, B.; Raabe, D. Interstitial atoms enable joint twinning and transformation induced plasticity in strong and ductile high-entropy alloys. *Sci. Rep.* **2017**, *7*, 40704. [CrossRef] [PubMed]
19. Saeed-Akbari, A.; Imlau, J.; Prahl, U.; Bleck, W. Derivation and variation in composition-dependent stacking fault energy maps based on subregular solution model in high-manganese steels. *Metall. Mater. Trans. A* **2009**, *40*, 3076–3090. [CrossRef]
20. Hielscher, R.; Schaeben, H. A novel pole figure inversion method: Specification of the MTEX algorithm. *J. Appl. Crystallogr.* **2008**, *41*, 1024–1037. [CrossRef]
21. Bachmann, F.; Hielscher, R.; Schaeben, H. Texture analysis with MTEX—Free and open source software toolbox. *Soild State Phenom* **2010**, *160*, 63–68. [CrossRef]
22. Bachmann, F.; Hielscher, R.; Schaeben, H. Grain detection from 2d and 3d EBSD data—Specification of the MTEX algorithm. *Ultramicroscopy* **2011**, *111*, 1720–1733. [CrossRef]

23. Haase, C.; Barrales-Mora, L.A.; Molodov, D.A.; Gottstein, G. Tailoring the mechanical properties of a twinning-induced plasticity steel by retention of deformation twins during heat treatment. *Metall. Mater. Trans. A* **2013**, *44*, 4445–4449. [CrossRef]
24. Haase, C.; Chowdhury, S.G.; Barrales-Mora, L.A.; Molodov, D.A.; Gottstein, G. On the relation of microstructure and texture evolution in an austenitic Fe-28Mn-0.28C TWIP steel during cold rolling. *Metall. Mater. Trans. A* **2013**, *44*, 911–922. [CrossRef]
25. Sathiaraj, G.D.; Ahmed, M.Z.; Bhattacharjee, P.P. Microstructure and texture of heavily cold-rolled and annealed fcc equiatomic medium to high entropy alloys. *J. Alloys Compd.* **2016**, *664*, 109–119. [CrossRef]
26. Sevsek, S.; Brasche, F.; Haase, C.; Bleck, W. Combined deformation twinning and short-range ordering causes serrated flow in high-manganese steels. *Mater. Sci. Eng. A* **2019**, *746*, 434–442. [CrossRef]
27. Hirsch, J.; Lücke, K.; Hatherly, M. Mechanism of deformation and development of rolling textures in polycrystalline F.C.C. metals—III. The influence of slip inhomogeneities and twinning. *Acta Metall.* **1988**, *36*, 2905–2927. [CrossRef]
28. Vercammen, S.; Blanpain, B.; De Cooman, B.C.; Wollants, P. Cold rolling behaviour of an austenitic Fe–30Mn–3Al–3Si TWIP-steel: The importance of deformation twinning. *Acta Mater.* **2004**, *52*, 2005–2012. [CrossRef]
29. Kusakin, P.; Belyakov, A.; Haase, C.; Kaibyshev, R.; Molodov, D.A. Microstructure evolution and strengthening mechanisms of Fe–23Mn–0.3C–1.5Al TWIP steel during cold rolling. *Mater. Sci. Eng. A* **2014**, *617*, 52–60. [CrossRef]
30. Saleh, A.A.; Haase, C.; Pereloma, E.V.; Molodov, D.A.; Gazder, A.A. On the evolution and modelling of brass-type texture in cold-rolled twinning-induced plasticity steel. *Acta Mater.* **2014**, *70*, 259–271. [CrossRef]
31. Stepanov, N.; Tikhonovsky, M.; Yurchenko, N.; Zyabkin, D.; Klimova, M.; Zherebtsov, S.; Efimov, A.; Salishchev, G. Effect of cryo-deformation on structure and properties of CoCrFeNiMn high-entropy alloy. *Intermetallics* **2015**, *59*, 8–17. [CrossRef]
32. Klimova, M.; Stepanov, N.; Shaysultanov, D.; Chernichenko, R.; Yurchenko, N.; Sanin, V.; Zherebtsov, S. Microstructure and mechanical properties evolution of the Al, C-containing CoCrFeNiMn-type high-entropy alloy during cold rolling. *Materials* **2018**, *11*, 53. [CrossRef]
33. Klimova, M.; Zherebtsov, S.; Stepanov, N.; Salishchev, G.; Haase, C.; Molodov, D.A. Microstructure and texture evolution of a high manganese TWIP steel during cryo-rolling. *Mater. Charact.* **2017**, *132*, 20–30. [CrossRef]
34. Gottstein, G. Annealing texture development by multiple twinning in F.C.C. crystals. *Acta Metall.* **1984**, *32*, 1117–1138. [CrossRef]
35. Haase, C.; Barrales-Mora, L.A.; Molodov, D.A.; Gottstein, G. Texture evolution of a cold-rolled Fe-28Mn-0.28C TWIP steel during recrystallization. *Mater. Sci. Forum* **2013**, *753*, 213–216. [CrossRef]
36. Haase, C.; Kühbach, M.; Barrales Mora, L.A.; Wong, S.L.; Roters, F.; Molodov, D.A.; Gottstein, G. Recrystallization behavior of a high-manganese steel: Experiments and simulations. *Acta Mater.* **2015**, *100*, 155–168. [CrossRef]
37. Haase, C.; Ingendahl, T.; Güvenç, O.; Bambach, M.; Bleck, W.; Molodov, D.A.; Barrales-Mora, L.A. On the applicability of recovery-annealed twinning-induced plasticity steels: Potential and limitations. *Mater. Sci. Eng. A* **2016**, *649*, 74–84. [CrossRef]
38. Haase, C.; Barrales-Mora, L.A.; Roters, F.; Molodov, D.A.; Gottstein, G. Applying the texture analysis for optimizing thermomechanical treatment of high manganese twinning-induced plasticity steel. *Acta Mater.* **2014**, *80*, 327–340. [CrossRef]
39. Berrenberg, F.; Haase, C.; Barrales-Mora, L.A.; Molodov, D.A. Enhancement of the strength-ductility combination of twinning-induced/transformation-induced plasticity steels by reversion annealing. *Mater. Sci. Eng. A* **2017**, *681*, 56–64. [CrossRef]
40. Yanushkevich, Z.; Belyakov, A.; Haase, C.; Molodov, D.A.; Kaibyshev, R. Structural/textural changes and strengthening of an advanced high-Mn steel subjected to cold rolling. *Mater. Sci. Eng. A* **2016**, *651*, 763–773. [CrossRef]
41. Yanushkevich, Z.; Belyakov, A.; Kaibyshev, R.; Haase, C.; Molodov, D.A. Effect of cold rolling on recrystallization and tensile behavior of a high-Mn steel. *Mater. Charact.* **2016**, *112*, 180–187. [CrossRef]
42. Sevsek, S.; Haase, C.; Bleck, W. Strain-rate-dependent deformation behavior and mechanical properties of a multi-phase medium-manganese steel. *Metals* **2019**, *9*, 344. [CrossRef]

43. Haase, C.; Zehnder, C.; Ingendahl, T.; Bikar, A.; Tang, F.; Hallstedt, B.; Hu, W.; Bleck, W.; Molodov, D.A. On the deformation behavior of κ-carbide-free and κ-carbide-containing high-Mn light-weight steel. *Acta Mater.* **2017**, *122*, 332–343. [CrossRef]
44. Haase, C.; Barrales-Mora, L.A.; Molodov, D.A.; Gottstein, G. Application of texture analysis for optimizing thermo-mechanical treatment of a high Mn TWIP steel. *Adv. Mater. Res.* **2014**, *922*, 213–218. [CrossRef]
45. Haase, C.; Kremer, O.; Hu, W.; Ingendahl, T.; Lapovok, R.; Molodov, D.A. Equal-channel angular pressing and annealing of a twinning-induced plasticity steel: Microstructure, texture, and mechanical properties. *Acta Mater.* **2016**, *107*, 239–253. [CrossRef]
46. Kies, F.; Köhnen, P.; Wilms, M.B.; Brasche, F.; Pradeep, K.G.; Schwedt, A.; Richter, S.; Weisheit, A.; Schleifenbaum, J.H.; Haase, C. Design of high-manganese steels for additive manufacturing applications with energy-absorption functionality. *Mater. Des.* **2018**, *160*, 1250–1264. [CrossRef]
47. Deng, Y.; Tasan, C.C.; Pradeep, K.G.; Springer, H.; Kostka, A.; Raabe, D. Design of a twinning-induced plasticity high entropy alloy. *Acta Mater.* **2015**, *94*, 124–133. [CrossRef]
48. He, J.Y.; Wang, H.; Huang, H.L.; Xu, X.D.; Chen, M.W.; Wu, Y.; Liu, X.J.; Nieh, T.G.; An, K.; Lu, Z.P. A precipitation-hardened high-entropy alloy with outstanding tensile properties. *Acta Mater.* **2016**, *102*, 187–196. [CrossRef]
49. Schuh, B.; Mendez-Martin, F.; Völker, B.; George, E.P.; Clemens, H.; Pippan, R.; Hohenwarter, A. Mechanical properties, microstructure and thermal stability of a nanocrystalline CoCrFeMnNi high-entropy alloy after severe plastic deformation. *Acta Mater.* **2015**, *96*, 258–268. [CrossRef]
50. Ewald, S.; Kies, F.; Hermsen, S.; Voshage, M.; Haase, C.; Schleifenbaum, J.H. Rapid alloy development of extremely high-alloyed metals using powder blends in laser powder bed fusion. *Materials* **2019**, *12*, 1706. [CrossRef] [PubMed]
51. Haase, C.; Tang, F.; Wilms, M.B.; Weisheit, A.; Hallstedt, B. Combining thermodynamic modeling and 3D printing of elemental powder blends for high-throughput investigation of high-entropy alloys—Towards rapid alloy screening and design. *Mater. Sci. Eng. A* **2017**, *688*, 180–189. [CrossRef]
52. Miracle, D.B.; Senkov, O.N. A critical review of high entropy alloys and related concepts. *Acta Mater.* **2017**, *122*, 448–511. [CrossRef]

metals

MDPI

Article

Processing Variants in Medium-Mn Steels

John Speer [1,*], Radhakanta Rana [1,2], David Matlock [1], Alexandra Glover [1], Grant Thomas [3] and Emmanuel De Moor [1]

[1] Advanced Steel Processing and Products Research Center, Colorado School of Mines, Golden, CO 80401, USA
[2] Tata Steel, Wenckebachstraat 1, 1970 CA IJmuiden, The Netherlands
[3] AK Steel, AK Steel Research and Innovation Center, 6180 Research Way, Middletown, OH 45005, USA
* Correspondence: jspeer@mines.edu; Tel.: +1-303-273-3897

Received: 10 June 2019; Accepted: 4 July 2019; Published: 10 July 2019

Abstract: This paper highlights some recent efforts to extend the use of medium-Mn steels for applications other than intercritically batch-annealed steels with exceptional ductility (and strengths in the range of about 1000 MPa). These steels are shown to enable a range of promising properties. In hot-stamping application concepts, elevated Mn concentration helps to stabilize austenite and to provide a range of attractive property combinations, and also reduces the processing temperatures and likely eliminates the need for press quenching. The "double soaking" concept also provides a wide range of attractive mechanical property combinations that may be applicable in cold-forming applications, and could be implemented in continuous annealing and/or continuous galvanizing processes where Zn-coating would typically represent an additional austempering step. Quenching and partitioning of steels with elevated Mn concentrations have exhibited very high strengths, with attractive tensile ductility; and medium-Mn steels have been successfully designed for quenching and partitioning using room temperature as the quench temperature, thereby effectively decoupling the quenching and partitioning steps.

Keywords: medium-Mn steel; hot-stamping; double soaking; continuous annealing; quenching and partitioning; high strength steel

1. Introduction

Medium-manganese steels with Mn concentrations in the range of about 5 to 10 wt. pct. have been explored over the past decade as potential options to provide "3rd Generation" Advanced High Strength Steel (AHSS) sheets for automobile bodies, with excellent combinations of strength and ductility. These steels are most commonly employed in the intercritically batch-annealed condition, where an ultra-fine microstructure is obtained, and extensive Mn partitioning to austenite helps to stabilize substantial amounts of austenite, providing effective transformation-induced plasticity (TRIP) and attractive tensile property combinations. Intercritical annealing may also be employed in the hot-band condition, to obtain the requisite Mn partitioning, while softening the microstructure before cold-rolling. These steels are of particular interest at tensile strength levels of about 1000 MPa, as the levels of Mn alloying are less than those employed in twinning-induced plasticity steels (and yield strengths are higher), and their tensile ductilities are greater than leaner-alloyed AHSS products employing alternative processing approaches. A variety of challenges are also associated with medium-Mn steels, such as very large yield-point elongations in some instances, and sensitivity of mechanical properties to small changes in annealing temperature.

In this overview, some processing routes for medium-Mn steels are considered that are different than intercritical batch-annealing. The applications involve press-hardening of steels subject to hot- stamping, and cold-formable steels at strength levels greater than are commonly achieved in intercritically-annealed medium-Mn steels. Recent work by the authors on these processing variants is highlighted in the sections below.

2. Hot-Forming of Medium-Mn Steels

Press-hardening of low-alloy B-added steels (e.g., 22MnB5) is commonly employed in high strength components, where a martensitic microstructure is obtained by die-quenching after forming. High strengths (above 1500 MPa) are common, beyond the levels typically associated with medium-Mn steels. The ductilities of these high strength components are limited, with values typical of low-carbon martensitic steels, and greater ductilities are of interest to enhance crashworthiness or "toughness" in some applications [1,2]. Medium-Mn steels offer some interesting possibilities to consider in press-hardening steels (PHS) subjected to hot-stamping. The austenite-stabilizing effect of Mn offers the potential to develop and employ retained austenite to enhance ductility, and also reduces the temperatures that would apply to full austenitizing or intercritical annealing prior to hot stamping [3–5]. These lower reheating temperatures provide additional benefits related to suppression of liquid-metal embrittlement during forming of Zn-coated PHS, and perhaps reduced oxidation/decarburization during heating [6]. In addition, the increased hardenability associated with an elevated Mn concentration could potentially eliminate the need for in-die heat extraction during forming (press quenching), enabling development of higher-productivity processing concepts.

To assess the hot-stamping response, a medium manganese steel containing 9.76 Mn, 0.16 C, 1.37 Al, 0.19 Si, 0.0018 S (wt. pct.), was subjected to thermal treatments simulating hot forming at different temperatures [3]. Prior to reheating for hot stamping, the material was received in the as-cold-rolled condition after hot-band annealing, and was then soaked for 3 min at temperatures between 650 and 800 °C (simulating approximately, for example, the influence of the annealing step in a continuous hot-dip coating process before reheating and hot-stamping). Figure 1a,b illustrate the mechanical property results for samples held 3 min at temperatures of 650 or 700 °C (simulating heating before hot forming), followed by rapid cooling. A variety of strength-ductility combinations resulted, with tensile strengths up to about 1450 MPa and total elongations in the range of about 16 to 43 pct.

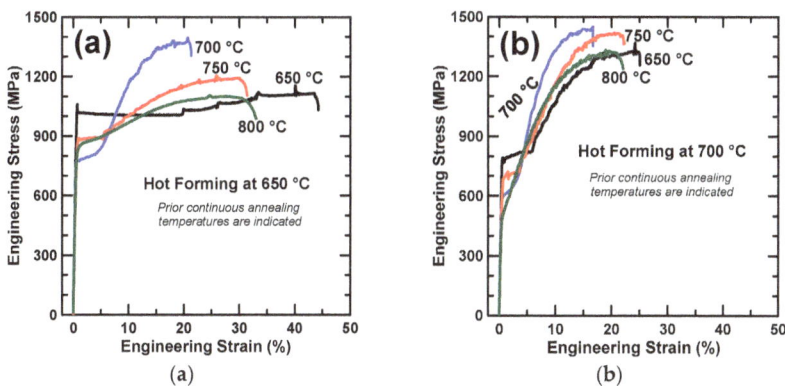

Figure 1. Representative room-temperature engineering stress-strain curves of a 9.76 Mn, 0.16 C, 1.37 Al (wt. pct.) steel subjected to hot-forming thermal simulation (without deformation) using reheating temperatures of (**a**) 650, or (**b**) 700 °C, preceded by continuous annealing at 650–800 °C for 3 min [3].

Without the continuous annealing simulation step applicable to the results in Figure 1, even higher strengths are obtainable, with properties after heating for hot stamping exceeding 1600 MPa tensile strength and 16 pct. ductility [3]. Retained austenite fractions exceeded about 20 pct. in the conditions of interest; less than is typical for intercritically batch-annealed medium-Mn steel, but sufficient to provide excellent ductility at high strength levels. The relatively high (10 wt. pct. Mn) concentration associated with the particular steel examined in this work is an exemplar and is not considered to represent an "optimized" steel composition for this processing path.

Other studies on hot-stamped medium-Mn steels have focused more on fully martensitic variants processed by full austenitizing at higher reheating temperatures [4,7], reporting high strength with enhanced ductility. In comparison, the intercritical heating approach as indicated in Figure 1 should provide further increases in ductility and toughness, due to the greater stabilization of austenite through Mn partitioning. The hot stamping process involves deformation before cooling, and aspects relating to that have not been considered in our work, although other authors have provided important information on the effects of warm deformation of the austenite, in terms of elevated temperature strengths as a function of deformation temperature and strain rate, as well as pressure effects on martensite, and the influence of warm deformation of austenite (i.e., thermomechanical treatment) on the final properties at room temperature [6,8]. Overall, research has shown that medium-Mn TRIP steels are promising candidate materials for hot-stamping following a variety of processing paths, not only due to the potential for achieving extraordinary properties, but also due to the additional benefits from enhanced hardenability and lower reheat temperatures.

3. "Double Soaking" of Medium-Mn Steels for Cold Forming Applications

De Moor and Glover have demonstrated concepts for achieving higher strengths in cold-formed medium-Mn steels, by designing "double soaking" (DS) treatments [9,10]. The DS concept aims to achieve substantial Mn-enrichment of austenite in a first soaking that could involve intercritical batch-annealing (typically after cold-rolling), followed by a second soaking step at a higher temperature. The goal of the second soaking step at a higher temperature is to replace some or all of the intercritical ferrite with austenite during heating and holding, and then transform it to martensite during final cooling, thus replacing the typical ferrite + austenite microstructure in medium-Mn steels with a higher strength TRIP microstructure, containing both martensite and austenite. In this concept, the austenitizing time in the second soaking step is envisioned to be relatively short, allowing substantial carbon redistribution but avoiding, as much as possible, redistribution of Mn. In this way, inhomogeneous austenite is obtained at a high temperature, with interlaced ultra-fine regions having higher and lower manganese concentrations, respectively. Retention of the higher-Mn regions should enable much greater austenite fractions to be maintained upon final cooling, while the lower-Mn regions readily transform to martensite (and perhaps ferrite). The second soaking step may be implemented during the annealing portion of a final continuous annealing or continuous galvanizing operation.

To explore this concept, thermal processing was conducted following the schematic illustration in Figure 2a of "discontinuous" double soaking. An industrially batch-annealed (first soak) 7.14 Mn, 0.14 C (wt. pct.) medium manganese steel was received and subjected to continuous annealing treatments (second soak) between 675 and 825 °C, for holding times between 10 and 120 s. These holding times are shorter than employed in the hot-stamping work discussed above, and resulted in greater levels of austenite retention in most instances (despite the lower Mn concentration), presumably due to reduced homogenization of manganese.

Resulting mechanical properties are illustrated in Figure 2b for 30 s holding times during the second soak at temperatures between 675 and 825 °C [10]. This range of "second soak" temperatures creates an interesting and substantial range of properties, with ultimate tensile strengths from about 1000 MPa to 1700 MPa, and corresponding tensile ductilities ranging from about 40 pct. down to about 15 pct. at the highest strength levels. The properties are nearly unchanged from the batch-annealed starting conditions after the second annealing soak for 30 s at 675 °C, while the strength increases substantially as the second soak temperature increases up to 800 °C. Tensile ductility correspondingly decreased as the second soak temperature increased, but the overall combinations of strength and elongation are very good across the full range of results exemplified in Figure 2b. The property changes are related to increasing martensite fractions and decreasing ferrite fractions (and to a lesser extent decreasing austenite fractions) after final cooling, associated with increasing the second soak temperature. These properties might be readily achievable by thermal processing characteristics of industrial continuous annealing or continuous galvanizing facilities.

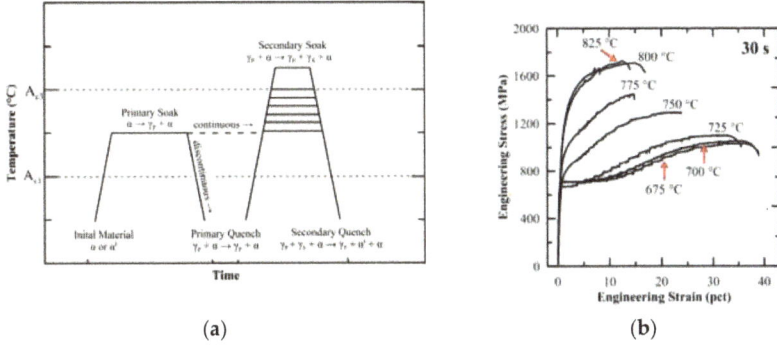

(a)   (b)

Figure 2. Schematic illustration of double soaking thermal treatments (**a**), and tensile properties (**b**) for a 7.14 Mn, 0.14 C (wt. pct.) steel after a second soak of 30 s at indicated temperatures [10].

The elevated Mn concentration in the steel, in conjunction with Mn partitioning in a first soak and the limited time for Mn redistribution in the second soak, provide sufficient austenite stabilization and retention, and an effective contribution of transformation-induced plasticity to the property balance. Microstructure evolution associated with double soaking is illustrated by the example in Figure 3. The intercritical batch anneal (first soak) leads to an ultra-fine mixture, as shown in Figure 3a, of equiaxed intercritical ferrite (α) and austenite (γ), along with a small amount of martensite (M). The double soaked condition in Figure 3b shows the microstructure after a 30 s second soak at 800 °C, consisting of a mixture of martensite and austenite. The austenite in the as-received condition largely remained after the second soak, but the equiaxed ferrite was largely replaced by martensite. The phase fractions vary with processing parameters; e.g., longer soaking times at high temperature result in more martensite, while the fractions of both ferrite and retained austenite are correspondingly reduced [10]. While refinement of the austenite was not a primary objective of this processing strategy, the rapid austenitization process was accomplished without substantial coarsening of the microstructure during the second soak at 800 °C [11].

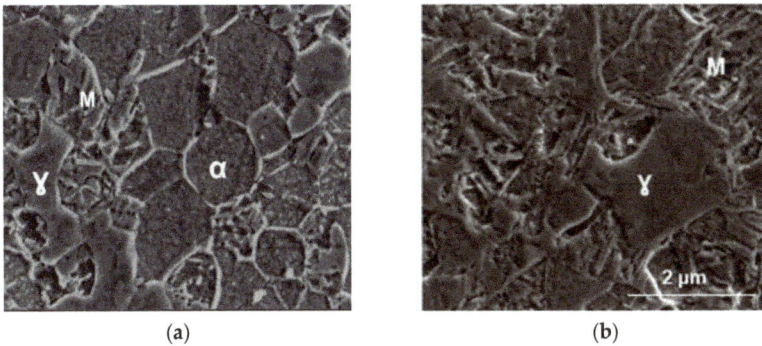

(a)   (b)

Figure 3. Example microstructures of a 7.14 Mn, 0.14 C (wt. pct.) steel (**a**) after intercritical batch annealing (first soak), and (**b**) after double soaking with a second soak of 30 s at 800 °C [10,11]. (Nital etch; SEM secondary electron images).

To further enhance the properties of medium-Mn steels processed by double soaking, additional tempering might be considered. The microstructure after cooling following the second soak contains a considerable amount of "fresh" martensite. Martensite tempering could improve tensile ductility, and even more likely, hole expansion performance, since hole expansion has been shown to benefit from reduced variation in the hardness between the various constituents in the microstructure [12]. Tempering might be achieved industrially by an overaging step in a continuous annealing process after quenching from the second soak temperature, or potentially in a continuous galvanizing facility with cooling capability, prior to the coating application. With more conventional hot-dip galvanizing line configurations, the coating application would constitute an austempering step. Tempering was included in the "multi-step partitioning" (MSP) work of Liu et al. [13] and in similar work by Heo et al. [14]; both of these studies also included an austenitizing and quenching step prior to the intercritical treatments, which leads to a more "lamellar" microstructure.

Tempering effects are being explored in ongoing research, following the double soaking thermal path achieved using molten salt baths with heating rates on the order of 80 °C/s [11,15]. Figure 4 illustrates the tensile property changes, showing engineering stress-strain behavior after the primary soak (intercritical batch anneal—BA), then after double-soaking for 30 s at 800 °C (DS), as well as after subsequent cooling to room temperature and then tempering for 300 s at 450 °C (DS-T) or after subsequent austempering (DS-A) for 300 s at 450 °C [11]. The results show that the tempering step decreases the work hardening rate and tensile strength while increasing ductility; austempering decreases the strength to a lesser degree while also enhancing ductility. While the tempering and austempering conditions were identical, the DS-T conditions involved tempering of martensite that was formed during the cooling step after the second soak (and potentially some carbon partitioning [13]), while the austempering step of the DS-A process was completed in advance of martensite formation, with dilation results indicating replacement of some austenite by ferrite.

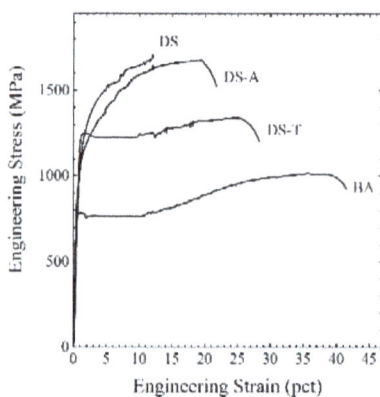

Figure 4. Tensile properties of a 7.14 Mn, 0.14 C (wt. pct.) steel after intercritical batch annealing (BA) first soak, after double soaking (DS) with a second soak of 30 s at 800 °C, and after double soaking, cooling to room temperature and tempering for 300 s at 450 °C (DS-T), and after double soaking followed by austempering for 300 s at 450 °C (DS-A) [11].

The double soaking approach is just one of the possible multistep thermal treatments that has been considered to control the fractions of martensite, ferrite and austenite, as well as the phase compositions, particularly the austenite where austenite stability during cooling or during deformation in service is sensitively related to its Mn and C concentrations. In this regard, some studies have suggested that multiple intercritical annealing steps may enhance the Mn and C partitioning response and properties, where a lower intercritical temperature is employed later, to enhance the final austenite stability [16,17].

4. Quenching and Partitioning of Steels Containing Elevated Manganese Additions

Quenching and partitioning (Q&P) is a heat treatment to stabilize austenite by carbon partitioning from martensite, formed during interrupted quenching after austenitizing or intercritical annealing, into untransformed austenite, leading to martensite-austenite mixtures with attractive combinations of strength and formability (e.g., [18–22]. Compositions similar to bainitic TRIP steels are often employed, with small Mn additions, typically on the order of 1.5 wt. pct., made to help avoid diffusional transformations during the quenching step. Even higher levels of Mn can be considered to enhance austenite retention, employing the austenite stabilizing effects of Mn to adjust the relative fractions of martensite and austenite [18,19]. The use of elevated Mn additions in Q&P steels has been successful in developing very high strength cold-formable sheet steels with tensile strengths in excess of 1500 MPa, in combination with total elongations exceeding 20 pct. [18], and numerous subsequent studies on Q&P processing of medium-Mn steels have also been reported by others. An example of a high strength Q&P microstructure developed through this approach is shown in Figure 5, for a 0.2C, 3Mn, 1.6Si (wt. pct.) steel quenched and partitioned after full austenitizing. The microstructure consists of martensite with a substantial quantity of interlath retained austenite.

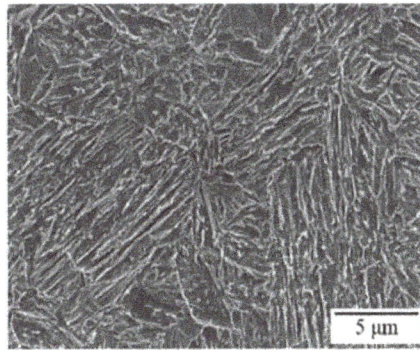

Figure 5. SEM micrograph of a 0.2 C, 3 Mn, 1.6 Si (wt. pct.) steel processed by quenching and partitioning after full austenitization [18]. (Nital etch; secondary electron image).

The quenching temperature employed in the Q&P process is normally well in excess of room temperature, requiring careful process control in the interrupted quenching step. Higher Mn concentrations have also been utilized to suppress the M_s temperature, and reduce the quench temperature in Q&P processing all the way down to room temperature. Use of an ambient "interrupted" quench temperature in these alloy designs allows decoupling of the quenching and partitioning steps. Such alloys were initially designed for fundamental studies of partitioning mechanisms [20,21], but later studies reported that attractive mechanical properties can also be obtained using this approach [22,23]. Ambient temperature quenching also provides additional process flexibility, and could be readily implemented, for example, in hot-rolled products or in a continuous annealing line with water quenching. Enhanced hardenability associated with elevated Mn concentrations can likely mitigate concerns related to cooling rates in thicker sections. In the context of hot stamping, Q&P can be achieved without the need for careful control of the quench temperature after hot forming [24].

As an example of properties obtained in a steel alloy designed for quenching to room temperature followed by partitioning (at a higher temperature), Figure 6a presents engineering tensile stress-strain curves for a steel containing 0.25 C, 8.23 Mn, 1.87 Si, 0.05 Ni, 0.24 Mo (wt. pct.). After quenching to room temperature, the steel was partitioned at different times at temperatures from 200 to 400 °C; the results in Figure 6a apply to 300 °C partitioning, and tensile strengths exceeding 1800 MPa were obtained with total elongations of about 15 pct. Even higher strengths with good tensile ductility were obtained at a partitioning temperature of 200 °C, and a similar approach was employed more recently to generate interesting properties for hot-stamping of a medium-Mn steel with 3% aluminum added to reduce density (0.3 C, 8 Mn, 3 Al, 0.2 V, in wt. pct.), and using a paint baking thermal treatment at 150 or 170 °C as the partitioning step after cooling to room temperature [25].

While the tensile ductility of the 0.25 C, 8.23 Mn (wt. pct.) steel in Figure 6a is outstanding for a steel with 1800 MPa tensile strength [26], some fascinating aspects of microstructure development were also observed in this work [22]. Specifically, the austenite fraction was found to increase with partitioning time and temperature in this steel, following quenching to room temperature, consuming martensite also, due to growth associated with migration of martensite-austenite interfaces. In contrast, the austenite fraction did not increase during partitioning of a high-Ni steel (0.3 C, 14 Ni, 1.5 Si, 0.25 Mo, in wt. pct.) similarly designed for room temperature quenching [22]. A clear understanding of the effects of Mn and/or Ni on partitioning response is still lacking, but the differences in evolution of the austenite fraction were hypothesized to relate potentially to variations in interface mobilities resulting from alloy-dependent interfacial structures associated with different γ/α' orientation relationships observed in the two steels [27].

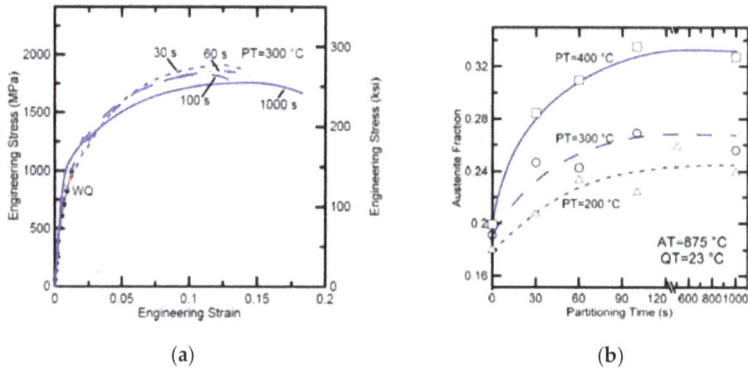

(a)

(b)

Figure 6. Quenching and partitioning (Q&P) response of 0.25 C, 8.23 Mn, 1.87 Si, 0.05 Ni, 0.24 Mo (wt. pct.) steel. (**a**) Tensile stress-strain results after quenching to room temperature and partitioning at 300 °C for times indicated. (**b**) Evolution of austenite fraction during partitioning at temperatures indicated [21].

5. Summary

Some atypical and potentially novel processing scenarios for medium-Mn steels were discussed, including (1) use of medium-Mn steels for hot-stamping, (2) "double soaked" medium-Mn steels for cold-formable applications at strength levels greater than are commonly achieved via intercritical annealing, and (3) medium-Mn steels processed by quenching and partitioning. For all processing histories emphasized here, increased strength levels are achieved by employing appreciable martensite fractions, with ductility enhancements resulting from TRIP behavior associated with Mn-enriched austenite. An apparent influence of Mn on austenite evolution during partitioning was also highlighted. The austenite stabilizing effects of Mn facilitate a variety of industrially important processing pathways, including room-temperature quenching temperatures. These processing opportunities and associated mechanisms of microstructure evolution remain fruitful areas for further research.

Author Contributions: Authors contributed to conceptualization (J.S., E.D.M., D.M., R.R., G.T., A.G.), methodology (J.S., E.D.M., D.M., R.R., G.T., A.G.), investigation (E.D.M., R.R., G.T., A.G.), formal analysis (J.S., E.D.M., D.M., R.R., G.T., A.G.), data curation (E.D.M., R.R., G.T., A.G.), original draft preparation (J.S.), review and editing (J.S., E.D.M., D.M., R.R., G.T., A.G.), supervision (J.S., E.D.M., D.M., R.R.), project administration (J.S., D.M.), and funding acquisition (J.S., E.D.M., D.M.).

Funding: The sponsors of the Advanced Steel Processing and Products Research Center (ASPPRC) at the Colorado School of Mines are gratefully acknowledged.

Acknowledgments: The advice and feedback of ASPPRC sponsors is acknowledged. AK Steel, POSCO, US Steel and Baosteel are acknowledged for provision of experimental materials.

Conflicts of Interest: The authors declare no conflict of interest.

References

1. Wang, J.; Liu, Y.; Lu, Q.; Pang, J.; Wang, Z.; Enloe, C.M.; Singh, J.P.; Horvath, C.D. Effect of microstructure on impact toughness of press hardening steels with tensile strength above 1.8 GPa. In Proceedings of the 6th CHS2 Conference, Atlanta, GA, USA, 4–7 June 2017; AIST: Warrendale, PA, USA, 2017; pp. 717–727.
2. Lu, Q.; Wang, J.; Liu, Y.; Wang, Z. Impact toughness of a medium-Mn steel after hot stamping. In Proceedings of the 6th CHS2 Conference, Atlanta, GA, USA, 4–7 June 2017; AIST: Warrendale, PA, USA, 2017; pp. 737–746.
3. Rana, R.; Carson, C.H.; Speer, J.G. Hot forming response of medium manganese transformation induced plasticity steels. In Proceedings of the 5th CHS2 Conference, Toronto, ON, Canada, 31 May–3 June 2015; AIST: Warrendale, PA, USA, 2015; pp. 391–399.
4. Han, Q.; Bi, W.; Jin, X.; Xu, W.; Wang, L.; Xiong, X.; Wang, J.; Belanger, P. Low temperature hot forming of medium-Mn steel. In Proceedings of the 5th CHS2 Conference, Toronto, ON, Canada, 31 May–3 June 2015; AIST: Warrendale, PA, USA, 2015; pp. 381–390.
5. Speer, J.; Matlock, D.K.; De Moor, E.; Thomas, G.A. Highlights of recent progress in automotive sheet development. In Proceedings of the Fifth Baosteel Biennial Academic Conference, Shanghai, China, 4–6 June 2013; pp. E59–E65.
6. Li, X.; Chang, Y.; Wang, C.; Hu, P.; Dong, H. Comparison of the hot-stamped boron-alloyed steel and the warm-stamped medium-Mn steel on microstructure and mechanical properties. *Mater. Sci. Eng. A* **2017**, *679*, 240–248. [CrossRef]
7. Hanamura, T.; Torizuka, S.; Sunahara, A.; Imagumba, M.; Takechi, H. Excellent total mechanical-properties-balance of 5% Mn, 30000 MPa% steel. *ISIJ Int.* **2011**, *51*, 685–697. [CrossRef]
8. Chang, Y.; Wang, C.Y.; Zhao, K.M.; Dong, H.; Yan, J.W. An introduction to medium-Mn steel: Metallurgy, mechanical properties and warm stamping process. *Mater. Des.* **2016**, *94*, 424–432. [CrossRef]
9. De Moor, E.; Speer, J.G.; Matlock, D.K. Heat treating opportunities for medium manganese steels. In Proceedings of the ICAS 2016 & HMnS 2016, Jeju, Korea, 5–10 September 2016; pp. 182–185.
10. Glover, A.; Speer, J.G.; De Moor, E. Double soaking of a 0.14C-7.14Mn steel. In Proceedings of the International Symposium on New Developments in Advanced High-Strength Sheet Steels, Keystone, CO, USA, 30 May–2 June 2017; AIST: Warrendale, PA, USA, 2017; pp. 189–197.
11. Glover, A. *Private Communication*; Colorado School of Mines: Golden, IL, USA, 2018.
12. Taylor, M.D.; Choi, K.S.; Sun, X.; Matlock, D.K.; Packard, C.E.; Xu, L.; Barlat, F. Correlations between nanoindentation hardness and macroscopic mechanical properties in DP980 steels. *Mater. Sci. Eng. A* **2014**, *597*, 431–439. [CrossRef]
13. Liu, S.; Xiong, Z.; Guo, H.; Shang, C.; Misra, R.D.K. The significance of multi-step partitioning: Processing-structure-property relationship in governing high strength-high ductility combination in medium-manganese steel. *Acta Mater.* **2017**, *124*, 159–172. [CrossRef]
14. Heo, Y.U.; Suh, D.W.; Lee, H.C. Fabrication of an ultrafine-grained structure by a compositional pinning. *Acta Mater.* **2014**, *77*, 236–247. [CrossRef]
15. Glover, A.; Gibbs, P.J.; Liu, C.; Brown, D.W.; Clausen, B.; Speer, J.G.; De Moor, E. Deformation behavior of a double soaked medium manganese steel with varied martensite strength. *Metals* **2019**, *7*, 761. [CrossRef]
16. Xie, Z.J.; Yuan, S.F.; Zhou, W.H.; Yang, J.R.; Guo, H.; Shang, C.J. Stabilization of retained austenite by the two-step intercritical heat treatment and its effect on the toughness of a low alloyed steel. *Mater. Des.* **2014**, *59*, 193–198. [CrossRef]
17. Xu, Y.; Hu, Z.; Zou, Y.; Tan, X.; Han, D.; Chen, S. Effect of two-step intercritical annealing on microstructure and mechanical properties of hot-rolled medium manganese TRIP steel containing δ-ferrite. *Mater. Sci. Eng. A* **2017**, *688*, 40–55. [CrossRef]
18. De Moor, E.; Kähkönen, J.; Speer, J.G.; Matlock, D.K. Quenching and partitioning of a 5 wt pct manganese steel. In Proceedings of the Ninth Pacific Rim International Conference Advanced Materials and Processing (PRICM9), Kyoto, Japan, 1–5 August 2016; Furuhara, T., Nishida, M., Miura, S., Eds.; Japan Institute of Metals and Materials: Sendai, Japan, 2016; pp. 167–170.

19. De Moor, E.; Speer, J.G.; Matlock, D.K.; Kwak, J.H.; Lee, S.B. Quenching and partitioning of CMnSi steels containing elevated manganese levels. *Steel Res. Int.* **2012**, *83*, 322–327. [CrossRef]

20. Thomas, G.; Speer, J.G.; Matlock, D.K.; De Moor, E.; Garza, L. Alloy design for fundamental study of quenched and partitioned steels. In Proceedings of the Materials Science & Technology (MS&T) 2011, Columbus, OH, USA, 16–20 October 2011; TMS: Plano, TX, USA, 2011; pp. 552–567.

21. Bigg, T.D.; Matlock, D.K.; Speer, J.G.; Edmonds, D.V. Dynamics of the quenching and partitioning (Q&P) process. *Solid State Phenom.* **2011**, *172–174*, 827–832.

22. Thomas, G.A.; De Moor, E.; Speer, J.G. Tensile properties obtained by Q&P processing of Mn-Ni steels with room temperature quench temperatures. In Proceedings of the International Symposium on New Developments in Advanced High-Strength Sheet Steels, Vail, CO, USA, 23–27 June 2013; AIST: Warrendale, PA, USA, 2013; pp. 153–165.

23. Hou, Z.R.; Zhao, X.M.; Zhang, W.; Liu, H.L.; Yi, H.L. A medium manganese steel designed for water quenching and partitioning. *Mater. Sci. Technol.* **2018**, *34*, 1168–1175. [CrossRef]

24. Yi, H.L.; Du, P.J.; Wang, B.G. A new invention of press-hardened steel achieving 1880 MPa tensile strength combined with 16% elongation in hot-stamped parts. In Proceedings of the 5th CHS2 Conference, Toronto, ON, Canada, 31 May–3 June 2015; AIST: Warrendale, PA, USA, 2015; pp. 725–734.

25. Pang, J.; Lu, Q.; Wang, J.; Enloe, C.; Wang, G.; Yi, H. A new low density press hardening steel with superior performance. In Proceedings of the 7th CHS2 Conference, Luleå, Sweden, 2–5 June 2019; AIST: Warrendale, PA, USA, 2019; pp. 123–130.

26. Hu, B.; Luo, H.; Yang, F.; Dong, H. Recent progress in medium-Mn steels made with new designing strategies, a review. *J. Mater. Sci. Technol.* **2017**, *33*, 1457–1464. [CrossRef]

27. Thomas, G.; Speer, J.G. Interface migration during partitioning of Q&P steel. *Mater. Sci. Technol.* **2014**, *30*, 998–1007.

metals

MDPI

Article

Accelerated Ferrite-to-Austenite Transformation During Intercritical Annealing of Medium-Manganese Steels Due to Cold-Rolling

Josh J. Mueller *, David K. Matlock, John G. Speer and Emmanuel De Moor *

Advanced Steel Processing and Products Research Center, Colorado School of Mines, Golden, CO 80401, USA
* Correspondence: joshmueller@mines.edu (J.J.M.); edemoor@mines.edu (E.D.M.);
  Tel.: +1-(303)-273-3624 (E.D.M.)

Received: 5 July 2019; Accepted: 16 August 2019; Published: 23 August 2019

Abstract: Prior cold deformation is known to influence the ferrite-to-austenite ($\alpha \rightarrow \gamma$) transformation in medium-manganese (Mn) steels that occurs during intercritical annealing. In the present study, a 7Mn steel with ultra-low residual carbon content and varying amounts of prior cold deformation was intercritically annealed using various heating rates in a dilatometer. The study was conducted using an ultra-low carbon steel so that assessments of austenite formation during intercritical annealing would reflect the effects of cold deformation on the $\alpha \rightarrow \gamma$ transformation and Mn partitioning and not effect cementite formation and dissolution or paraequilibrium partitioning induced austenite growth from carbon. Increasing prior cold deformation was found to decrease the A_{c1} temperature, increase austenite volume fraction during intercritical annealing, and increase the amount of austenite nucleation sites. Phase field simulations were also conducted in an attempt to simulate the apparent accelerated $\alpha \rightarrow \gamma$ transformation with increasing prior cold deformation. Mechanisms for accelerated $\alpha \rightarrow \gamma$ transformation explored with phase field simulations included an increase in the amount of austenite nucleation sites and an increased Mn diffusivity in ferrite. Simulations with different amounts of austenite nucleation sites and Mn diffusivity in ferrite predicted significant changes in the austenite volume fraction during intercritical annealing.

Keywords: intercritical annealing; medium manganese steel; phase field simulation

---

1. Introduction

Intercritical annealing of medium-manganese (Mn) steels, containing 4–10 weight percent (wt%) Mn, is an advanced high-strength steel concept primarily intended for automobile components. Due to the high hardenability of medium-Mn steels, a microstructure primarily consisting of martensite is generally observed upon cooling from hot-rolling [1]. Prior to cold-rolling, tempering is often employed to soften the steel, resulting in a microstructure consisting of deformed ferrite/tempered martensite immediately prior to intercritical annealing. Upon intercritical annealing, austenite forms, and carbon (C) and Mn partitioning to austenite occurs, allowing all or some of the austenite to be stabilized to ambient temperatures upon cooling [1–3]. The mechanical performance of intercritically annealed medium-Mn steels is dependent on the austenite content, composition, and stability [3]. To develop medium-Mn steel grades of various strength levels, it is crucial to control the austenite content and composition resulting from intercritical annealing heat treatments.

It has been shown that increasing the amount of cold deformation prior to intercritical annealing accelerates the ferrite-to-austenite ($\alpha \rightarrow \gamma$) transformation [4]. Studies of austenite formation upon heating and intercritical annealing of conventional Fe-Mn-C alloys with leaner Mn content suggest that interplay between ferrite recrystallization and austenite formation affects the austenite amount and dispersion formed during intercritical annealing [5–7]. Carbon bearing alloys have additional complexity associated with austenite nucleation and growth during intercritical annealing relative to binary Fe-Mn alloys due to the presence of cementite. Studies on austenite formation during intercritical annealing have reported austenite to form at ferrite boundaries [5,8] as well as at cementite-ferrite interfaces [9]; it is reasonable to expect that austenite formation under these different conditions would not be associated with identical growth rates or compositions. Considerable Mn enrichment of cementite during intercritical annealing has also been reported [8]. Mn enrichment in cementite likely influences the cementite dissolution behavior and therefore the amount of Mn and C solute available to partition to austenite. In the absence of carbon, austenite formation assessments should enable a clearer understanding of the effects of cold-rolling on austenite nucleation and growth during intercritical annealing.

The work presented herein pertains to a two part study on intercritical annealing of a 7Mn steel with an ultra-low residual carbon concentration; the study is comprised of an experimental section that includes austenite formation assessments and a simulation section that explores some possible mechanisms for the accelerated $\alpha \rightarrow \gamma$ transformation upon intercritical annealing due to increased prior cold deformation. For the experimental portion of this study, three different conditions of the 7Mn steel were produced, each with a different amount of prior cold deformation, so that the effects of prior cold deformation on austenite formation during intercritical annealing may be distinguished.

For the simulation portion of this study, phase field simulations were conducted for the prediction of microstructural evolution during intercritical annealing of a 7Mn steel. Phase field simulations give a visual output for the translation of domain boundaries over space and time. For the purpose of simulating microstructural evolution, domains are specified to be specific phases or constituents present in the microstructure. Each domain in a phase field simulation is designated a unique order parameter, which is a function of position and time. When the order parameters are plotted as a function of their positions, the output is analogous to a micrograph. The evolution of the order parameter distribution over time is motivated by the minimization of the local free energy of the system. A derivation for the free energy functional of the system and how order parameter transitions are derived from the minimization of the local free energy can be found in a review of the phase field method by Steinbach et al. [10]. The temporal evolution of the order parameter distribution is solved by a set of coupled differential equations commonly referred to as the 'phase field equation'. Additionally, solute concentration fields and diffusivities can also be coupled to the phase field equation to influence the migration of the order parameter domain boundaries. For more information on the application of phase field method to modeling of microstructure evolution in steels, the reader is referred to a review article by Militzer [11].

Simulations in this study were conducted using the phase field simulation software MICRESS®. Various simulations were conducted using different values for select parameters in an attempt to incorporate mechanisms that may lead to accelerated $\alpha \rightarrow \gamma$ transformation due to prior cold deformation. The mechanisms considered include an increase in the number of austenite nucleation sites due to increased ferrite grain boundary area and an increase in the Mn diffusivity in ferrite due to the high defect density anticipated in cold-worked ferrite.

2. Materials and Methods

The composition of the steel used in this study was 0.0005C-7.19Mn-0.25Si wt% The as-received material was cast from a vacuum melt, reheated to 1230 °C, hot-rolled to a thickness of 2.87 mm and furnace-cooled from 650 °C, before being cold-rolled to a thickness of 1.42 mm (50 pct reduction). Three different conditions of the 7Mn steel were used in this study. The condition denoted from here on as 'CR50' is the material in its as-received, cold-rolled condition. Some of the CR50 material was further cold-rolled to a thickness of 0.97 mm (66 pct total reduction); these samples are denoted as 'CR66'. Additionally, to provide a uniform initial microstructure for comparison to the cold-rolled conditions, samples were cut from the CR50 sheet, austenitized under vacuum at 850 °C for 300 s and

---

quenched to 30 °C at 100 °C/s with Helium (He) gas before being placed in liquid nitrogen for 300 s; these samples are denoted as 'AQ' (as-quenched).

Rectangular samples measuring 4.0 mm × 10.0 mm of each material condition were machined with the long axis parallel to the rolling direction and processed under vacuum in a DIL 805A dilatometer (TA Instruments, New Castle, DE, USA). For in situ austenite volume fraction assessments during intercritical annealing, samples were heated to 650 °C at rates ranging from 0.05 to 96 °C/s and held for 1000 s. Subsequently, the samples were heated to 850 °C at a rate of 10 °C/s.

Samples for metallography were either heated to 650 °C and quenched or heated to 500 °C and held for 10,000 s and quenched. The A_{c1} temperature was determined via dilatometry by applying a linear offset corresponding to one volume pct austenite formation from the heating portion of the dilation data prior to the A_{c1} temperature. The A_{c1} temperature was determined to be the temperature where the experimental dilation curve intersected the linear offset. For samples with heating rates exceeding 3 °C/s, heating continued to 900 °C to ensure the A_{c1} temperature was realized upon heating. A_{c1} temperature measurements were taken on three samples for each condition and heating rate. The austenite volume fraction during intercritical annealing was assessed by application of a lever rule to dilation measurements [6,12]. Figure 1 shows exemplary dilatometry data for a sample heated to 650 °C and held for 1000 s followed by heating to 850 °C. Overlaid on the data are dashed lines that extrapolate thermal expansion of the sample corresponding to regions when the sample is fully ferritic/tempered martensite and fully austenitic, labeled α and γ, respectively. Another overlaid line, vertical at 650 °C and bounded by points A and D, corresponds to the relative dilation of the sample corresponding to complete α→γ transformation. The length of line A-B divided by the length of line A-D corresponds to the austenite volume fraction formed upon heating, the length of line B-C divided by the length of line A-D corresponds to the austenite fraction formed during isothermal holding, the length of line A-C divided by the length of line A-D corresponds to total austenite fraction formed during intercritical annealing, and the length of line C-D divided by the length of line A-D corresponds to the fraction of additional austenite formed upon heating to 850 °C, after intercritical annealing.

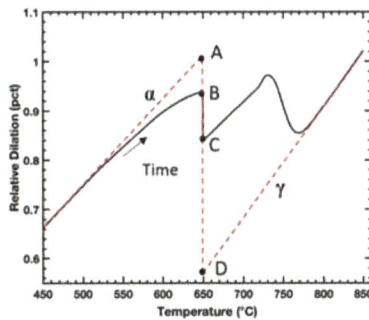

Figure 1. Schematic depicting method for in situ austenite volume fraction assessments from dilatometry during intercritical annealing. Overlaid on the dilation data are dashed lines that extrapolate thermal expansion of the sample corresponding to regions when the sample is fully ferritic/tempered martensite and fully austenitic labeled α and γ, respectively. The line length fraction AC/AD corresponds to austenite volume fraction formed during heating and isothermal holding.

For metallography, samples were sectioned with the cross-section normal perpendicular to the rolling direction. Samples were then mounted, polished, and etched with a one pct Nital solution and analyzed with a JOEL 7000 field-emission scanning electron microscope (FESEM) (JOEL USA, Peabody, MA, USA) using a 15 kV accelerating voltage and a working distance of 10 mm.

Simulations of microstructural evolution during intercritical annealing were conducted using the phase field simulation software MICRESS® (version 7.117, ACCESS e.V., Aachen, Germany) utilizing the TQ module which allows MICRESS® to interface with Thermo-Calc® Software databases; the

TCFE9 Steels/Fe-alloys database was used for all MICRESS® simulations. All MICRESS® simulations conducted were for intercritical annealing at 650 °C and a heating rate of 96 °C/s for a steel with a composition of 7.19Mn-0.25Si wt%. The 5 μm × 5 μm area simulated had an initial microstructure consisting of elongated ferrite which contained a specified amount of stored strain energy; this initial microstructure was constructed with the intent of representing cold-rolled ferrite. As time progressed in each simulation, the elongated ferrite was consumed concurrently by nucleation and growth of both strain-free ferrite and austenite. Strain-free ferrite nucleation was specified to nucleate within the initial elongated ferrite grains while austenite was specified to nucleate at ferrite grain boundaries, including grain boundaries between strain-free and deformed ferrite. The number of austenite nucleation sites was limited to a specified number of nucleation events. To reduce simulation time, all simulations began with heating from 400 °C. Table 1 lists values used for the initial simulation relating to interfaces, austenite nucleation, Mn diffusivity in ferrite, and stored strain energy of ferrite. Interfacial energy values are identical to those used by Zhu and Militzer for a phase field simulation developed for predicting the microstructural evolution during intercritical annealing of dual-phase steels [13]. The α-γ interface mobility pre-factor, M_0, and activation energy, ΔG^*, used by Zhu and Militzer were initially adopted for all interfaces in the simulation. The mobility pre-factor was adjusted to produce simulations that predicted stable austenite growth at 650 °C. The number of austenite nucleation sites was chosen to reflect qualitative metallography observations. The diffusivity pre-factor and activation energy for Mn in ferrite were adopted from De Cooman and Speer [14]. The stored strain energy in ferrite was chosen to be greater than that use by Zhu and Militzer (2 J/cm^3) by approximately a factor of six and was specified as a range to reflect inhomogeneity in the amount of stored strain energy in cold-rolled ferrite.

Table 1. Parameters for Initial MICRESS® Simulation.

Parameter	Value
Interface Energy (J/cm^2)	-
α-α	8.0×10^{-5}
γ-γ	5.0×10^{-5}
α-γ	7.0×10^{-5}
Interface Mobility (α-α/γ-γ/α-γ)	-
M_0 (cm^4/J s)	0.05
ΔG^* (J/mol)	140,000
Number of γ Nucleation Sites	20
Mn Diffusivity in α	-
D_0 (cm^2/s)	0.756
Q (J/mol)	224,500
Stored Strain Energy in α (J/cm^3)	10–15

Additional simulations were conducted using variations of select input parameters in an attempt to incorporate mechanisms that may cause accelerated $\alpha \rightarrow \gamma$ transformation due to prior cold deformation; these parameter variations are listed in Table 2. The number of austenite nucleation sites was varied from 10–40 to reflect a potential increase in ferrite grain boundary area, which serves as an austenite nucleation site. The diffusivity pre-factor for Mn in ferrite was varied from 0.756–7560 cm^2/s to reflect a potential increase in the effective bulk diffusivity of Mn through ferrite due to pipe diffusion in deformed ferrite. Initially, the stored strain energy in ferrite was also varied in order to reflect a potential increase in driving force due to stored strain energy in deformed ferrite, however, it was discovered that MICRESS® does not consider the stored strain energy in ferrite to contribute to the driving force for the $\alpha \rightarrow \gamma$ transformation. Stored strain energy is apparently only considered for like-phase grain boundary migration (e.g., recrystallization).

Table 2. Parameter Variations for MICRESS® Simulations.

Parameter	Value
Number of γ Nucleation Sites	10, 20, 30, 40
Mn Diffusivity Pre-Factor in α-D_0 (cm^2/s)	0.756, 7.56, 75.6, 7560

3. Results

Samples of each condition of the steel were intercritically annealed at 650 °C in a dilatometer using various heating rates to observe austenite formation during the intercritical anneal. The A_{c1} temperatures for each condition and heating rate were also determined using dilatometry. Some samples were cross-sectioned and metallographically prepared for FESEM analysis to assess the austenite formation upon heating to 650 °C or heating to 500 °C and holding for 10,000 s. Phase field simulations were conducted that incorporated variation of some parameters in an attempt to simulate $\alpha \rightarrow \gamma$ transformation due to prior cold deformation.

3.1. Dilatometry

Austenite formed during intercritical annealing was found to transform to martensite upon cooling. Figure 2 shows the relative dilation response for the CR50 condition upon heating to 900 °C at 12 °C/s followed by cooling to 30 °C at 100 °C/s. Upon heating it is evident that the sample passes through A_{c1} and A_{c3}, while upon cooling the martensite start (M_s) temperature is distinguished at approximately 380 °C. Figure 3 shows the relative dilation response produced from the CR50 condition upon heating to 650 °C and holding for 50,000 s followed by quenching to 30 °C at 100 °C/s. Austenite growth is evident from the contraction during the isothermal hold; the M_s temperature can be distinguished at approximately 230 °C. The lower M_s temperature observed for the sample isothermally held at 650 °C suggests that Mn partitioning has occurred in this sample, stabilizing the austenite to a lower temperature.

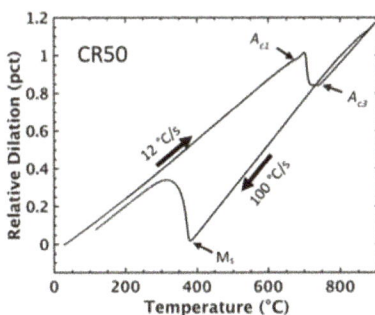

Figure 2. Dilation response of the CR50 condition during heating to 900 °C followed by quenching to 30 °C at 100 °C/s.

Figure 4 shows the variation in A_{c1} temperature observed via dilatometry for each condition and heating rate. Increasing prior cold deformation is shown to decrease the A_{c1} temperature, suggesting that cold deformation accelerates austenite formation upon heating. This is consistent with the findings of Azizi-Alizamini et al. [15], where A_{c1} temperatures observed upon heating a plain low-carbon steel were lower for an as-cold-rolled condition relative to an as-hot-rolled condition. Figure 5 shows austenite volume fraction assessments during intercritical annealing for each condition and for heating rates of 0.05 °C/s (Figure 5a), 3 °C/s (Figure 5b), and 96 °C/s (Figure 5c). The time shown on the x-axis has been adjusted in each figure so that 0 s in the figures corresponds to the time when the sample reached 400 °C. The shaded portion of each plot corresponds to the isothermal holding portion of each intercritical anneal. Austenite formation prior to reaching the isothermal holding temperature of 650 °C is consistent with the A_{c1} measurements in Figure 4; considerable austenite formation is detected in the samples heated at 0.05 °C/s, a small amount of austenite is detected in the samples heated at 3 °C/s, and virtually no austenite is detected prior to reaching the isothermal holding temperature for the samples heated at 96 °C/s. For each heating rate, increasing prior cold deformation corresponds to increasing the $\alpha \rightarrow \gamma$ transformation rate.

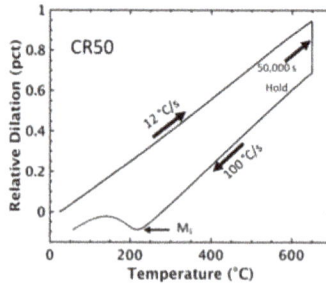

Figure 3. Dilation response of the CR50 condition during heating and isothermal holding at 650 °C for 50,000 s followed by quenching to 30 °C at 100 °C/s.

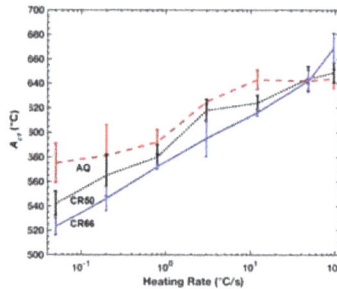

Figure 4. A_{c1} temperatures assessed via dilatometry for each condition and heating rate.

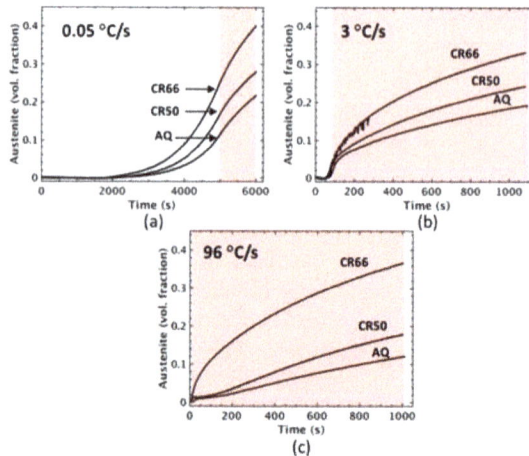

Figure 5. In situ austenite volume fraction assessments from dilatometry for intercritical annealing at 650 °C for each condition and heating rates of 0.05 °C/s (**a**), 3 °C/s (**b**), and 96 °C/s (**c**). Shaded areas indicate approximate duration of isothermal hold at 650 °C.

3.2. Metallography

Figure 6 shows two magnifications of the microstructure of the CR50 condition after being heated to 650 °C at 0.05 °C/s followed by quenching to 30 °C at 100 °C/s. Considering the corresponding austenite volume fractions shown in Figure 5a and the dilation response indicating a martensite transformation upon cooling from a 650 °C isothermal hold shown in Figure 3, the dark etching phase is identified as martensite which was previously austenite that formed upon heating. The microstructure indicates that austenite formed as a fine dispersion amongst the deformed ferrite grains. Austenite appears to have formed predominantly at deformed ferrite grain boundaries. In some instances, elongated regions of martensite (previously austenite) are distinguished that have grown preferentially along the deformed ferrite grain boundaries. Other areas of the microstructure evidence more equiaxed growth of austenite.

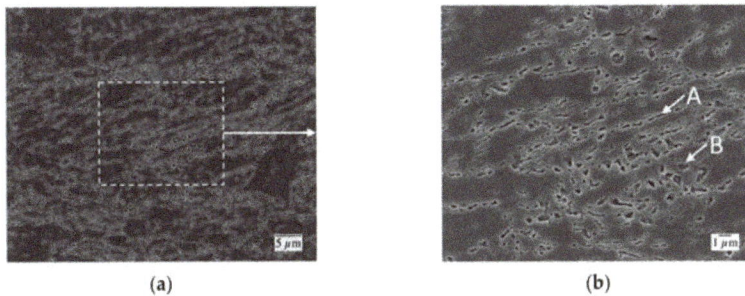

(a)   (b)

Figure 6. Field-emission scanning electron microscope (FESEM) images of the CR50 condition after heating to 650 °C at 0.05 °C/s followed by quenching to 30 °C at 100 °C/s. Higher magnification of the boxed region in (**a**) is shown in (**b**). Elongated regions of martensite (previously austenite) at deformed ferrite grain boundaries are labeled A. More equiaxed regions of martensite (previously austenite) are labeled B. One pct Nital Etch.

Figure 7 shows microstructures for each condition after being heated to 500 °C at 12 °C/s and isothermally held for 10,000 s followed by quenching to 30 °C at 100 °C/s. Based on the A_{c1} temperatures shown in Figure 4, austenite in these samples is anticipated to have formed isothermally at 500 °C. The dark etching areas are identified as either martensite or austenite, corresponding to austenite that formed during the isothermal hold, however, there may be more retained austenite in these samples due to the greater Mn enrichment in austenite that is anticipated with lower intercritical annealing temperatures. For the AQ condition, shown in Figure 7a, austenite has formed predominantly on prior austenite grain boundaries. Alternatively, for the CR50 and CR66 conditions, austenite has formed predominantly on deformed ferrite grain boundaries. It can be distinguished that a greater number of austenite grains had formed in the CR66 condition than the CR50 condition, and that a far greater number of austenite grains had formed in both cold-rolled conditions relative to the AQ condition, reflecting that increasing prior cold deformation increases the number of austenite nucleation sites.

3.3. MICRESS® Phase Field Simulation

Figure 8 shows microstructural evolution results predicted by MICRESS®. The simulation was developed to simulate concurrent austenite growth and ferrite recrystallization. As time progresses in the simulation, the deformed ferrite in the initial microstructure is replaced by strain-free recrystallized ferrite and austenite. Recovery phenomena are not considered in the simulation; stored strain energy is only reduced by growth of a strain-free grain into the ferrite with stored strain energy. Recrystallizing ferrite nucleates homogenously and austenite nucleates at ferrite grain boundaries. Austenite forms interfaces with both deformed and recrystallized ferrite grains. Austenite growth occurs with an enriched Mn concentration, while the Mn content in the ferrite is depleted.

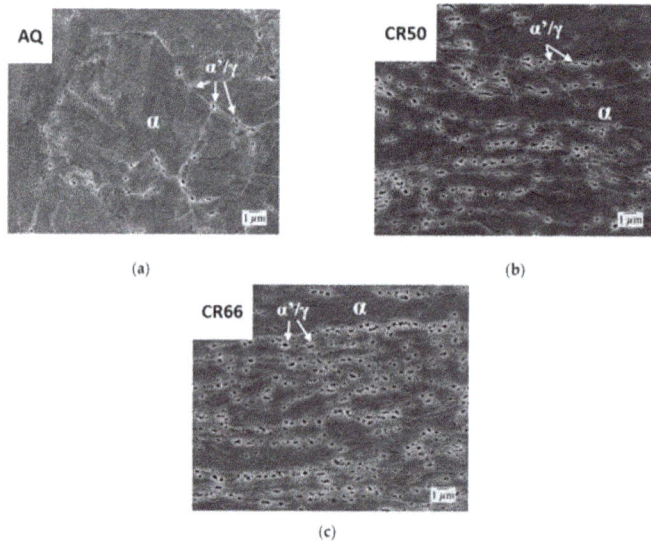

Figure 7. FESEM images of the (**a**) as-quenched (AQ), (**b**) CR50, and (**c**) CR66 conditions after heating to 500 °C at 12 °C/s and isothermal holding for 10,000 s followed by quenching to 30 °C at 100 °C/s. Regions of martensite (previously austenite) are labeled α'/γ. Regions of ferrite are labeled α. One pct Nital Etch.

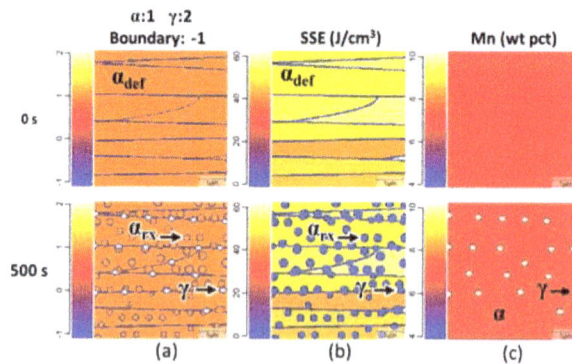

Figure 8. Exemplary MICRESS® simulation results of intercritical annealing with an isothermal holding temperature of 650 °C and a heating rate of 96 °C/s. Timesteps are shown at 0 and 500 s. In the phase maps: (**a**) The simulation is shown to initially contain only deformed ferrite; after 500 s the phase map shows the presence of homogenously nucleated recrystallized ferrite and ferrite grain boundary nucleated austenite. The stored strain energy maps (**b**) shows that the deformed ferrite has a distribution of stored strain energy and that recrystallized ferrite and austenite have no stored strain energy. The medium-manganese (Mn) distribution map (**c**) shows the simulation initially has a homogenous distribution of Mn and that austenite growth occurs with an enriched Mn concentration.

Similar to the argument for increased Mn diffusivity in martensite [16,17], the high density of defects in deformed ferrite likely increases the effective bulk diffusivity of Mn in deformed ferrite relative to recrystallized ferrite. Thus, it is considered that austenite may grow more rapidly into deformed ferrite than recrystallized ferrite due to this increased diffusivity of Mn to the α-γ interface. Diffusivity (*D*) of a solute is described by the following Arrhenius-type relation:

$$D = Do \; e^{(-Q/RT)} \tag{1}$$

where D_0 is the diffusivity pre-factor, Q is the activation energy, and RT takes its usual meaning of thermal energy per mole [18]. Figure 9 shows austenite volume fraction plotted against time for MICRESS® simulations incorporating the different diffusivity pre-factors for Mn diffusivity in ferrite listed in Table 2. Increasing the diffusivity pre-factor for Mn in ferrite is shown to increase the predicted austenite volume fraction. However, the amount of increase in the austenite volume fraction decreases with increasing Mn diffusivity. It should be noted that the greatest diffusivity pre-factor is two orders of magnitude greater than the second greatest, while the other diffusivity pre-factors are only one order of magnitude difference.

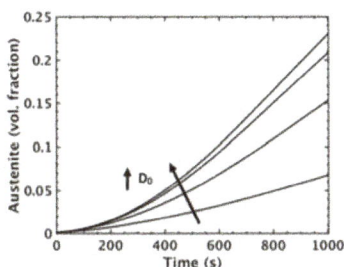

Figure 9. MICRESS® simulation results for austenite volume fraction during intercritical annealing for simulations incorporating different Mn diffusivity pre-factors for Mn diffusivity in ferrite (D_0).

Figure 10 shows predictions for the Mn distributions across α-γ interfaces at 500 s for the MICRESS® simulations incorporating different Mn diffusivity pre-factors in ferrite. The interface width for the simulations was 0.06 μm; the center of the interface is overlaid in the appropriate locations on the figure. For the two lowest Mn diffusivity pre-factors in ferrite the Mn concentration in ferrite was predicted to decrease approaching the interface, while the simulations for the two greater Mn diffusivity pre-factors predict virtually no gradient in the Mn concentration in the ferrite. The difference in Mn gradients in ferrite ahead of the advancing interface corresponds to austenite growth transitioning from more diffusion controlled to more interface controlled. It is also interesting to note that the austenite is predicted to grow with a greater Mn enrichment for simulations with greater Mn diffusivity pre-factors. This suggests that there is driving force for a range of austenite compositions for the α → γ transformation and that the Mn concentration of the growing austenite can be altered by the Mn diffusivity in ferrite.

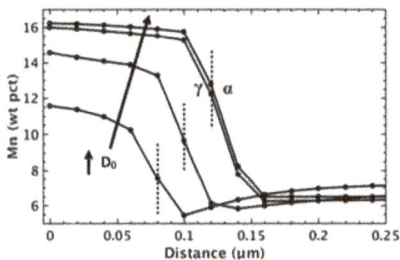

Figure 10. MICRESS® predictions for Mn distributions across the α-γ interface after a 500 s isothermal hold at 650 °C for simulations incorporating different Mn diffusivity pre-factors (D_0) for Mn diffusivity in ferrite.

An increased number of austenite nucleation sites due to an increase in grain boundary area in cold-rolled ferrite was considered as a factor contributing to accelerated $\alpha \rightarrow \gamma$ transformation. MICRESS® simulations incorporating the different amounts of austenite nucleation sites listed in Table 2 were conducted to assess the change in the predicted austenite volume fraction. Figure 11 shows the predicted austenite volume fraction plotted against time for these simulations. As expected, increasing the number of austenite nucleation sites increased the predicted austenite volume fraction during intercritical annealing.

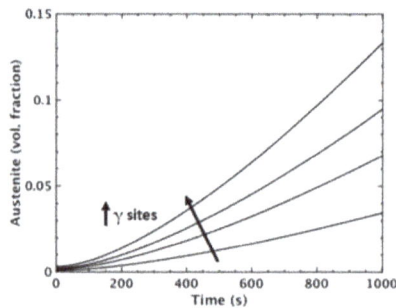

Figure 11. MICRESS® simulation results for austenite volume fraction during intercritical annealing for simulations incorporating different amounts of austenite nucleation sites (γ sites).

4. Discussion

4.1. Martensite Formation Upon Cooling

Generally, the intent of intercritical annealing for medium-Mn steels is for the austenite to become enriched in C and/or Mn to the extent that it is stabilized to room temperature. However, the intercritical annealing heat treatments applied in this study led to martensite transformation upon cooling. Comparing the dilation curves upon cooling in Figures 2 and 3, it is evident that the austenite formed during isothermal holding at 650 °C underwent martensite transformation at a considerably lower temperature than the austenite that formed upon heating to 900 °C (above the A_{c3} temperature). In the absence of carbon, the lower M_s temperature exhibited by the intercritically annealed sample is interpreted to reflect Mn partitioning to austenite during intercritical annealing.

4.2. A_{c1} Temperature

Increasing prior cold deformation was found to decrease the A_{c1} temperature for low heating rates. It is suggested that the decrease in A_{c1} temperature may be attributed to increased ferrite grain boundary area caused by cold rolling, and/or an increased driving force for austenite nucleation due to stored strain energy in ferrite. The prior mechanism is consistent with a previously formulated austenite nucleation mechanism during intercritical annealing of so-called manganese-partitioning dual-phase steels proposed by Navara et al. [8]. They proposed that austenite preferentially forms in a Mn enriched zone behind a migrating ferrite grain boundary, and not at cementite-ferrite interfaces. Migration of the ferrite boundary was suggested to be caused by diffusion-induced grain boundary migration which leaves a solute enriched zone behind the migrating boundary [19]. In the present study, it was found that austenite is capable of ample nucleation in the absence of cementite. With increasing cold deformation there will be increased ferrite grain boundary area. With a greater amount of ferrite grain boundary area serving as possible austenite nucleation sites, it would be expected that a greater number of austenite nuclei would form and that austenite detection upon heating would therefore be realized at lower temperatures.

The latter mechanism mentioned can be visualized by the schematic shown in Figure 12, where the deformed ferrite (α_{Def}) Gibbs free energy curve is raised relative to the undeformed ferrite (α)

curve by an amount equivalent to the stored strain energy induced by cold deformation (ΔG_{SSE}). Using the criteria for driving force for nucleation of a phase with a different composition [20] it is shown that for the arbitrary temperature and alloy composition C_0 in the schematic, there is driving force for austenite nucleation from deformed ferrite whereas there is no driving force for austenite nucleation from ferrite without stored strain energy. Considering this schematic, austenite should form upon heating at lower temperatures from ferrite with greater amounts of stored strain energy.

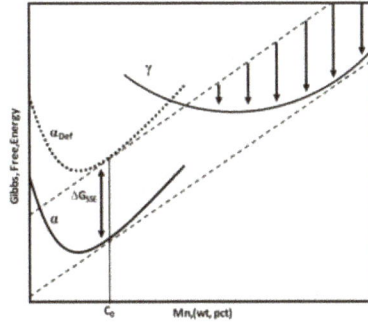

Figure 12. Schematic of Gibbs free energy curves for ferrite, deformed ferrite, and austenite showing driving force for austenite nucleation only from deformed ferrite.

4.3. Austenite Growth

Consistent with Miller's work [4], the $\alpha \rightarrow \gamma$ transformation during intercritical annealing was shown in Figure 6 to increase with increasing prior cold deformation. Possible mechanisms for the increased austenite growth rate are discussed herein which were also incorporated into phase field simulations. The first mechanism discussed is essentially identical to the argument for decreased A_{c1} temperatures due to an increased number of austenite nucleation sites resulting from increased ferrite grain boundary area produced from prior cold deformation. The microstructures shown in Figure 7 indicate that increasing prior cold deformation results in a greater number of austenite grains upon intercritical annealing. Additionally, phase field simulations with variations in the amount of austenite nucleation sites show that increasing the number of austenite nucleation sites is predicted to accelerate the $\alpha \rightarrow \gamma$ transformation; the predicted austenite volume fraction after a 1000 s isothermal hold at 650 °C is increased approximately 330 pct by increasing the amount of austenite nucleation sites from 10 to 40.

Another potential mechanism for the observed increase in the $\alpha \rightarrow \gamma$ transformation rate is based on how an increase in Mn diffusivity in ferrite may affect the α-γ interface velocity. As the intent of intercritical annealing of medium-Mn steels is to partition Mn to austenite, intercritical annealing should be conducted at temperatures that require austenite to grow with an enriched Mn concentration, as opposed to a partitionless $\alpha \rightarrow \gamma$ transformation (e.g., massive). Partitioning of Mn after austenite growth has been predicted to require very long isothermal hold times due to the low diffusivity of Mn in austenite [21]. Austenite growth that occurs with an enriched Mn concentration requires the transport of Mn through ferrite to the α-γ interface. Figures 8 and 10 indicate that the MICRESS® simulations predict the growth of austenite to occur with an enriched Mn concentration. Under this condition, the migration of the α-γ interface may be described as being under mixed control from diffusion of Mn to the interface and the mobility of the interface. Figure 13 depicts a Mn concentration gradient in ferrite ahead of an advancing α-γ interface for mixed control of interface migration as well as the two extreme cases of diffusion-controlled and interface-controlled [22], where X_0, X_i, and X_α are the nominal Mn concentration of the steel, the actual Mn concentration at the interface, and the equilibrium Mn concentration in ferrite. The Mn gradient in ferrite should be expected to reside closer to the diffusion controlled growth curve if there is low Mn diffusivity in ferrite or the interface is highly mobile, whereas the Mn gradient in the ferrite would be expected to reside near the interface controlled line if the Mn diffusivity in ferrite is great or the interface mobility is low.

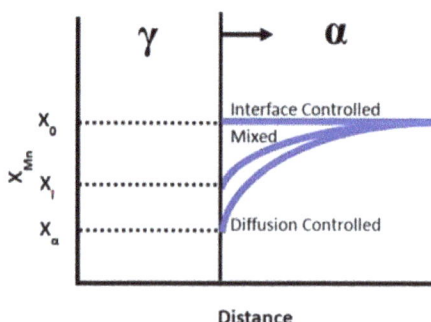

Figure 13. Schematic showing the Mn concentration in ferrite ahead of an advancing α-γ boundary.

The consideration of increased Mn diffusivity through ferrite due to the high defect density in deformed ferrite was incorporated into MICRESS® simulations by input of a range of different diffusivity pre-factors for Mn in ferrite. Increasing the Mn diffusivity pre-factor in simulations corresponded to an increase in the $\alpha \rightarrow \gamma$ transformation rate. The amount of increase in the transformation rate decreased with increasing Mn diffusivity as the predicted α-γ interface migration transitioned from more diffusional controlled to more interface controlled; which is evidenced by the predicted Mn distributions in ferrite ahead of the advancing α-γ interface.

5. Conclusions

In situ austenite growth assessments indicate that increasing prior cold deformation is associated with accelerated $\alpha \rightarrow \gamma$ transformation during intercritical annealing for an ultra-low carbon medium-Mn steel. A_{c1} temperatures were also observed to decrease with increasing prior cold deformation at low heating rates. Metallography indicates that upon intercritical annealing austenite forms readily along deformed ferrite boundaries for the CR50 and CR66 conditions and along prior austenite grain boundaries for the AQ condition. An increase in the number of austenite grains formed during intercritical annealing was observed for samples with increasing prior cold deformation.

MICRESS® simulations for the prediction of microstructural evolution during intercritical annealing were conducted with the intent of incorporating mechanisms that may contribute to accelerated $\alpha \rightarrow \gamma$ transformation during intercritical annealing. Simulations with variation in the number of austenite nucleation sites predicted an accelerated $\alpha \rightarrow \gamma$ transformation with increasing number of austenite nucleation sites. Consistent with metallography and in situ austenite volume fraction assessments during intercritical annealing for conditions with varying amounts of prior cold deformation, simulations suggest that an increased austenite nucleation site density is a significant factor in accelerating $\alpha \rightarrow \gamma$ transformation caused by increased prior cold deformation. Simulations including variations of the diffusivity pre-factor of Mn in ferrite were conducted based on the expectation that increasing prior cold deformation will increase the dislocation density in ferrite and thus provide accelerated transport of Mn through ferrite. The $\alpha \rightarrow \gamma$ transformation rate was predicted to increase with increasing diffusivity pre-factor for Mn in ferrite, suggesting that changes in Mn diffusivity in ferrite due to prior cold deformation could be significant in accelerating $\alpha \rightarrow \gamma$ transformation during intercritical annealing. The predicted increase in the $\alpha \rightarrow \gamma$ transformation rate diminished as the Mn diffusivity in ferrite increased due to the predicted α-γ interface migration transitioning from diffusion controlled to interface controlled.

Author Contributions: Author contributions for this article are as follows: conceptualization, J.J.M., J.G.S., D.K.M., and E.D.M.; methodology, J.J.M.; formal analysis, J.J.M., J.G.S., D.K.M., and E.D.M.; investigation, J.J.M., J.G.S.,

D.K.M., and E.D.M.; writing—original draft preparation, J.J.M.; writing—review and editing, J.J.M., J.G.S., D.K.M., and E.D.M.; supervision, E.D.M.;

Funding: The authors gratefully acknowledge the support of Pacific Northwest National Laboratory, the sponsors of the Advanced Steel Processing and Products Research Center at the Colorado School of Mines, and the ASM Material Genome Toolkit. This work was funded by the Department of Energy's Office of Vehicle Technologies under the Automotive Lightweighting Materials Program.

Conflicts of Interest: The authors declare no conflicts of interest.

References

1. Merwin, M.J. Microstructure and Properties of Cold Rolled and Annealed Low-Carbon Manganese TRIP Steels. *Iron Steel Technol.* **2008**, *5*, 66–84.
2. De Moor, E.; Matlock, D.K.; Speer, J.G.; Merwin, M.J. Austenite Stabilization Through Manganese Enrichment. *Scr. Mater.* **2011**, *64*, 185–188. [CrossRef]
3. Gibbs, P.J.; De Moor, E.; Merwin, M.J.; Clausen, B.; Speer, J.G.; Matlock, D.K. Austenite stability effects on tensile behavior of manganese-enriched- austenite transformation-induced plasticity steel. *Metall. Mater. Trans. A* **2011**, *42*, 3691–3702. [CrossRef]
4. Miller, R.L. Ultrafine-Grained Microstructures and Mechanical Properties of Alloy Steels. *Metall. Trans.* **1972**, *3*, 905–912. [CrossRef]
5. Yang, D.Z.; Brown, E.L.; Matlock, D.K.; Krauss, G. Ferrite Recrystallization and Austenite Formation in Cold-Rolled Intercritically Annealed Steel. *Metall. Trans. A* **1985**, *16*, 1385–1392. [CrossRef]
6. Huang, J.; Poole, W.J.; Militzer, M. Austenite Formation During Intercritical Annealing. *Metall. Mater. Trans. A* **2004**, *35*, 3363–3375. [CrossRef]
7. Chbihi, A.; Barbier, D.; Germain, L.; Hazotte, A.; Gouné, M. Interactions Between Ferrite Recrystallization and Austenite Formation in High-Strength Steels. *J. Mater. Sci.* **2014**, *49*, 3608–3621. [CrossRef]
8. Navara, E.; Bengtsson, B.; Easterling, K.E.; Easterling, K.E. Austenite Formation in Manganese-Partitioning Dual-Phase Steel. *Mater. Sci. Technol.* **1986**, *2*, 1196–1201. [CrossRef]
9. Lee, S.; De Cooman, B.C. Influence of Carbide Precipitation and Dissolution on the Microstructure of ultra-fine-grained intercritically annealed medium manganese steel. *Metall. Mater. Trans. A* **2016**, *47*, 3263–3270. [CrossRef]
10. Steinbach, I.; Pezzolla, F.; Nestler, B.; Seeßelberg, M.; Prieler, R.; Schmitz, G.J.; Rezende, J.L.L. A Phase Field Concept for Multiphase Systems. *Phys. D* **1996**, *94*, 135–147. [CrossRef]
11. Militzer, M. Phase Field Modeling of Microstructure Evolution in Steels. *Curr. Opin. Solid State Mater. Sci.* **2011**, *15*, 106–115. [CrossRef]
12. ASTM Int. *Standard Practice for Quantitative Measurement and Reporting of Hypoeutectoid Carbon and Low-Alloy Steel Phase Transformations*; ASTM International: West Conshohocken, PA, USA, 2004.
13. Zhu, B.; Militzer, M. Phase-Field Modeling for Intercritical Annealing of a Dual-Phase Steel. *Metall. Mater. Trans. A* **2014**, *46*, 1073–1084. [CrossRef]
14. De Cooman, B.C.; Speer, J.G. *Fundamentals of Steel Product Physical Metallurgy*; Association for Iron & Steel Technology: Materials Park, OH, USA, 2011; ISBN 978-1-935117-16-2.
15. Azizi-Alizamini, H.; Militzer, M.; Poole, W.J. Austenite Formation in Plain Low-Carbon Steels. *Mater. Sci. Eng. A* **2011**, *42*, 1544–1557. [CrossRef]
16. Mueller, J.J.; De Moor, E. Austenite growth and retention simulations in intercritically annealed medium manganese steels. In Proceedings of the Materials Science and Technology Conference and Exhibition 2017, Pittsburgh, PA, USA, 8–12 October 2017; Material Science and Technology: Warrendale, PA, USA, 2017.
17. Dmitrieva, O.; Ponge, D.; Inden, G.; Millán, J.; Choi, P.; Sietsma, J.; Raabe, D. Chemical Gradients Across Phase Boundaries Between Martensite and Austenite in Steel Studied by Atom Probe Tomography and Simulation. *Acta Mater.* **2011**, *59*, 364–374. [CrossRef]
18. Shewmon, P. *Diffusion in Solids*; The Minerals, Metals & Materials Society: Pittsburgh, PA, USA, 1989; ISBN 9780873391054.
19. Balluffi, R.W.; Cahn, J.W. Mechanism for Diffusion Induced Grain Boundary Migration. *Acta Metall.* **1980**, *29*, 493–500. [CrossRef]

20. Hillert, M. *Phase Equilibria, Phase Diagrams and Phase Transformations: Their Thermodynamic Basis*; Cambridge University Press: Cambridge, UK, 2007.
21. De Moor, E.; Kang, S.; Speer, J.G.; Matlock, D.K. Manganese Diffusion in Third Generation Advanced High Strength Steels. In Proceedings of the International Conference on Mining, Materials and Metallurgical Engineering, Prague, Czech Republic, 11–12 August 2014. Keynote Lecture II.
22. Porter, D.A.; Easterling, K.E. *Phase Transformations in Metals and Alloys*; CRC Press: Boca Raton, FL, USA, 1992.

![metals logo] *metals*

MDPI

Article

Processing–Microstructure Relation of Deformed and Partitioned (D&P) Steels

Li Liu [1,2], Binbin He [1,2] and Mingxin Huang [1,2,*] ![ORCID]

[1]   Department of Mechanical Engineering, The University of Hong Kong, Hong Kong 999077, China;
        liliu@hku.hk (L.L.); hebinbin@hku.hk (B.H.)
[2]   Shenzhen Institute of Research and Innovation, The University of Hong Kong, Shenzhen 518057, China
[*]   Correspondence: mxhuang@hku.hk; Tel.: +852-28597906; Fax: +852-28585415

Received: 22 May 2019; Accepted: 19 June 2019; Published: 20 June 2019

Abstract: An ultrastrong and ductile deformed and partitioned (D&P) steel developed by dislocation engineering has been reported recently. However, the microstructure evolution during the D&P processes has not yet been fully understood. The present paper aims to elucidate the process–microstructure relation in D&P process. Specifically, the evolution of phase fraction and microstructure during the corresponding D&P process are captured by means of X-ray diffraction (XRD) and electron backscatter diffraction (EBSD). Subsequently, the effect of partitioning temperature on dislocation density and mechanical properties of D&P steel is investigated with the assistance of uniaxial tensile tests and synchrotron X-ray diffraction. It is found that a heterogeneous microstructure is firstly realized by hot rolling. The warm rolling is crucial in introducing dislocations, while deformation-induced martensite is mainly formed during cold rolling. The dislocation density of the D&P steel gradually decreases with the increase of partitioning temperature, while the high yield strength is maintained owing to the bake hardening. The ductility is firstly enhanced while then deteriorated by increasing partitioning temperature due to the strong interaction between dislocation and interstitial atoms at higher partitioning temperatures.

Keywords: D&P steel; processing; microstructure; phase transformation; dislocation density; mechanical properties

---

1. Introduction

Strong and ductile metallic materials are ideal to develop high-performance yet energy-efficient structural components for many applications [1,2]. Unfortunately, strength and ductility are, in general, mutually exclusive in metallic materials [3,4]. Sustained effort has been paid to overcome the strength–ductility trade-off of structural materials. Alloying by the addition of cobalt and titanium is an effective way to simultaneously improve the strength and ductility. Nevertheless, this strategy is not cost-efficient for the industrial application and not sustainable for the limited resources on earth [5]. Developing novel thermal-mechanical process to tune the microstructure is an alternative way to resolve such trade-off. For example, quenching and partitioning (Q&P) treatment enables a dual-phase microstructure of retained austenite embedded and martensite matrix has been successfully applied to medium Mn steel, austenitic stainless steel, and martensitic stainless steel [6–8]. The retained austenite can provide transformation-induced plasticity (TRIP) effect to improve strain hardening rate, delaying the onset of necking and leading to the high strength and good ductility [9]. Recently, a novel dislocation engineering concept was employed to introduce intensive dislocations in a deformed and partitioned (D&P) steel that achieves an ultrahigh yield strength (~2.2 GPa) without compromising ductility (uniform elongation is up to ~16%) [10]. In general, strength tends to increase as dislocation density increases by resisting dislocation motion. However, ductility, which requires the glide of dislocations, is deteriorated by abundant dislocations [11]. Opposite to such traditional view, the D&P

steel demonstrates that a high dislocation density can simultaneously introduce ultrahigh yield strength as well large ductility [10]. The D&P steel shows a dual-phase heterogeneous microstructure with metastable retained austenite embedded in a highly dislocated martensitic matrix [10,12]. The retained austenite provides TRIP effect upon deformation [9]. Intensive dislocations not only provide dislocation hardening for high strength but promote ductility by glides of mobile dislocations [10,13].

The excellent mechanical properties are tailored by the corresponding D&P process. However, the evolution of microstructure during the D&P process has not yet been fully understood, which are crucial to the industrialization of the D&P steel. Meanwhile, the effect of partitioning parameters on the properties of D&P steel has not yet been discussed. Therefore, the present study investigates the contribution of different processing step of the D&P process, especially the partitioning temperature, on microstructure evolution and consequently mechanical properties of the D&P steel.

2. Materials and Methods

The investigated D&P steel has a chemical composition of Fe-10Mn-0.47C-0.7V-2Al (in wt %). Figure 1 is a schematic illustration on the thermal-mechanical processing route to produce the D&P steel. The as-received ingot is firstly homogenized at 1150 °C for 2.5 h, followed by hot rolling (HR) down to a thickness of 4 mm. The hot rolled sheet is further warm rolled (WR) at 750 °C with a total thickness reduction of 50% and is then intercritical annealed (IA) at 620 °C for 5 h. Afterwards, the sheet is further subjected to cold rolling (CR) with a thickness reduction of 30%, giving the final thickness of about 1.4 mm. The specimens subjected to HR, WR, IA, and CR are named as "Deformed" samples for brevity. Finally, Deformed samples are tempered at various temperatures for carbon partitioning from martensite to austenite to optimize austenite stability. Specimens tempered at 200 °C, 300 °C, and 400 °C are referred to D&P200, D&P300, and D&P400, respectively.

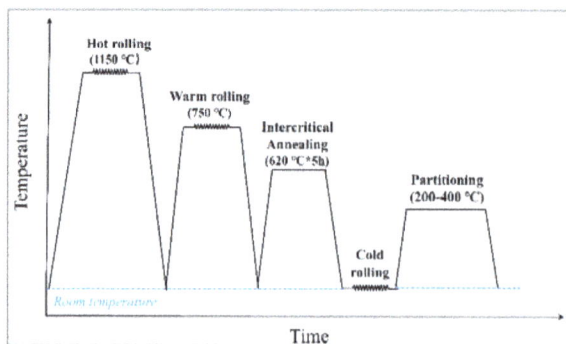

Figure 1. The schematic illustration on the thermal-mechanical processing route to produce the deformed and partitioned (D&P) steel.

The evolution of phase fraction during the whole thermomechanical process is monitored by X-ray diffraction (XRD) carried out on a Rigaku diffractometer (Rigaku Corporation, Tokyo, Japan) operating in the reflection mode with Cu Kα radiation (wavelength = 1.542 Å). Scanning is performed from 40 to 100 degree at a counting rate of $0.01°\cdot s^{-1}$. Diffraction peaks including $(110)\alpha$, $(200)\alpha$, $(211)\alpha$, $(200)\gamma$, $(220)\gamma$, and $(311)\gamma$ are selected to determine the phase fraction of austenite and martensite. The cross-section along rolling direction (RD) of samples is mechanically polished down to 1 µm followed by electrical polishing in a solution of 20% perchloric acid and 80% acetic acid (vol %) for XRD measurement and further microstructure observation. The microstructure is captured by electron backscatter diffraction (EBSD) with an LEO 1530 FEG SEM (Zeiss, Oberkochen, Germany) operated at 20 kV with a step size of 0.25 µm. Vickers hardness measurements are conducted on the electropolished samples with a peak load of 200 gf at ambient temperature. Dog-bone-shaped tensile specimens with a gauge length of 12 mm, thickness of 1.4 mm, and width of 4 mm are machined from the tempered sheets. Tensile tests are performed using a universal tensile testing machine under a strain rate of 10^{-3} s^{-1} at room temperature. To determine the dislocation density of the martensitic matrix of the D&P steel, synchrotron X-ray diffraction with a wavelength of 0.0688 nm, and a two-dimensional (2D) detector are performed on the electropolished samples at the BL14B beamline of Shanghai Synchrotron Radiation Facility (SSRF) (Shanghai, China). The 2D X-ray diffraction pattern is converged to intensity-2theta profiles by Fit2d for further calculation. The modified Williamson–Hall method is employed to obtain the dislocation density from the synchrotron X-ray profiles [14].

3. Results

The evolution of phase fraction during thermal-mechanical processing of the D&P steel process is demonstrated in Figure 2. The specimens had almost fully austenitic microstructure after HR and WR (Figure 2a). The volume fraction of austenite was slightly decreased after intercritical annealing, indicating that ferrite transformation is negligible during IA. This can be explained by the strong hardenability caused by high Mn content (10%) (Figure 2a). A large amount of austenite grains transformed to martensite during the subsequent cold rolling process. Around 18% of austenite was retained after cold rolling and it basically remained unchanged during subsequent partitioning process, confirming that almost no phase transformations took place during low-temperature partitioning process (Figure 2b).

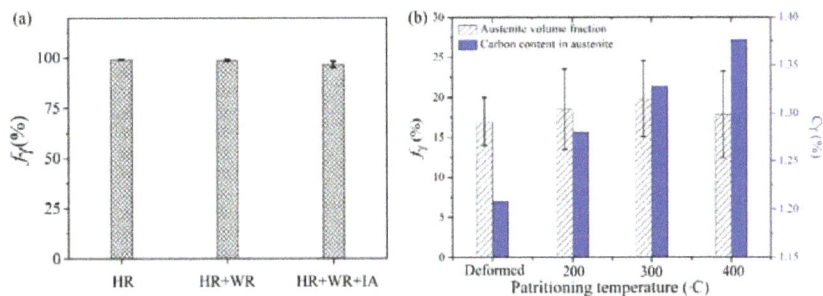

Figure 2. (**a**) The evolution of austenite phase fraction during HR (hot rolling), WR (warm rolling), and IA process (intercritical annealed). (**b**) The phase fraction and increment of carbon content of austenite in Deformed, D&P200, D&P300, and D&P400 samples.

Although the partitioning process had a less-significant effect on the austenite fraction, it influenced the carbon content and, consequently, the stability of austenite. The carbon content of austenite, which can be estimated from the peak shift of XRD profiles [15,16], was increased by 0.072, 0.12, and 0.168% after partitioning at 200, 300, and 400 °C, respectively (Figure 2b).

The typical microstructure of HR, HR + WR, and D&P specimens were characterized by EBSD, as shown in Figure 3. The HR specimen had an austenitic microstructure with prior austenite grain boundaries (PAGBs) decorated with granular submicron ferrite grains (Figure 3a1,a2). The fraction of austenite obtained from EBSD microstructure was quantitively consistent with the XRD results. The coarse austenite grains (5–30 μm) dominated the microstructure, while some recrystallized austenite grains (~0.5 μm) were found at PAGBs (Figure 3a1,a2).

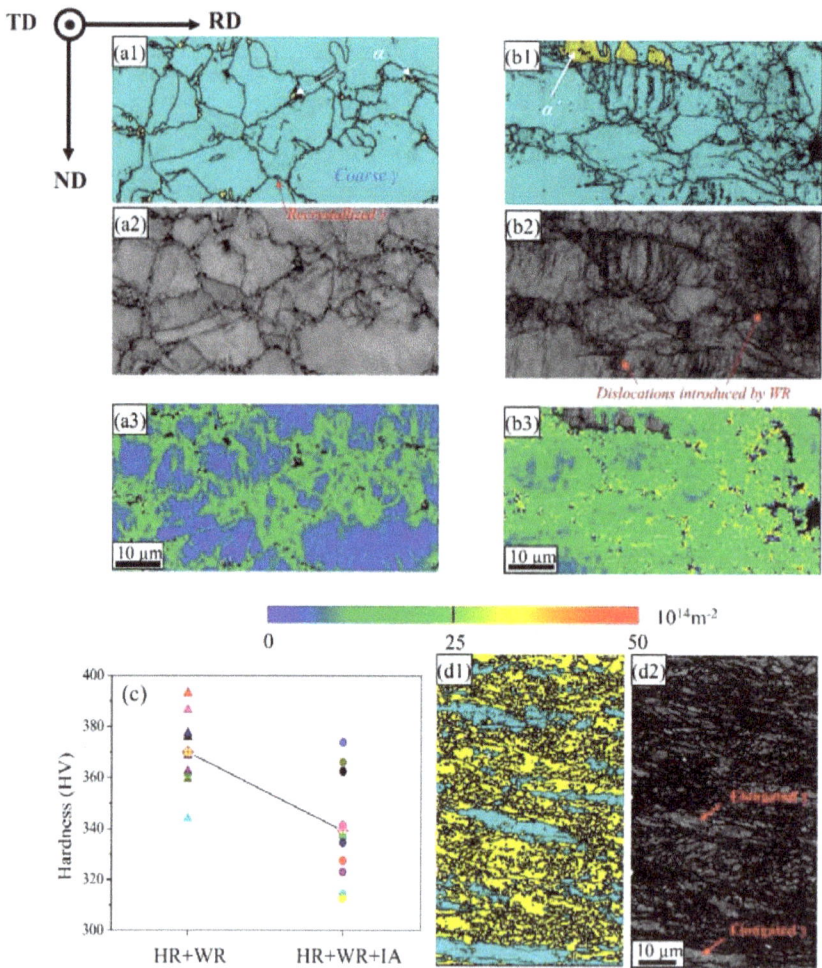

Figure 3. The microstructure of (**a1,a2,a3**) HR sample and (**b1,b2,b3**) HR+WR sample. (**c**) The Vickers hardness (HV) of HR+WR sample and HR+WR+IA sample. The microstructure of (**d1,d2**) D&P samples partitioned at 300 °C for 6 min. (a1, b1 and d1 are electron backscatter diffraction (EBSD) phase maps, wherein yellow and blue colors represent martensite/ferrite and austenite, respectively; a2, b2 and d2 are EBSD band contrast maps; a3 and b3 are the distribution of geometrically necessary dislocations (GND) densities of austenite matrix estimated based on the kernel average misorientation (KAM)). RD: Rolling direction; ND: Normal direction; TD: Transverse direction.

The dominated austenitic microstructure was maintained after warm rolling (Figure 3b1). To distinguish the difference of dislocation density between HR and HR + WR samples, the geometrically necessary dislocation (GND) densities were further estimated based on the kernel average misorientation (KAM), which represents the average misorientation between the measured point and the nearest neighbor points in the EBSD measurement (Figure 3b3). It was found that intensive dislocations were generated in austenite grains by warm rolling (Figure 3b3). Moreover, prior austenite grains were substantially elongated along the rolling direction or even fragmented into subgrains, resulting in obvious grain refinement.

The HR + WR specimen was further softened by means of intercritical annealing to facilitate subsequent deformation at room temperature. The Vickers hardness of specimens, before and after intercritical annealing, are summarized in Figure 3c. The hardness measurements were conducted in different regions through the thickness of the sample. This leads to the dispersion of hardness values due to the edge of the steel sheet which are stronger than the center resulting from warm rolling. The slight decrease of the hardness implies that some dislocations in the HR + WR specimen were annealed after IA.

Cold rolling greatly promoted the deformation-induced martensitic transformation, leading to a dual-phase lamella microstructure with austenite grains embedded in a highly dislocated martensite matrix in D&P steel (Figure 3d1,d2). The prior austenite grain boundaries (PAGB) exist before the martensitic transformation and can be retained after the formation of martensite. The martensitic matrix possesses a heterogeneous microstructure, consisting of large lenticular grains and small martensite lath. Retained austenite grains also exhibit heterogeneous morphologies and bimodal distribution. The large austenite grains with a length of 30 μm or above were elongated along the rolling direction and constituted most of the austenite phase, while fine granular austenite grains can be observed at PAGB. Abundant dislocations were introduced to the martensitic matrix by cold rolling, as showed by the low image quality of martensite phase in Figure 3d2. In contrast, it seems that the austenite grains, especially the coarse elongated ones, still contained relatively lower defects than the martensite phase (Figure 3d2).

The mechanical properties of the D&P steel processed by varying partitioning temperatures are summarized in Figure 4a. The Deformed specimen is brittle, although it possesses an ultrahigh yield strength of up to 2200 MPa. The ductility of the D&P steel was greatly enhanced after low-temperature partitioning (200 and 300 °C) without a noticeable decrease of the yield strength. However, very different deformation behaviors were obtained after partitioning at a relatively high temperature (400 °C). Necking was found to proceed upon yielding for the D&P400 specimen. A typical yield drop phenomenon, which is induced by the un-locking of dislocations from the interstitial C atoms [17], was observed in all specimens. It was interesting to find that Lüders deformation is also tailored by the partitioning conditions (Figure 4b). Lüders deformation dominated in D&P300 specimen and accounted for 65.5% of the total elongation. The highest ductility was obtained in D&P300 specimen, which possessed the largest Lüders strain, followed by work hardening at large plastic deformation regime (Figure 4a).

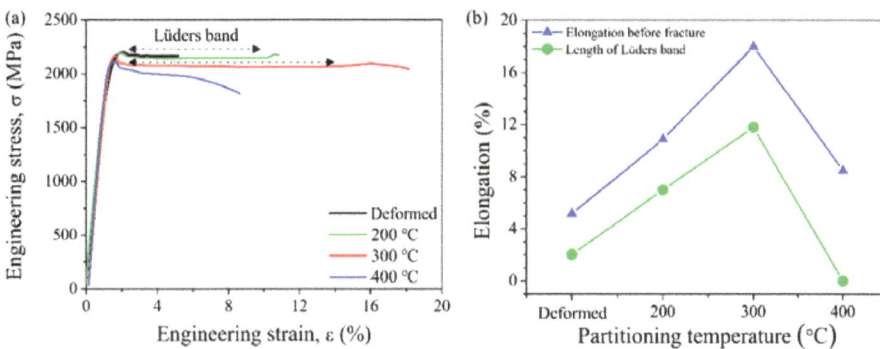

Figure 4. (**a**) The engineering stress–strain curves of D&P steels. (**b**) Effect of partitioning temperature on the ductility of D&P steels.

To understand the difference in the mechanical behaviors of various D&P specimens, 2D synchrotron X-ray diffraction measurements were carried out and the modified Williamson–Hall method was employed to study the evolution of dislocation density during the partitioning process.

2D synchrotron X-ray diffraction can provide better intensity and statistics for quantitative analysis and phase identification, especially for samples with texture, large grain size, and small quantity [18]. The broadening of diffraction peaks depends on the dislocation density (ρ), the average crystallite size (d'), and the faulting probability (β) by the following equation [14]:

$$\Delta K = 0.9/d' + (\pi A^2 b^2/2) \cdot \rho^{1/2} \cdot K^2 \overline{C} + \beta W_{hkl} + o(K^4 \overline{C}^2) \tag{1}$$

where $\Delta K = \cos\theta_B (\Delta 2\theta)/\lambda$ and $K = 2\sin\theta_B/\lambda$, θ_B is the diffraction angle at certain Bragg position, $\Delta 2\theta$ is the full width at half-maximum (FWHM) of the diffraction peak at θ_B, and λ is the wavelength of the X-ray. A is a constant which can be determined by the effective outer cut-off radius of dislocation, b is the Burgers vector of dislocations and W_{hkl} is the scale factor representing the peak broadening induced by a fault at exact {hkl} deflection. Here, the faulting probability (β)-induced peak broadening was not considered because twin martensite was barely founded in the present material. $\overline{C}$ is the average dislocation contrast factor determined by the empirical equations. Six martensite diffraction rings and four austenite diffraction rings were obtained from 2D synchrotron X-ray (Figure 5a). The 2D diffraction patterns are further converted to peak profiles to calculate ΔK and K (Figure 5b). Five martensite peaks, including (110), (200), (211), (220), and (310) were selected to calculate dislocation density of martensitic matrix. Dislocation density (ρ) and d' were subsequently determined by considering the best linear fitting between ΔK and $K^2 \overline{C}$.

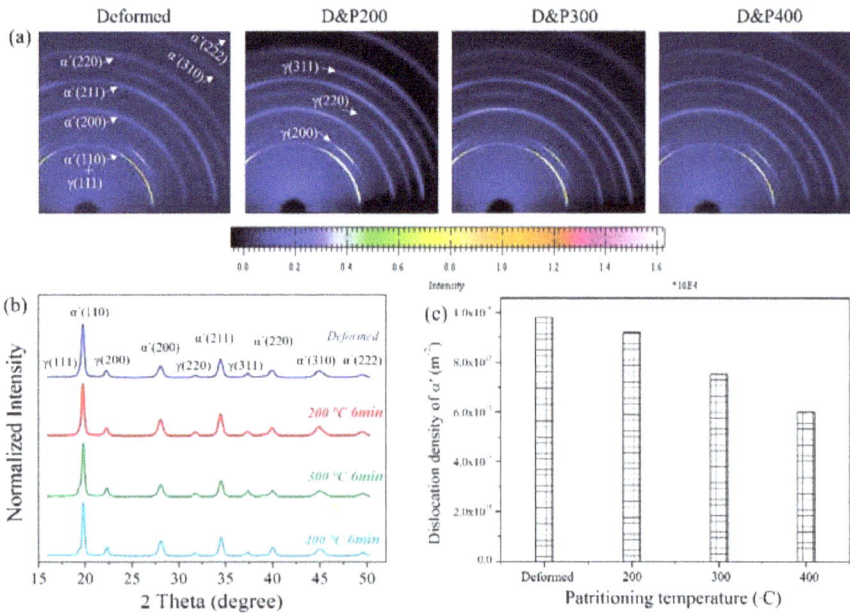

Figure 5. (a) 2D synchrotron X-ray diffraction patterns of D&P steels processed at different partitioning temperatures. Images have been colorized with high contrast to highlight the Debye diffraction rings. (b) 1D synchrotron X-ray diffraction profiles (intensity-2 theta curves) converged by 2D diffraction patterns. (c) The dislocation density of the martensitic matrix in D&P steels.

As shown in Figure 5c, the dislocation density of martensitic matrix was around 10^{16} m^{-2} after cold rolling, which is two-orders of magnitude larger than conventional thermally transformed martensite (6×10^{13} m^{-2}) and cold-rolled martensite (5.39×10^{14} m^{-2}) [19,20]. The ultrahigh dislocation density resulted from the deformation (hot rolling, warm rolling, and cold rolling) and displacive shear deformation. The dislocation density decreased with an increase of partitioning temperature and reduced to 6×10^{15} m^{-2} after partitioning at 400 °C for 6 min.

4. Discussion

A heterogeneous microstructure with bimodal grain distribution was firstly developed by hot rolling (Figure 3a1). The subsequent warm rolling further enhanced the inhomogeneity and introduced intensive dislocations in austenite grains (Figure 3b1–3b3). These dislocations were slightly recovered during the intercritical annealing, while the majority can be inherited by the martensite formed during cold rolling. Note that the intercritical annealing between warm rolling and cold rolling process plays a key role in relieving the residual internal stress. Sudden cracking may take place before reaching the targeted cold rolling reduction if the HR + WR sheet is not sufficiently annealed. The amount of ferrite transformed during warm rolling and intercritical annealing were very low (Figure 2a). The sluggish transformation kinetics of ferrite was due to the high alloying contents, especially the high Mn (10%) contents, of the present steel [21]. Cold rolling enables the occurrence of deformation-induced martensitic transformation. Consequently, substantial amount of dislocations was introduced during the cold rolling process owing to the displacive shear transformation and deformation of early transformed martensite grains (Figure 3d2). In a word, although the majority of martensite was formed during cold rolling, the previous hot rolling, warm rolling, and intercritical annealing were of great importance in developing heterogeneous and dislocated microstructure in the final martensite matrix.

The partitioning process was applied to tailor the austenite stability, dislocation density, and consequently, mechanical behaviors of the D&P steel. The carbon content of austenite only increased slightly after partitioning (Figure 2b), which could be because of the formation of abundant Cottrell atmosphere, leading to a retarded carbon diffusion from martensite to austenite. These results suggest that the stability of retained austenite is not the key factor controlling the mechanical behaviors of D&P steel.

The high yield strength of D&P steels is mainly contributed by the presence of high dislocation density (Figure 4a). The contribution of the dislocation density to the yield strength can be estimated by the Taylor hardening law [11]:

$$\sigma = M\alpha\mu b \sqrt{\rho_t} \qquad (2)$$

where M represents the Taylor factor and is taken as 2.9, $\alpha = 0.23$ is an empirical constant for martensite with dislocation cell structure, $\mu = 85$ GPa is shear modulus, and $b = 0.25$ nm is Burgers vector [10,11]. Therefore, the contribution of the dislocation density was estimated to be 1400, 1360, 1230, and 1100 MPa for Deformed, D&P200, D&P300, and D&P400 specimens, respectively. Therefore, the decreased yield strength of D&P400 should be around 300 MPa owing to the dislocation recovery. However, the upper yield stress of deformed sample was decreased slightly from 2204 to 2115 MPa after partitioning at 400 °C for 6 min. Therefore, other strengthening mechanisms may take place to compensate the reduction of strength caused by dislocation recovery. The reduction of dislocation density without obviously sacrificing the strength may be ascribed to the bake hardening. The carbon atoms tend to diffuse into adjacent dislocations in martensitic matrix to minimize the strain energy during the partitioning process, resulting in the formation of "Cottrell atmosphere" [17]. Dislocations surrounded by atmosphere were immobilized, leading to an increase of the upper yield stress. The increase of yield stress realized by the combination of pre-strain and strain-aging is known as bake hardening [22]. Therefore, bake hardening to some extent compensates the loss of dislocation hardening, beneficial to the high yield strength of the D&P steel. However, dislocations can break away the Contrell atmosphere and become mobile when a sufficiently large stress is applied. This causes the occurrence of yield drop as observed in some D&P specimens (Figure 4a) [17].

The propagation of Lüders band after a yield drop depends on the work hardening ability of the specimen [23,24]. The tangled dislocation structure makes the slip of dislocation very difficult to proceed. Hence, the deformed specimen has low work hardening rate and very limited ductility

(Figure 4). After low-temperature partitioning (200 °C), D&P steel was recovered but not yet sufficiently. The D&P200 sample had improved ductility but still fractured during Lüders deformation (Figure 4). Further increasing the partitioning temperature greatly improved the ductility of samples. The D&P300 specimen showed a completed propagation of Lüders band, followed by work hardening which could be provided by the TRIP effect [10]. Although some reports on austenitic steels suggest that the TRIP effect is the reason for the propagation of Lüders band [25], it was found that the Lüders band is still present in medium Mn steels during relatively high temperature tensile deformation where the TRIP effect is completely suppressed [26]. Therefore, the TRIP-induced Lüders band mechanism is not favorable in explaining the deformation behaviors of the D&P steels with both of high dislocation density and high carbon content. The discontinuous yielding in D&P400 sample results in localized necking which is not able to propagate owing to the low work hardening ability, leading to the disappearance of Lüders deformation. This can be explained in two aspects. The annihilation of dislocation was promoted at high temperature, which greatly reduces the mobile dislocation density in the specimen. Moreover, the diffusion of interstitial carbon atoms to dislocations was energetically favored at high temperature. Namely, dislocations were more likely to be locked by carbon atoms, leading to a further reduction of the mobile dislocations. Consequently, necking proceeded quickly after the yield drop in D&P400 specimen (Figure 4).

5. Conclusions

The present work investigates the phase transformation and microstructure evolution during the D&P process. The effect of partitioning temperatures on mechanical properties of D&P steel is also addressed. The present results show that the hot rolling process can result in an initial single-phase heterogeneous microstructure. The following warm rolling is highly effective in introducing dislocations. Intercritical annealing reduces the residual stress by partially recovering dislocations, which facilitates the cold rolling at room temperature. The subsequent cold rolling enables the deformation induced martensitic transformation. Partitioning tunes the mechanical behaviors by influencing the dislocation density and austenite stability. The increase of partitioning temperature slightly changes the austenite stability because the strong interaction between dislocations and carbon atoms restricts the carbon partitioning from martensite to austenite grains. Bake hardening compensates the reduction of strength after partitioning process, enables the ultrahigh yield strength of D&P steels.

Author Contributions: M.H. proposed the project and supervised all the work. L.L. performed the experimental testing and characterization. B.H. made the material. The paper was written by L.L., B.H. and M.H.

Funding: This research was funded by the National Natural Science Foundation of China (grant number U1764252, U1560204), Research Grants Council of Hong Kong (grant number 17255016, 17203014 and 17210418), and seed fund for Basic Research of HKU (grant number 201711159029).

Acknowledgments: Authors acknowledge the experimental support from the BL14B beamline at Shanghai Synchrotron Radiation Facility, Shanghai, China.

Conflicts of Interest: The authors declare no conflict interest.

References

1. Wei, Y.; Li, Y.; Zhu, L.; Liu, Y.; Lei, X.; Wang, G.; Wu, Y.; Mi, Z.; Liu, J.; Wang, H. Evading the strength–ductility trade-off dilemma in steel through gradient hierarchical nanotwins. *Nat. Commun.* **2014**, *5*, 3580–3587. [CrossRef] [PubMed]
2. Lu, K. The future of metals. *Science* **2010**, *328*, 319–320. [CrossRef] [PubMed]
3. Li, Z.; Pradeep, K.G.; Deng, Y.; Raabe, D.; Tasan, C.C. Metastable high-entropy dual-phase alloys overcome the strength-ductility trade-off. *Nature* **2016**, *534*, 227–230. [CrossRef] [PubMed]
4. Ritchie, R.O. The conflicts between strength and toughness. *Nat. Mater.* **2011**, *10*, 817–822. [CrossRef] [PubMed]
5. Li, X.; Lu, K. Improving sustainability with simpler alloys. *Science* **2019**, *364*, 733–734. [CrossRef] [PubMed]

6. Seo, E.J.; Cho, L.; De Cooman, B.C. Application of quenching and partitioning processing to medium Mn steel. *Metall. Mater. Trans. A* **2015**, *46*, 27–31. [CrossRef]

7. Wendler, M.; Ullrich, C.; Hauser, M.; Krüger, L.; Volkova, O.; Weiß, A.; Mola, J. Quenching and partitioning (Q&P) processing of fully austenitic stainless steels. *Acta Mater.* **2017**, *133*, 346–355.

8. Mola, J.; De Cooman, B.C. Quenching and partitioning (Q&P) processing of martensitic stainless steels. *Metall. Mater. Trans. A* **2013**, *44*, 946–967.

9. Bleck, W.; Guo, X.; Ma, Y. The TRIP effect and its application in cold formable sheet steels. *Steel Res. Int.* **2017**, *88*, 1700218. [CrossRef]

10. He, B.; Hu, B.; Yen, H.; Cheng, G.; Wang, Z.; Luo, H.; Huang, M. High dislocation density-induced large ductility in deformed and partitioned steels. *Science* **2017**, *357*, 1029–1032. [CrossRef]

11. Taylor, G.I. The mechanism of plastic deformation of crystals. Part I—Theoretical. *Proc. R. Soc. London, Ser. A* **1934**, *145*, 362–387. [CrossRef]

12. Liu, L.; He, B.; Huang, M. The Role of Transformation-Induced Plasticity in the Development of Advanced High Strength Steels. *Adv. Eng. Mater.* **2018**, *20*, 1701083. [CrossRef]

13. Huang, M.; He, B. Alloy design by dislocation engineering. *J. Mater. Sci. Technol.* **2018**, *34*, 417–420. [CrossRef]

14. Ungár, T.; Ott, S.; Sanders, P.; Borbély, A.; Weertman, J. Dislocations, grain size and planar faults in nanostructured copper determined by high resolution X-ray diffraction and a new procedure of peak profile analysis. *Acta Mater.* **1998**, *46*, 3693–3699. [CrossRef]

15. Liu, L.; He, B.; Cheng, G.; Yen, H.; Huang, M. Optimum properties of quenching and partitioning steels achieved by balancing fraction and stability of retained austenite. *Scr. Mater.* **2018**, *150*, 1–6. [CrossRef]

16. Van Dijk, N.; Butt, A.; Zhao, L.; Sietsma, J.; Offerman, S.; Wright, J.; Van der Zwaag, S. Thermal stability of retained austenite in TRIP steels studied by synchrotron X-ray diffraction during cooling. *Acta Mater.* **2005**, *53*, 5439–5447. [CrossRef]

17. Cottrell, A.H.; Bilby, B. Dislocation theory of yielding and strain ageing of iron. *Proc. Phys. Soc., London, Sect. A* **1949**, *62*, 49–62. [CrossRef]

18. He, B.B.; Preckwinkel, U.; Smith, K.L. Comparison between conventional and two-dimensional XRD. *Adv. X-Ray Anal.* **2003**, *46*, 37–42.

19. Pešička, J.; Kužel, R.; Dronhofer, A.; Eggeler, G. The evolution of dislocation density during heat treatment and creep of tempered martensite ferritic steels. *Acta Mater.* **2003**, *51*, 4847–4862. [CrossRef]

20. Shintani, T.; Murata, Y. Evaluation of the dislocation density and dislocation character in cold rolled Type 304 steel determined by profile analysis of X-ray diffraction. *Acta Mater.* **2011**, *59*, 4314–4322. [CrossRef]

21. Coates, D. Diffusional growth limitation and hardenability. *Metall. Trans.* **1973**, *4*, 2313–2325. [CrossRef]

22. Soenen, B.; De, A.; Vandeputte, S.; De Cooman, B. Competition between grain boundary segregation and Cottrell atmosphere formation during static strain aging in ultra low carbon bake hardening steels. *Acta Mater.* **2004**, *52*, 3483–3492. [CrossRef]

23. Johnston, W.G. Yield points and delay times in single crystals. *J. Appl. Phys.* **1962**, *33*, 2716–2730. [CrossRef]

24. Schwab, R.; Ruff, V. On the nature of the yield point phenomenon. *Acta Mater.* **2013**, *61*, 1798–1808. [CrossRef]

25. Gao, S.; Bai, Y.; Zheng, R.; Tian, Y.; Mao, W.; Shibata, A.; Tsuji, N. Mechanism of huge Lüders-type deformation in ultrafine grained austenitic stainless steel. *Scripta Mater.* **2019**, *159*, 28–32. [CrossRef]

26. Wang, X.-G.; Huang, M.-X. Temperature dependence of Lüders strain and its correlation with martensitic transformation in a medium Mn transformation-induced plasticity steel. *J. Iron Steel Res. Int.* **2017**, *24*, 1073–1077. [CrossRef]

![metals logo] *metals*

MDPI

Article

Austenite Reversion Tempering-Annealing of 4 wt.% Manganese Steels for Automotive Forging Application

Alexander Gramlich*[ID], Robin Emmrich[ID] and Wolfgang Bleck

Steel Institute of RWTH Aachen University, 52072 Aachen, Germany;
robin.emmrich@iehk.rwth-aachen.de (R.E.); bleck@iehk.rwth-aachen.de (W.B.)
* Correspondence: alexander.gramlich@iehk.rwth-aachen.de; Tel.: +49-241-80-95786

Received: 25 April 2019; Accepted: 14 May 2019; Published: 17 May 2019

Abstract: New medium Mn steels for forged components, in combination with a new heat treatment, are presented. This new annealing process implies air-cooling after forging and austenite reversion tempering (AC + ART). This leads to energy saving compared to other heat treatments, like quenching and tempering (Q + T) or quenching and partitioning (Q + P). Furthermore, the temperature control of AC + ART is easy, which increases the applicability to forged products with large diameters. Laboratory melts distinguished by Ti, B, Mo contents have been casted and consecutively forged into semi-finished products. Mechanical properties and microstructure have been characterized for the AC and the AC + ART states. The as forged-state shows YS from 900 MPa to 1000 MPa, UTS from 1350 MPa to 1500 MPa and impact toughness from 15 J to 25 J. Through the formation of nanostructured retained metastable austenite an increase in impact toughness was achieved with values from 80 J to 100 J dependent on the chemical composition.

Keywords: medium-manganese; forging; austenite reversion; mechanical properties; microstructure

---

1. Introduction

In the last decade several material concepts were designed in order to substitute the classic quench and tempering (Q + T) steels. These materials required high yield strength by keeping applicable levels of toughness and ductility to prevent brittle fracture [1]. Besides the Q + T steels, precipitation hardening ferritic pearlitic (PHFP) steels are also widely used in the automotive industry. These steels do not accomplish the balance of mechanical properties of the Q + T steels but achieve their final mechanical properties through air-cooling out of the forging heat. In order to improve the mechanical properties of these steels micro-alloying concepts where applied on the PHFP steels in order to increase the yield strength [2]. Recently, steels with a ductile bainitic microstructure (HDB) have been developed and are now widely used [3,4]. These steels offer comparable toughness to the Q + T steels but still not a sufficient yield strength level. The gap in properties could finally be closed by using steels, which contain a small fraction of retained austenite in their matrix. Through the retained austenite strength and toughness can be further increased as a consequence of the transformation induced plasticity (TRIP) [5]. Alternatively, the quenching and partitioning steels (Q + P) have been designed, with the material quenched into the range between the martensite start temperature (M_s) and the martensite finish temperature (M_f) [6]. After the quenching these materials are directly tempered in order to achieve carbon and to some extend manganese partitioning into the retained austenite, in order to increase its stability. These materials achieve good combinations of strength and ductility but require a complex processing that shows a high sensibility of the achieved properties on the process parameters. As the narrow parameter windows during these processes limits the possibilities for industrial application, the current focus is shifted again to materials, which achieve their final properties through air-cooling. Air-hardening medium manganese (LHD) steels can be used for these purposes. Through the high amounts of manganese, these materials develop a martensitic microstructure via air-cooling, by keeping reasonable levels of toughness [7,8]. The increase in toughness in the as-forged state can be reached through the alloying of grain boundary strengthening elements like boron [9,10] or molybdenum [11,12]. Alternatively, through an addition of aluminium, retained austenite can be stabilized to room temperature in order to increase the ductility of these alloys [13]. These air hardening medium Mn steels combine good mechanical properties with a simple process route, which leads to cost and CO_2-emission reduction.

Recently, it was shown that this alloy group shows a further increase in toughness by keeping applicable levels of strength when the materials are intercritically annealed [14]. Intercritical annealing of medium manganese steels leads to an stabilization of the newly formed austenite until room temperature through the partition of alloying elements into the austenite. This mechanism is know as austenite reverted transformation (ART) annealing [15,16]. Figure 1 compares the new air-cooling + austenite reversion tempering process (AC + ART) which standard heat treatments. Most papers using the ART effect deal with applications for sheet products, as the necessary heat treatment can be best tailored by continuous strip annealing treatments. Thus, the composition requirement and the process parameters have been studied for a wide range of medium manganese steels [17–20]. As the time-temperature profiles of sheet products cannot be applied on forging parts because of the complex geometries and great cross sections, only simple and robust heat treatments can be used to modify the microstructure. Therefore, the annealing in the intercritical area and the influence on the mechanical properties need further investigation in order to reach the full potential of these alloys.

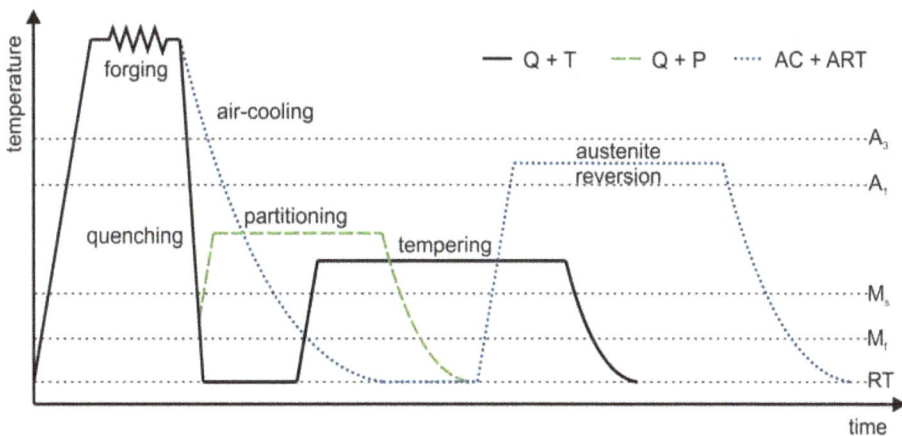

Figure 1. Comparison of the new air-cooling + austenite reversion tempering process (AC + ART) compared to the standard quenching + tempering (Q + T) and quenching + partitioning (Q + P) heat treatments. The time axis of this diagram is schematic and does not reflect the real tempering times.

2. Materials and Methods

Three laboratory melts were ingot-casted in an ALD vacuum induction furnace (80 kg, 140 mm × 140 mm × 500 mm). The melts differ in the type and concentration of their grain boundary strengthening elements B and Mo. Alloy L1 and L2 are alloyed with boron in different amounts and titanium for nitrogen control. Alloy L3 is alloyed with molybdenum. Additionally, all melts are alloyed with niobium in order to control the austenite grain size during the forging process. The resulting chemical compositions can be seen in Table 1. After casting the blocks were homogenized at 1200 °C for 5 h and consecutively forged into rods (60 × 60 mm² base area). The samples for the mechanical tests were manufactured from the transition region between corner and center of the rod.

For tensile tests, round tensile samples with a diameter of 6 mm and a test length of 30 mm and for impact tests standard Charpy-V-notch samples were produced. The quasi-static tensile test have been performed with an elongation rate of 0.0005 s^{-1}, for the impact test a 300 J hammer was used.

Table 1. Chemical composition of the laboratory melts. All concentrations are given in wt.%.

Alloy	C *	Si	Mn	P	S *	Al	Mo	Ti	Nb	B	N
L1	0.19	0 50	4.02	0.008	0.011	0.031	0.02	0.020	0.035	0.0016	0.011
L2	0.17	0.50	3.99	0.010	0.009	0.025	0.02	0.020	0.033	0.0057	0.010
L3	0.15	0.49	4.02	0.011	0.009	0.027	0.20	<0.003	0.035	<0.0005	0.010

* C, S determined with Leco-combustion analyses.

As it can be seen in Figure 2, the intercritical range for the present alloys is between 585 °C and 750 °C, with a maximum carbon enrichment in the austenite at 650 °C. The annealing was carried out between $T_{ART, min}$ temperature of 600 °C and $T_{ART, max}$ temperature of 675 °C in order to investigate the influence of different carbon concentrations and therefore different austenite stabilities on the mechanical properties. The thermodynamic simulations have been carried out using MATCALC6 commercial software and the MC-Fe database. The ART-annealing was done in a salt bath (THERMCONCEPT GmbH, Bremen, Germany) for 1 h at 600 °C, 625 °C, 650 °C and 675 °C.

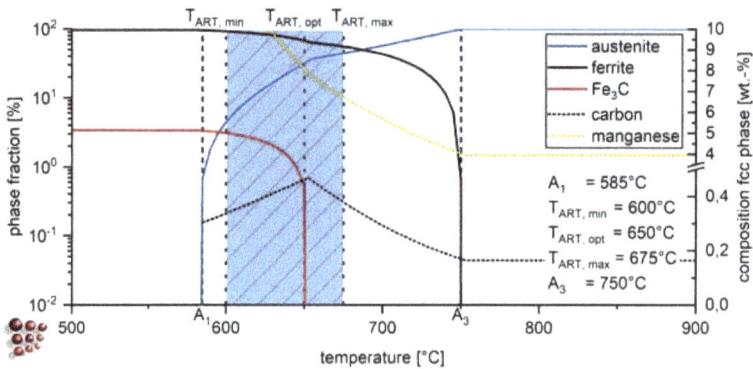

Figure 2. Thermodynamic equilibrium calculations of the temperature dependency of the different phases and the equilibrium C and Mn contents of the fcc phase. The investigated temperature range for ART annealing is indicated.

Preparation for electron backscatter diffraction (EBSD) measurements was done by mechanical grinding with SiC paper up to 4000 grit followed by mechanical polishing using 3 µm and 1 µm diamond slurry. Subsequently, the samples were electropolished at 37 V for 15 s using an electrolyte consisting of 700 mL ethanol (C_2H_5OH), 100 mL butyl glycol ($C_6H_{14}O_2$) and 78 mL perchloric acid (60%)($HClO_4$). For the scanning electron microscopy (SEM) experiments a Zeiss Sigma SEM (Carl Zeiss Microscopy GmbH, Jena, Germany) was used. The fracture surfaces were analyzed by taking secondary electron (SE) images at an acceleration of 20 kV. Chemical analyses of the fracture surfaces were done using Energy Dispersive X-ray Spectroscopy (EDS). EBSD measurements were done using an acceleration voltage of 20 kV, a step size of 50 nm and a working distance of 18 mm. Post-processing of the dataset was done using the MATLAB-based MTEX toolbox.

3. Results

3.1. Mechanical Properties

The materials in the as-forged state are characterized by high yield strengths between 930 MPa and 1000 MPa, tensile strengths from 1370 MPa to 1510 MPa, small uniform elongations from 3.7% to 4.3% and total elongations from 12.2% to 12.4%, respectively. The Charpy impact energy at this state is in the range between 15 and 30 J. After the annealing treatment, these properties change as it can be seen in Figure 3. The yield stress decreases continuously while the ultimate tensile stress has a minimum at 650 °C and increases at 675 °C. The yield ratio YS/UTS consequently increases from 0.66 in the as-forged state to 0.82 at 650 °C and decreases at 675 °C to 0.54. The uniform elongation increases continuously up to 18%, while the total elongation shows a maximum of 27% and 24% at 650 °C for the alloys L2 and L3 respectively but continuously increases up to 27% for alloy L1. The Charpy impact energy shows a maximum at 650 °C and reaches in the case of alloy L3 values around 100 J. An overview of the impact energy of all three alloys after different heat treatments can be seen in Table 2. The stress strain curves and the strain-hardening curves of alloy L2 are shown in Figure 4. The as-forged state shows in comparison to the annealed states a much higher strain hardening. The at 600 °C, 625 °C and 650 °C annealed samples have a very similar strain behavior, which is characterized by a clearly visible maximum at 0.2 true strain. The at 675 °C annealed sample shows a different strain hardening behavior with clearly visible serrations in the engineering stress-strain curve. These serrations represent inhomogeneities in the strain hardening behavior, due to pronounced differences in local plastic deformation. As can be concluded from Figure 3 all investigated steels show, in principle, the same temperature dependency in their mechanical properties irrespective of their specific chemical composition.

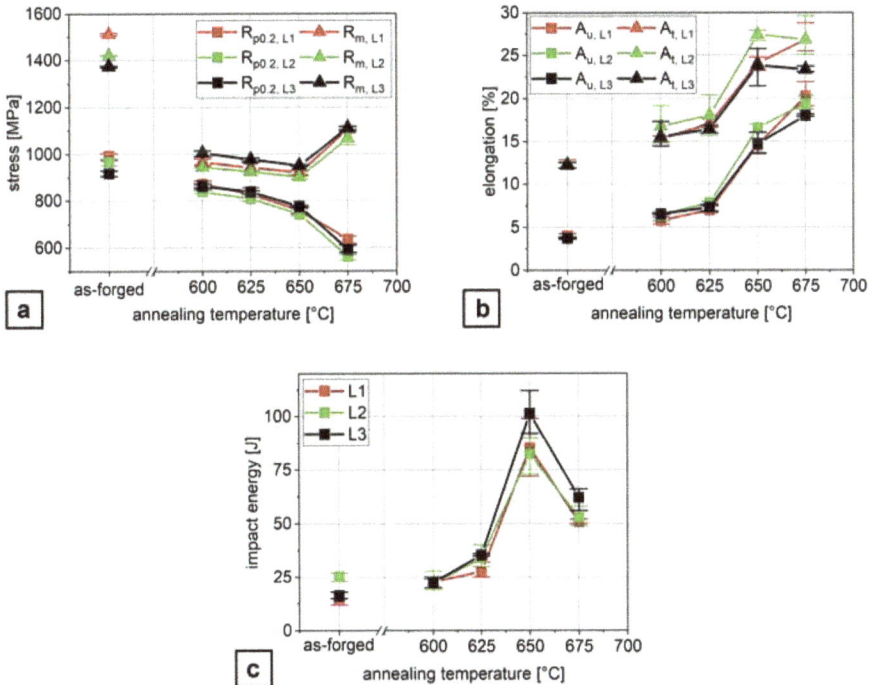

Figure 3. Tempering charts for the alloys L1–L3. The yield strength (YS) and ultimate tensile strength (UTS) (**a**), uniform A_u and total elongation A_t (**b**) and the impact energy (**c**) are shown.

Table 2. Impact energy at room temperature after different annealing treatments.

Alloy	As-Forged	600 °C	625 °C	650 °C	675 °C
L1	15 J	23 J	28 J	85 J	51 J
L2	25 J	22 J	34 J	82 J	53 J
L3	16 J	22 J	35 J	101 J	62 J

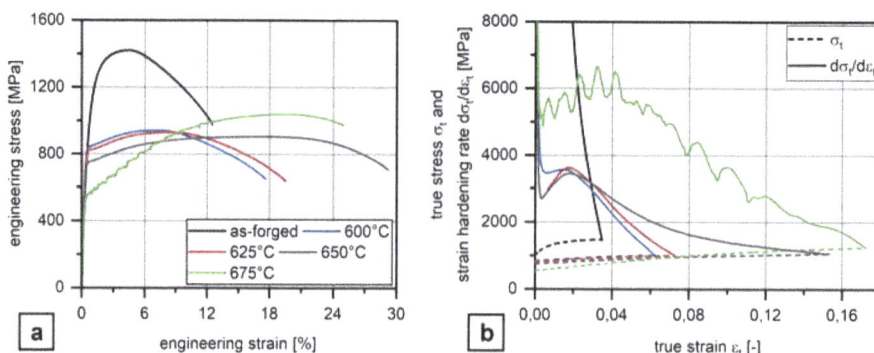

Figure 4. Annealing influence on the stress-strain-curve (**a**) and strain hardening curves (**b**).

3.2. Microstructure

The microstructure observed with light optical microscopy (LOM) shows marginal differences after the heat-treatment (Figure 5). The as-forged sample develops a fully martensitic microstructure after air cooling. After annealing, all samples show weakly rounded and widened martensite laths in comparison to the as-forged state. In the as-forged state small carbides below 1 μm can be found, which are hardly distinguishable from the matrix. The samples annealed at 600 °C and 625 °C show carbides, which grow with increasing annealing temperature. As a consequence of the smaller amount of dissolved carbon the etching behavior changes, which can be concluded from the differently colored martensitic matrix. After annealing at 650 °C, no carbides were found.

Figure 5. Microstructure observable with light optical microstructure (LOM) for the as-forged state and the at 600 °C, 650 °C and 675 °C annealed samples. Exemplary alloy L2 is shown in this figure.

Investigations of the fracture surface of the Charpy-V-notch samples have been performed by scanning electron microscopy (SEM). Figure 6 shows exemplary the evolution of the fracture morphology from 600 °C up to 675 °C. At 600 °C and 625 °C most of the fracture surface shows mainly intergranular cleavage fracture and only small regions with ductile pores. At 650 °C cleavage fracture still dominates but a second phase is observable on the fracture surface. These particles show a round morphology with a radius of around 2 μm. The chemical composition of these particles was measured using energy dispersive X-ray spectroscopy. In comparison to the matrix an enrichment of manganese up to 8 wt.% was measured, as well as an enrichment of carbon of from 3 wt.% to 4 wt.%. The Mn content corresponds well with the simulated data in Figure 2. The C-level increase corresponds as well while the absolute values measured are unrealistically high due to the detection restriction of light elements with EDX. If the annealing temperature is increased up to 675 °C the particles are coarser up to 5 μm and in some cases form a film at the grain boundaries. In all samples a mixture between intergranular and intragranular fracture was observed. The intergranular fracture takes place along the prior austenite grain boundaries of the martensite matrix. The exposed planes in the fracture surfaces have a diameter of 10 μm to 40 μm.

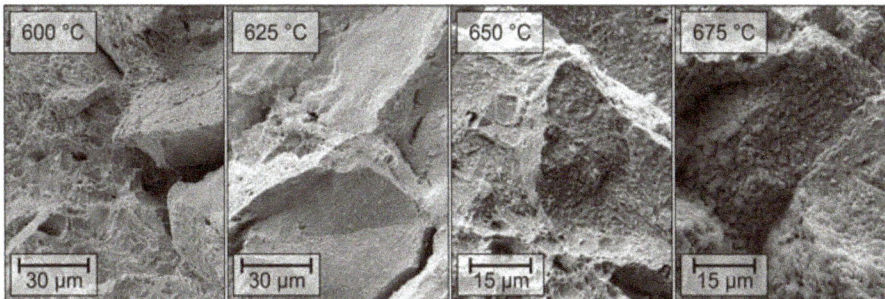

Figure 6. Scanning electron microscopy (SEM) micrographs of the fracture surfaces of Charpy-V-notch impact samples, annealed at different temperatures. Exemplary alloy L2 is shown in this figure.

As a result from the previous observations, EBSD-measurements were carried out to analyze the distribution of austenite in the ferritic matrix. Figure 7 shows the resulting phase maps exemplary for the laboratory melt L3, as the results for melt L1 and L3 are similar. The size of the martensitic laths seems to be in the same range for all tested samples, reaching from below 1 μm up to a maximum thickness of approximately 5 μm. For lower annealing temperatures (600 °C and 625 °C) a pure ferritic matrix has been found. Just some regions on boundaries haven been indexed as cementite or have not been indexed at all. If the temperature is increased to 650 °C small austenite particles can be observed on prior austenite grain boundaries or in between martensite laths. At 675 °C the austenite is present as a film at these grain boundaries, in some cases decorated with small cementite precipitates. Apart from the differences in morphology, the volume fraction of the austenite regions also increased in comparison to 650 °C. For both states it was found, that the austenite is not distributed homogeneously in the matrix but forms all colonies of austenite islands within a prior austenite grain or islands and films along prior austenite grain boundaries or at martensite boundaries. Besides ferrite, austenite and cementite, ε-martensite was also included in the EBSD-measurement but not detected.

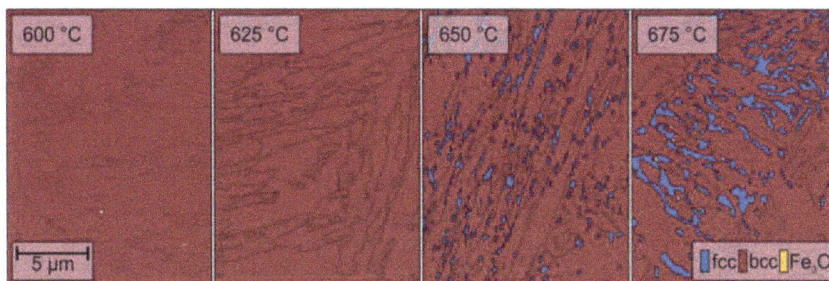

Figure 7. Electron backscatter diffraction (EBSD) phasemaps of austenite distribution after annealing for 1 h with different annealing temperatures. As the austenite is not homogenously distributed in the matrix the volume fraction can just be estimated. For 600 °C and 625 °C less than 1 vol.% was detected. For 650 °C the approximated volume fraction is around 10 vol.% and for 675 °C 20 vol.% have been detected.

4. Discussion

Mechanical properties, microstructure and Charpy impact energy indicate the usual response of hardened steels to tempering for annealing temperatures up to 625 °C but a significant change after annealing at 650 °C and 675 °C. Comparing the annealing temperatures of 650 °C and 675 °C the ultimate tensile strength shows an increase of 160 MPa while the yield stress decreases by 180 MPa with rising temperature. Comparing the annealing temperature of 625 °C, 650 °C and 675 °C, the Charpy impact energy is drastically improved at 650 °C but drops again at 675 °C. The uniform elongation increases with annealing temperature but the total elongation shows a maximum of 27% at 650 °C in comparison to 12% in the as-forged state for the alloy L2. The pronounced increase of Charpy impact energy indicates the presence of a second phase, which is supported by the observed particles on the fracture surfaces of the Charpy samples, as well as by evidence of the austenite phase and the carbides by the EBSD measurement. The observed change of Charpy impact energy can be explained by the TRIP effect on one hand and the impact of grain boundary segregation on the other hand.

The austenite phase is enriched with manganese and carbon, which can result in the TRIP effect during deformation contributing to improved ductility and toughness. But, if the austenite phase exceeds a range of about 10 vol.% the local enrichment gets smaller leading to a different stability of the austenite [21–23]. Both the larger volume fraction and the smaller stability result in discontinuous TRIP behavior that is reflected by the serrated stress-strain curve.

The austenite was always found at interfaces, primarily at prior austenite grain boundaries or at martensite lath boundaries. Due to the enrichment of manganese at these boundaries [21,24], austenite will form earlier in these regions during the heat treatment. When the nucleation of austenite starts, the newly formed austenite acts as a diffusion sink for manganese and carbon at the surrounding grain boundary, as the main driving force for manganese is the formation of austenite close to the thermodynamic equilibrium [25]. As the diffusivity at the grain boundary is high, the manganese segregation at grain boundaries will be reduced by austenite islands. This consecutively leads to a strengthening of the grain boundaries. Together with the effect of the grain boundary strengthening elements boron and molybdenum [11,12] the annealing treatment increases the Charpy impact energy drastically at the expense of yield strength and tensile strength. This loss in strength is caused by the absorption of the matrix carbon through the austenite. As the carbon is dissolved in the newly formed fcc-particles, it can no longer be used for strengthening the matrix through small carbides or through solid solution strengthening. This effect is already observable after annealing at 600 °C.

These results can be used to explain the Charpy impact energy alteration after annealing at different temperatures, which have been reported earlier [8]. Figure 8 shows the schematic development of the microstructure in dependence of the annealing temperature. In the as-forged state the material shows a martensitic microstructure with small carbides. Low amounts of retained austenite can also be found. After annealing above the dissolution temperatur of the carbides the matrix gets enriched with carbon while manganese diffuses along prior austenite grain boundaries. Through the enrichment of manganese and carbon in this region nanostructured austenite forms and get stabilized. If the optimum temperature (650 °C) is exceeded, austenite forms thin films on grain-boundaries, which lowers the Charpy impact energy drastically.

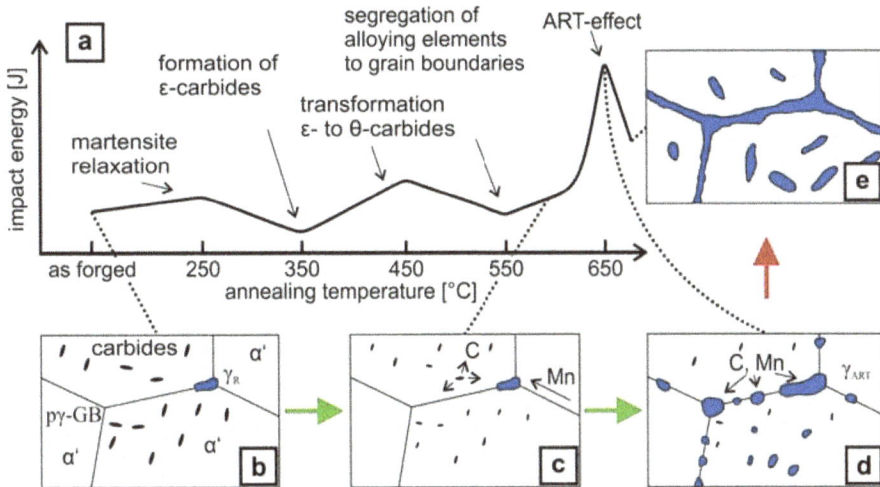

Figure 8. Correlation of the Charpy impact energy alteration [8] with austenite morphology obtained through ART annealing (**a**). The (**b**–**e**) describe the formation of the microstructure and are explained in detail in the text. α' is representing the martensitic matrix, γ is representing austenite in the retained (γ_R) and ART-annelead (γ_{ART}) condition, while pγ-GB stands for prior austenite grain boundary.

5. Conclusions

The proposed air cooling + austenite reversion tempering (AC + ART) treatment is a robust process that can be applied for forgings with various dimensions. From this study, the following conclusions can be drawn:

- Air-hardening martensitic forging steels develop high ultimate tensile strengths from 1370 MPa to 1510 MPa but low room temperature Charpy impact energy of below 20 J.
- Microalloying with boron and molybdenum only slightly increases the Charpy impact energy after air cooling.
- Through ART-annealing, finely distributed retained austenite islands are obtained that can provide Charpy impact energy values above 80 J.
- Obviously, there exist an optimum austenite content as by austenite fractions above about 10 vol.% Charpy impact energy decreases again together with the development of serrated plastic flow behavior in the tensile test.
- The Charpy impact energy improvement is due to the occurrence of a continuous TRIP effect and less Mn segregation at grain boundaries.

Author Contributions: Conceptualization, A.G.; methodology, A.G.; validation, A.G. and R.E.; formal analysis, A.G. and R.E.; investigation, A.G. and R.E.; data curation, A.G.; writing—original draft preparation, A.G.; writing—review and editing, A.G. and W.B.; visualization, A.G.; supervision and discussion, W.B.; project administration, W.B.; funding acquisition, W.B.

Funding: Results presented here are from the research project IGF 27 EWN. Funding was provided by the German Federal Ministry of Economics and Energy via the German Federation of Industrial Cooperative Research Associations "Otto von Guericke" (AiF) in the program to encourage the industrial Community research by an resolution of the German Bundestag and the Steel Forming Research Society (FSV).

Conflicts of Interest: The authors declare no conflict of interest. The funders had no role in the design of the study; in the collection, analyses, or interpretation of data; in the writing of the manuscript, or in the decision to publish the results.

Abbreviations

The following abbreviations are used in this manuscript:

AC + ART	Air cooling and austenite reversion tempering
Q + T	Quenching and tempering
Q + P	Quenching and partitioning
ART	Austenite reverted transformation
EDS	Energy dispersive X-ray spectroscopy
PHFP	Precipitation hardening ferritic pearlitic
EBSD	Electron backscattering diffraction
SE	Secondary electron
SEM	Scanning electron microscopy
LOM	Light optical microscopy
HDB	High ductile bainite
LHD	Air hardening ductile (German abbreviation)

References

1. Bleck, W.; Bambach, M.; Wirths, V.; Stieben, A. Microalloyed Engineering Steels with Improved Performance—An Overview. *HTM J. Heat Treatm. Mat.* **2017**, *72*, 346–354. [CrossRef]
2. Bleck, W.; Keul, C.; Zeislmair, B. Entwicklung eines höherfesten mikro legierten ausscheidungshärtenden ferritisch/perlitischen Schmiedestahls AFP-M. [The development of a high-strength, microalloyed, precipitation hardening, ferritic-perlitic formed steel AFP-M]. *Schmiede J.* **2010**, *3*, 42–44.
3. Keul, C.; Wirths, V.; Bleck, W. New bainitic steels for forging. *Arch. Civ. Mech. Eng.* **2012**, *12*, 119–125. [CrossRef]
4. Keul, C.; Urban, M.; Back, A.; Hirt, G.; Bleck, W. Entwicklung eines hochfesten duktilen bainitischen (HDB) Stahls für hochbeanspruchte Schmiedebauteile. [The Development of a High-Strength, Ductile, Bainitic (HDB) Steel for Heavy Duty Forged Components]. *Schmiede J.* **2010**, *9*, 28–31.
5. Wirths, W.; Wagener, R.; Bleck, W.; Melz, T. Bainitic Forging Steels for Cyclic Loading. *Adv. Mat. Res.* **2014**, *922*, 813–818. [CrossRef]
6. Edmonds, D.V.; He, K.; Rizzo, F.C.; De Cooman, B.C.; Matlock, D.K.; Speer, J.G. Quenching and partitioning martensite—A novel steel heat treatment. *Mater. Sci. Eng. A* **2006**, *438–440*, 25–34. [CrossRef]
7. Stieben, A.; Bleck W.; Schönborn, S. Lufthärtender duktiler Stahl mit mittlerem Mangangehalt für die Massivumformung. [Air hardening ductile steel with medium manganese content for forging]. *massivUmformung* **2016**, *9*, 50–55.
8. Stieben, A. Eigenschaften lufthärtender martensitischer Schmiedestähle mit Mangangehalten von 3–10 Gew.-%. [Properties of Air-hardening Martensitic Forging Steels with Manganese Concentrations of 3–10 wt.-%]. Ph.D. Thesis, RWTH Aachen University, Aachen, Germany, 16 March 2018.
9. Hwang, S.K.; Morris, J.W. The use of a boron addition to prevent intergranular embrittlement in Fe-12Mn *Metall. Trans. A* **1980**, *11*, 1197–1206. [CrossRef]
10. Kuzmina, M.; Ponge, D.; Raabe, D. Grain boundary segregation engineering and austenite reversion turn embrittlement into toughness: Example of a 9 wt.% medium Mn steel. *Acta Mater.* **2015**, *86*, 182–192. [CrossRef]
11. Song, S.; Faulkner, R.G.; Flewitt, P.E.J. Quenching and tempering-induced molybdenum segregation to grain boundaries in a 2.25Cr–1Mo steel. *Mater. Sci. Eng. A* **2000**, *281*, 23–27. [CrossRef]
12. Bolton, J.D.; Petty, E.R.; Allen, G.B. The mechanical properties of α-phase low carbon Fe-Mn alloys. *Metall. Trans.* **1971**, *2*, 2915–2923. [CrossRef]
13. Suh, D.W.; Ryu, J.H.; Joo, M.S.; Yang, H.S.; Lee, K.; Bhadeshia, H.K.D.K. Medium-Alloy Manganese-Rich Transformation-Induced Plasticity Steels. *Metall. Mater. Trans. A* **2013**, *44A*, 286–293. [CrossRef]
14. Gramlich, A.; Emmrich, R.; Bleck, W. Annealing of 4 wt.-% Mn forging steels for automotive application. In Proceedings of the 4th International Conference on Medium and High Manganese Steels, Aachen, Germany, 1–3 April 2019; pp. 283–286.
15. Miller, R.L. Ultrafine-grained microstructures and mechanical properties of alloy steels. *Metall. Mater. Trans. B* **1972**, *3*, 905–912. [CrossRef]

16. Ma, Y. Medium-manganese steels processed by austenite-reverted-transformation annealing for automotive applications. *Mater. Sci. Tech.* **2017**, *33*, 1713–1727. [CrossRef]

17. Kaar, S.; Schneider, R.; Krizan, D.; Béal, D.; Sommitsch, C. Influence of the Quenching and Partitioning Process on the Transformation Kinetics and Hardness in a Lean Medium Manganese TRIP Steel. *Metals* **2019**, *9*, 353. [CrossRef]

18. Grajcar, A.; Kilarski, A.; Kozlowska, A. Microstructure–Property Relationships in Thermomechanically Processed Medium-Mn Steels with High Al Content. *Metals* **2018**, *8*, 929. [CrossRef]

19. Liu, C.; Peng, Q.; Xue, Z.; Deng, M.; Wang, S.; Yang, Y. Microstructure-Tensile Properties Relationship and Austenite Stability of a Nb-Mo Micro-Alloyed Medium-Mn TRIP Steel. *Metals* **2018**, *8*, 615. [CrossRef]

20. Sevsek, S.; Haase, C.; Bleck, W. Strain-Rate-Dependent Deformation Behavior and Mechanical Properties of a Multi-Phase Medium-Manganese Steel. *Metals* **2019**, *9*, 344. [CrossRef]

21. Luo, H.; Shi, J.; Wang, C.; Cao, W.; Sun, X.; Dong, H. Experimental and numerical analysis on formation of stable austenite during the intercritical annealing of 5Mn steel. *Acta Mater.* **2011**, *59*, 4002–4014. [CrossRef]

22. Chen, J.; Lv, M.; Liu, Z.; Wang, G. Combination of ductility and toughness by the design of fine ferrite/tempered martensite-austenite microstructure in a low carbon medium manganese alloyed steel plate. *Mater. Sci. Eng. A* **2015**, *648*, 51–56. [CrossRef]

23. Twardowski, R.; Prahl, U. Microstructure based flow curve modeling of high-Mn steels with TWIP and TRIP effect. *Comput. Methods Mater. Sci.* **2012**, *12*, 130–136.

24. Chen, J.; Lv, M.; Liu, Z.; Wang, G. Influence of Heat Treatments on the Microstructural Evolution and Resultant Mechanical Properties in a Low Carbon Medium Mn Heavy Steel Plate. *Metall. Mater. Trans. A* **2016**, *47A*, 2300–2312. [CrossRef]

25. Nasim, M.; Edwards, B.C.; Wilson, E.A. A study of grain boundary embrittlement in an Fe-8%Mn alloy. *Mater. Sci. Eng. A* **2000**, *281*, 56–67. [CrossRef]

Article

Strain-Rate-Dependent Deformation Behavior and Mechanical Properties of a Multi-Phase Medium-Manganese Steel

Simon Sevsek * [ID], **Christian Haase** [ID] and **Wolfgang Bleck**

Steel Institute, RWTH Aachen University, 52072 Aachen, Germany; christian.haase@iehk.rwth-aachen.de (C.H.); bleck@iehk.rwth-aachen.de (W.B.)
* Correspondence: simon.sevsek@iehk.rwth-aachen.de; Tel.: +49-241-809-0138

Received: 3 March 2019; Accepted: 15 March 2019; Published: 18 March 2019

Abstract: The strain-rate-dependent deformation behavior of an intercritically annealed X6MnAl12-3 medium-manganese steel was analyzed with respect to the mechanical properties, activation of deformation-induced martensitic phase transformation, and strain localization behavior. Intercritical annealing at 675 °C for 2 h led to an ultrafine-grained multi-phase microstructure with 45% of mostly equiaxed, recrystallized austenite and 55% ferrite or recovered, lamellar martensite. In-situ digital image correlation methods during tensile tests revealed strain localization behavior during the discontinuous elastic-plastic transition, which was due to the localization of strain in the softer austenite in the early stages of plastic deformation. The dependence of the macroscopic mechanical properties on the strain rate is due to the strain-rate sensitivity of the microscopic deformation behavior. On the one hand, the deformation-induced phase transformation of austenite to martensite showed a clear strain-rate dependency and was partially suppressed at very low and very high strain rates. On the other hand, the strain-rate-dependent relative strength of ferrite and martensite compared to austenite influenced the strain partitioning during plastic deformation, and subsequently, the work-hardening rate. As a result, the tested X6MnAl12-3 medium-manganese steel showed a negative strain-rate sensitivity at very low to medium strain rates and a positive strain-rate sensitivity at medium to high strain rates.

Keywords: medium-manganese steel; TRIP; strain-rate sensitivity; Lüders band; serrated flow; in-situ DIC tensile tests

1. Introduction

Increasing demands for fuel-efficient vehicles led to the development of Advanced High-Strength Steels (AHSS). First generation AHSS usually possess a ferritic matrix and employs phase fractions of harder phases like martensite, bainite, and metastable retained austenite to increase the product of ultimate tensile strength (UTS) and elongation to fracture ε_f < 20 GPa% [1,2]. In search of further improvement of this so-called ECO-Index, the second generation of AHSS with fully austenitic microstructures was developed. Reaching an ECO-Index of >50 GPa%, the remarkable mechanical properties result from very high dislocation densities in combination with the activation of additional deformation mechanisms in addition to dislocation slip which increase the work-hardening rate (WHR) [3,4]. The critical parameter governing the activation of those secondary deformation mechanisms, such as deformation twinning (TWinning-Induced Plasticity–TWIP) and deformation-induced phase transformation (TRansformation-Induced Plasticity–TRIP), is the material's stacking fault energy (SFE), which is primarily a function of the chemical composition [5–8]. As the main alloying element of the second generation AHSS is Mn with alloying contents <35 wt.%, those steels are also termed high-manganese steels (HMnS).

While the total energy-absorption potential of HMnS makes them an attractive option for crash-relevant parts in the automotive industry, several aspects have hindered widespread industrial application. Due to the design criteria of automotive parts, the available deformation volume is rather low, emphasizing the importance of a high yield strength (YS), which HMnS do not possess compared to their UTS [9,10]. Additionally, during quasistatic tensile tests, strain localization phenomena have frequently been observed that result in the propagation of deformation bands in the sample and subsequent serrated flow curves [11–14]. The third generation of AHSS has drawn significant interest, as they have reduced Mn contents of 3–12 wt.% and are described as medium-manganese steels (MMnS). Intercritical annealing (IA) in the two-phase region of austenite (γ) and ferrite (α) is conventionally employed to develop an ultrafine-grained (UFG) multi-phase microstructure consisting of γ, α, and martensite (α'), which increases the YS considerably, while retaining good ductility [1,2]. However, a discontinuous elastic-plastic transition during quasistatic tensile testing has frequently been reported [15].

In the past, there have been numerous efforts to analyze the impact of annealing parameters during the IA on the initial microstructure and the partitioning of alloying elements on the individual phases [1,16–22]. Subsequent studies have dealt with the activation of secondary deformation mechanisms in γ and the strain partitioning in the microstructural constituents of different hardness [23–26]. However, most of the work on the analysis of mechanical properties has been performed within a small range of strain rates in the quasistatic tensile testing regime. On the other hand, for application in crash-relevant automotive parts, understanding the deformation behavior at very high strain rates is critical. It has been observed in the past that the UTS of HMnS and MMnS increases with increasing strain rate at high-speed tensile tests while also retaining good ductility. While microstructural investigations have revealed that adiabatic heating effects during high-speed tensile tests partially suppress the TWIP and TRIP effect, the strain partitioning behavior in multi-phase MMnS has primarily been investigated during quasistatic tensile testing. [2,4,22,23,27–29]

However, the microstructural response in combination with the strain partitioning behavior is especially interesting in the case of MMnS. This is because of possible differences in the response to the plastic deformation at high strain rates of the individual microstructural constituents of the multi-phase microstructure. In this work, the strain rate-dependent mechanical properties and deformation behavior of a X6MnAl12-3 MMnS are investigated. To this end, the strain partitioning behavior on a microstructural scale is investigated by means of interrupted tensile tests at different strain rates coupled with electron backscatter diffraction (EBSD) experiments. Additionally, in-situ digital image correlation (DIC) methods during tensile testing are employed to analyze the strain localization behavior in the specimen with respect to the occurrence of deformation bands.

2. Materials and Methods

The X6MnAl12-3 alloy was ingot-cast in an ALD vacuum induction furnace with ingot dimensions of 140 mm × 140 mm × 500 mm. Afterward, the MMnS was hot-forged at 1150 °C in 3 forging passes to a cross section of 160 mm × 40 mm. Subsequent hot rolling to a sheet thickness of 2.5 mm was performed at 1150 °C with reheating between passes. The MMnS was then homogenization annealed at 1100 °C for 2 h, followed by water quenching and an additional austenitization annealing at 850 °C for 10 min with subsequent water quenching to reduce eigenstrains. Finally, the MMnS was cold-rolled (CR) to a final thickness of 1.25 mm (50% thickness reduction).

Before intercritical annealing (IA) at 675 °C for 2 h in a salt bath furnace (THERMCONCEPT GmbH, Bremen, Germany), tensile test specimens with 30 mm initial gauge length and 6 mm gauge width were machined after cold rolling using water jet cutting. The tensile axis (TA) was perpendicular to the rolling direction (RD). The temperatures for the austenitization annealing and the IA were selected based on previous studies for a MMnS with a very similar chemical composition [22,23]. Additionally, dilatometer measurements were performed after cold rolling to determine A_{c1} and A_{c3}, ensuring that the IA was performed in the α-γ two-phase region and that the temperature for the austenitization was higher than A_{c3}. A_{c1}, A_{c3}, and the nominal chemical composition of the MMnS tested in this work are summarized in Table 1. The processing route, including the final heat treatment, is summarized in Figure 1.

Table 1. A_{c1}, A_{c3}, and nominal chemical composition of the X6MnAl12-3 MMnS.

| Designation | Nominal Chemical Composition [wt.%] | | | A_{c1} [°C] | A_{c3} [°C] |
	C	Mn	Al		
X6MnAl12-3	0.064	11.7	2.9	503	825

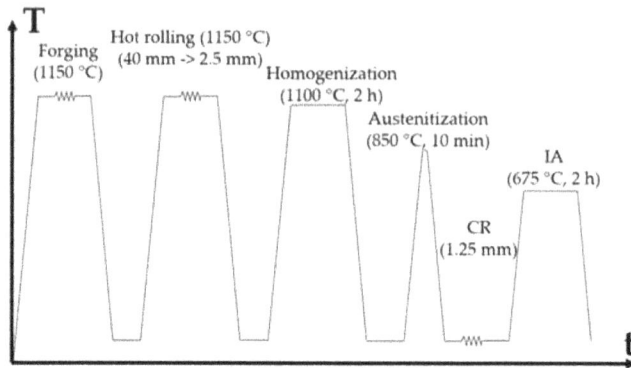

Figure 1. The processing route for the X6MnAl12-3 MMnS.

Tensile testing was performed on two different devices. A minimum number of three tensile tests per strain rate were performed and a representative test result was selected for the diagrams in this work. The tensile tests at constant strain rates $\dot{\varepsilon}$ of 0.00001 1/s, 0.0001 1/s, 0.001 1/s, 0.01 1/s, and 0.1 1/s were conducted using a Z250 tensile testing machine by ZwickRoell (Ulm, Germany). Strains were measured with the contact-type extensometer MultiXtens by ZwickRoell. The Aramis 12 Megapixel DIC system (GOM GmbH, Braunschweig, Germany) was employed during the quasistatic tensile tests to analyze the local $\dot{\varepsilon}$ and the strain distribution in the specimens. For the higher $\dot{\varepsilon}$ of 1.0 1/s, 11 1/s, 140 1/s, and 230 1/s, dynamic tensile test samples with 20 mm gauge length and 6 mm width were used. The dynamic tensile tests were conducted at room temperature on a servo-hydraulic HTM2012 testing machine by ZwickRoell (Ulm, Germany). Strain was measured using a high-speed opto-electronic device. A piezo-electric type load cell was used to measure the applied load. At 140 1/s and 230 1/s, additional strain gauges were attached to the sample shoulders linked in a Wheatstone bridge circuit to reduce the oscillation of the input load signal, which occurs at high testing speeds. In order to determine the flow curves, the force signal was smoothed with a polynomial spline function to eliminate the remaining oscillations.

Optical microscopy was conducted using a 5% Nital etchant. Metallographic preparation for electron backscatter diffraction (EBSD) measurements was performed by means of mechanical grinding with SiC paper up to 4000 grit with subsequent mechanical polishing using 3 µm and 1 µm diamond suspension. Electropolishing was then performed at 28 V for 20 s using an electrolyte containing 700 mL ethanol (C_2H_5OH), 100 mL butyl glycol ($C_6H_{14}O_2$), and 78 mL perchloric acid (60%) ($HClO_4$). EBSD analyses were performed on the sheet surface for two different kinds of specimens. The measurements of the initial microstructure and after fracture were performed on A_{30} tensile test specimens. For the measurements after fracture, the measured area was selected as close to the fracture surface as possible. Interrupted tensile tests of specimens with 12 mm gauge length and 2 mm gauge width were used to analyze the microstructural evolution after 2%, 10%, and 20% elongation, respectively. A Zeiss Sigma scanning electron microscope (SEM) (Carl Zeiss Microscopy

GmbH, Jena, Germany) was used for all SEM experiments. Secondary electron (SE) images were taken at an acceleration voltage of 20 kV and coupled with Energy Dispersive X-ray Spectroscopy (EDS) for line-scans with 20 nm step size. The chemical composition of γ was determined via EDS and was used for the calculation of the SFE, based on a subregular solution thermodynamic model that takes the impact of temperature into account too [7]. For larger-area EBSD measurements before deformation, after 20% uniform elongation at 0.001 1/s and after fracture of the 0.00001 1/s and 0.001 1/s strain rates, an acceleration voltage of 15 kV and a step size of 100 nm were selected. For the other high-resolution (HR) EBSD measurements, an acceleration voltage of 10 kV and a step size of 70 nm was used. The phase fractions given in this manuscript are based on those HR EBSD measurements, however, the phase fractions determined by the large-area EBSD measurements did not deviate significantly. The working distance was between 18 mm and 24 mm and a post-processing routine employing the HKL Channel 5 software (version 5.12j, Oxford Instruments, Abingdon-on-Thames, UK) was utilized, as well as the MATLAB-based MTEX toolbox [30,31]. Noise reduction was employed by removing wild spikes and considering at least 5 neighboring data points.

3. Results

The microstructures of the MMnS after cold-rolling prior to IA and after intercritical annealing at 675 °C for 2 h are shown in Figure 2. After cold-rolling (cf. Figure 2a), a very fine-grained lamellar microstructure was present. Prior γ grain boundaries (GB) were still visible in optical micrographs and separated regions of lamellar martensite. After intercritical annealing (cf. Figure 2c), a high fraction of the lamellar morphology was retained. However, in some areas (highlighted in Figure 2c), a more equiaxed recrystallized grain morphology was present. The SE image in Figure 2b with the corresponding EDS analysis in Figure 2d indicates that elemental partitioning of Al and Mn was present after intercritical annealing. Al-enriched regions showed approximately 3.4 wt.% Al and corresponded with Mn-depleted regions with about 10.5 wt.% Mn. Mn-rich regions showed about 14.0 wt.% Mn and corresponded with Al-depleted regions, where only approximately 2.5 wt.% Al was present. Comparing the microstructure in Figure 2b with the phase fractions from EBSD measurements shown in Figure 4a revealed that the Mn-rich regions correspond to austenite while the Al-rich regions correspond to ferrite and α'-martensite.

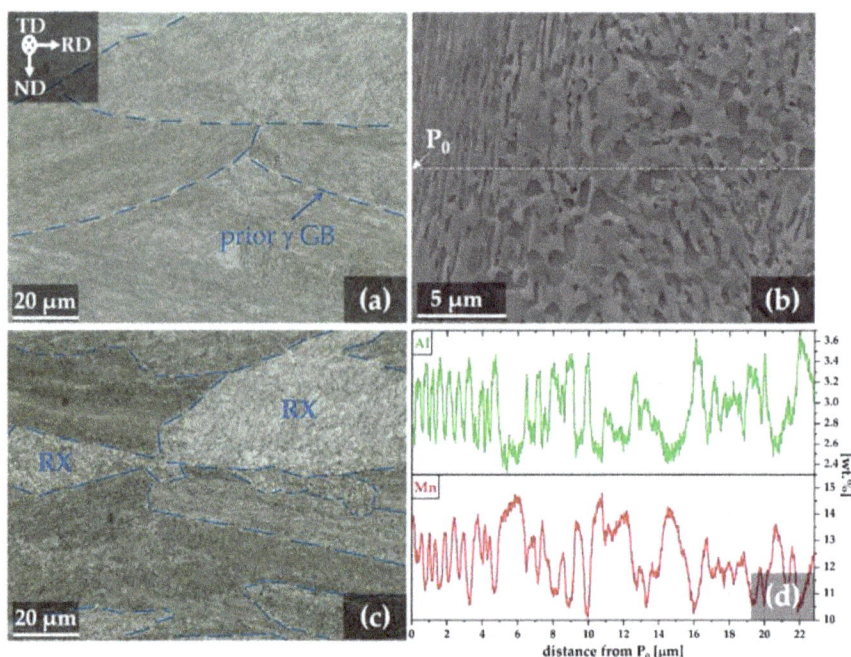

Figure 2. (**a**,**c**) Optical micrographs of the X6MnAl12-3 MMnS (**a**) after 50% cold rolling and (**c**) after intercritical annealing. Easily identifiable prior γ grain boundaries in (**a**) and grain boundaries of recrystallized grains in (**c**) are highlighted in blue. (**b**) Secondary electron (SE) image of the microstructure after intercritical annealing, and (**d**) the corresponding Energy Dispersive X-ray Spectroscopy (EDS) analysis of the local Al and Mn contents are shown.

The strain-rate-dependent mechanical properties as determined by tensile tests are shown in Figure 3, in addition to the strain-rate-dependent YS and stress values corresponding to engineering strains ε_{eng} of 5% and 10%. The mechanical response of the tested X6MnAl12-3 MMnS to an increase of $\dot{\varepsilon}$ can be divided into strain rate changes in the regime of low, medium, and high strain rates, as shown in Table 2. In general, the ductility was affected the most, as the uniform elongation (UE) and ε_f decreased from approximately UE = ε_f = 40% at 0.00001 1/s to about UE = 18% and ε_f = 30% at 1.0 1/s respectively. During dynamic tensile testing ($\dot{\varepsilon} \geq 1.0$ 1/s), the UE was not affected significantly, while ε_f increased slightly to 35% and 32% at 140 1/s and 230 1/s. The YS only decreased slightly from 720 MPa at 0.00001 1/s to 690 MPa at 230 1/s, which equals a variation in YS of less than 5%. The UTS decreased from 875 MPa at 0.00001 1/s to 790 MPa at 1.0 1/s. During the dynamic tensile tests, the UTS increased up to 835 MPa at 230 1/s. At strain rates of 0.0001 1/s and 0.00001 1/s, serrated flow occurred, as shown in Figure 3b. Additionally, Figure 3a shows that at <1.0 1/s, a discontinuous elastic-plastic transition was observed which extended up to approximately 2% ε_{eng}. For the high-speed dynamic tensile tests, the necessary application of the smoothing function for the force data did not allow for accurate observation of the transition phase. However, at 1.0 1/s, the discontinuous elastic-plastic transition was also observed in the raw data of the force signal.

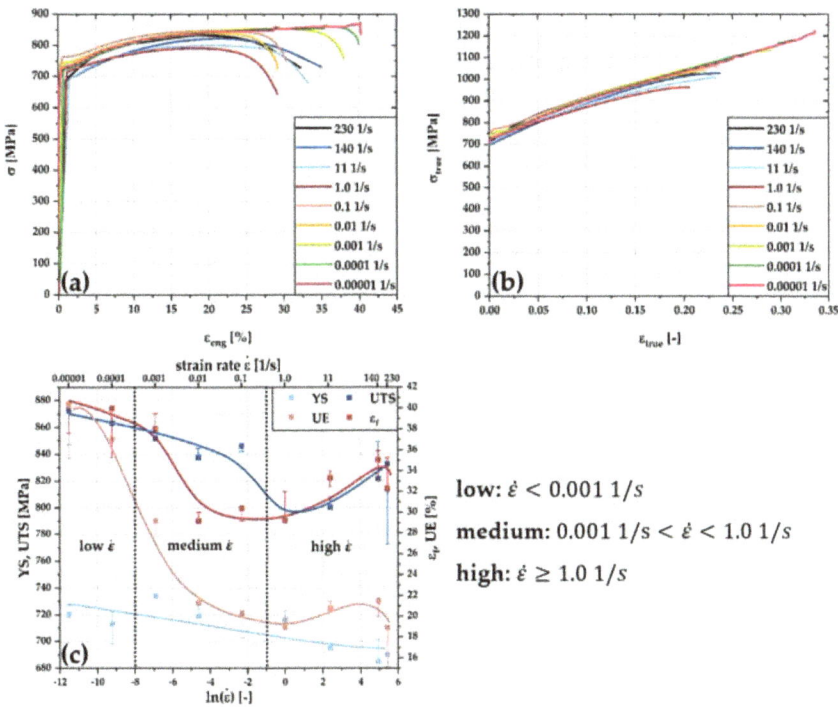

Figure 3. (a) Engineering stress-strain and (b) true stress-true strain curves of the X6MnAl12-3 alloy tested at different strain rates. (c) Overview of the strain-rate-dependent mechanical properties.

Table 2. Impact of an increase in strain rate for low, medium, and high strain rates (cf. Figure 3).

Mechanical Property	Low $\dot{\varepsilon}$	Medium $\dot{\varepsilon}$	High $\dot{\varepsilon}$
YS	↓	↓	↓
UTS	↓	↓	↑
UE	↓	↓	~
ε_f	~	↓	↑

In Figure 4, the initial austenite (γ), ferrite (α), and martensite (α') phase fractions after IA and the evolution of austenite phase fraction during quasistatic tensile testing at 0.001 1/s are shown, as measured by EBSD experiments. After IA at 675 °C for 2 h (cf. Figure 4a), the initial microstructure was lath-like with phase fractions of approximately 45% γ (face-centered cubic) and 55% α and/or recovered α' (body-centered cubic). At the end of the discontinuous elastic-plastic transition after approximately 2% elongation (cf. Figure 4b), no change in the γ phase fraction was detected. During straining at 0.001 1/s to ε_{eng} of 10% (Figure 4c) and 20% (Figure 4d), a significant fraction of γ underwent a martensitic $\gamma \to \alpha'$ phase transformation, resulting in a detected γ phase fraction of 35% and 15% respectively. No ε-martensite was detected.

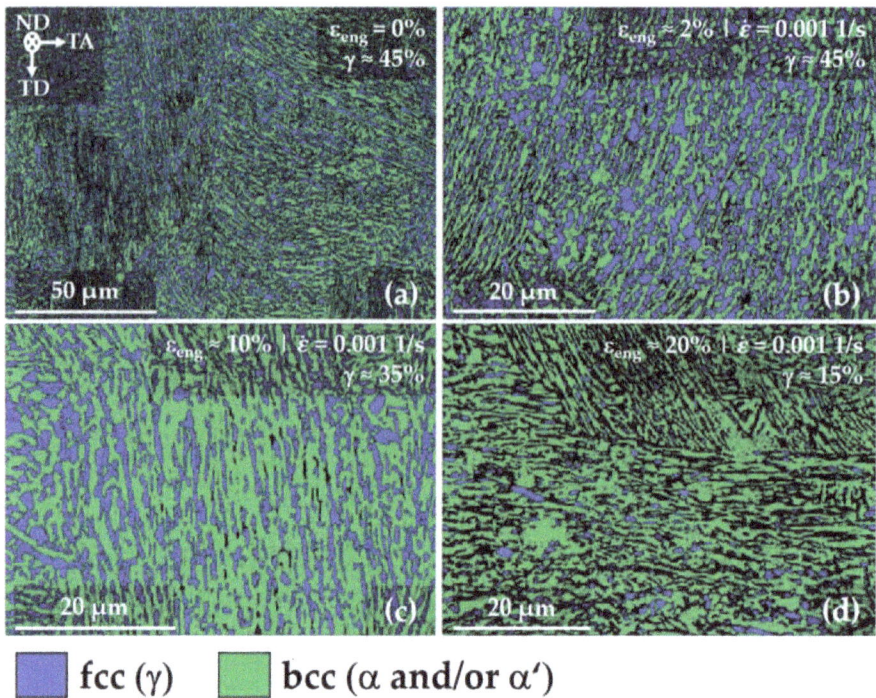

Figure 4. Electron backscatter diffraction (EBSD) phase maps of the investigated MMnS (a) before deformation and after straining to ε_{eng} of (b) 2%, (c) 10%, and (d) 20% at 0.001 1/s. Green areas indicate a body-centered cubic crystal structure, α and/or α', whereas blue areas indicate face-centered cubic austenite, γ.

To investigate the impact of strain rate on the phase transformation, Figure 5 compares the microstructure after tensile testing at different $\dot{\varepsilon}$ until fracture. While the medium $\dot{\varepsilon}$ of 0.001 1/s led to a nearly complete $\gamma \rightarrow \alpha\prime$ transformation at fracture (cf. Figure 5b), both very high strain rates of 1 1/s and 230 1/s (cf. Figure 5c,d) as well as a very low $\dot{\varepsilon}$ of 0.00001 1/s (cf. Figure 5a) facilitated significantly higher retained γ phase fractions of 29% and 25% after fracture, respectively.

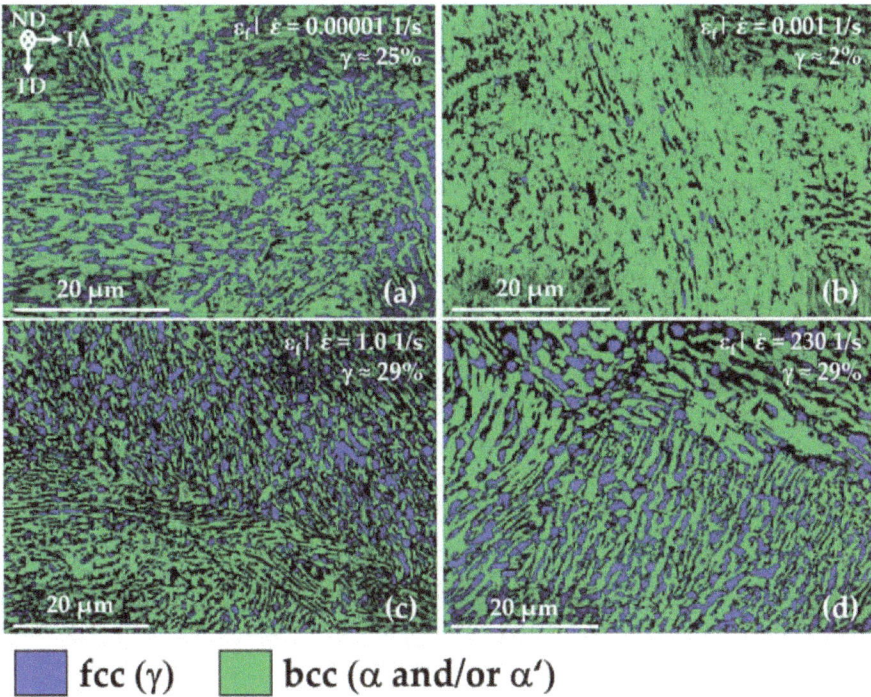

Figure 5. γ phase fractions (blue) as determined by EBSD measurements after tensile testing to fracture at (a) 0.00001 1/s, (b) 0.001 1/s, (c) 1.0 1/s, and (d) 230 1/s. Green areas indicate a body-centered cubic crystal structure, α and/or α'.

To highlight the onset of serrated flow at 0.00001 1/s, Figure 6 compares the flow curves and WHR of the tensile tests at 0.00001 1/s, 0.001 1/s, 1.0 1/s, and 230 1/s. While the flow curves at quasistatic strain rates of 0.001 1/s and 0.00001 1/s were almost identical, ε_f at the slower $\dot{\varepsilon}$ of 0.00001 1/s was significantly higher. The WHR during the initial stages of deformation were similar, as both curves dropped significantly during the Lüders band propagation. Afterwards the WHR increased and evolved in a similar way. However, at a true strain of approximately 0.25, the onset of the serrated flow at 0.00001 1/s was highlighted by a peak in the WHR curve, whereas the WHR at 0.001 1/s decreased to the flow curve, and necking was initiated. Similarly, the second serration in the flow curve at ε_{true} of approximately 0.33 marked the onset of the second peak in the WHR curve of the MMnS tested at 0.00001 1/s. Overall, the significantly more pronounced decrease in austenite phase fraction for 0.001 1/s did not result in a higher WHR compared to testing at 0.00001 1/s. Furthermore, tensile testing at 1.0 1/s led to a decrease of the WHR below those of 0.001 1/s and 0.00001 1/s, resulting in a reduction of uniform elongation. Increasing $\dot{\varepsilon}$ from 1.0 1/s to 230 1/s led to a small increase in the WHR to slightly lower levels than for the quasistatic tensile tests, which resulted in an increase in UTS, UE, and ε_f.

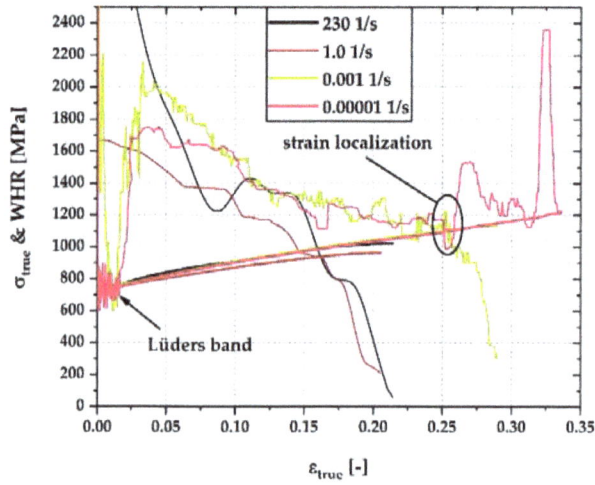

Figure 6. Flow curve and work-hardening rate (WHR) of the X6MnAl12-3 alloy tested at 0.00001 1/s, 0.001 1/s, 1.0 1/s, and 230 1/s, respectively.

Figure 7 shows the EBSD inverse pole figure (IPF) maps and grain orientation spread (GOS) maps for α and α', as well as γ and the kernel average misorientation (KAM) map. The GOS for α and γ was usually below 1.0 in equiaxed grains, indicating recrystallized grains, whereas the GOS values in lamellar morphologies were 2.0 or higher (cf. Figure 7c,d). The IPF maps in Figure 7 show that the orientation of the γ, α and α' grains within the lamellar structures did not vary significantly. On the other hand, the regions in the microstructure, where granular grains occurred frequently, did not show any preferred orientations. Those regions also showed significantly lower KAM values (cf. Figure 7e).

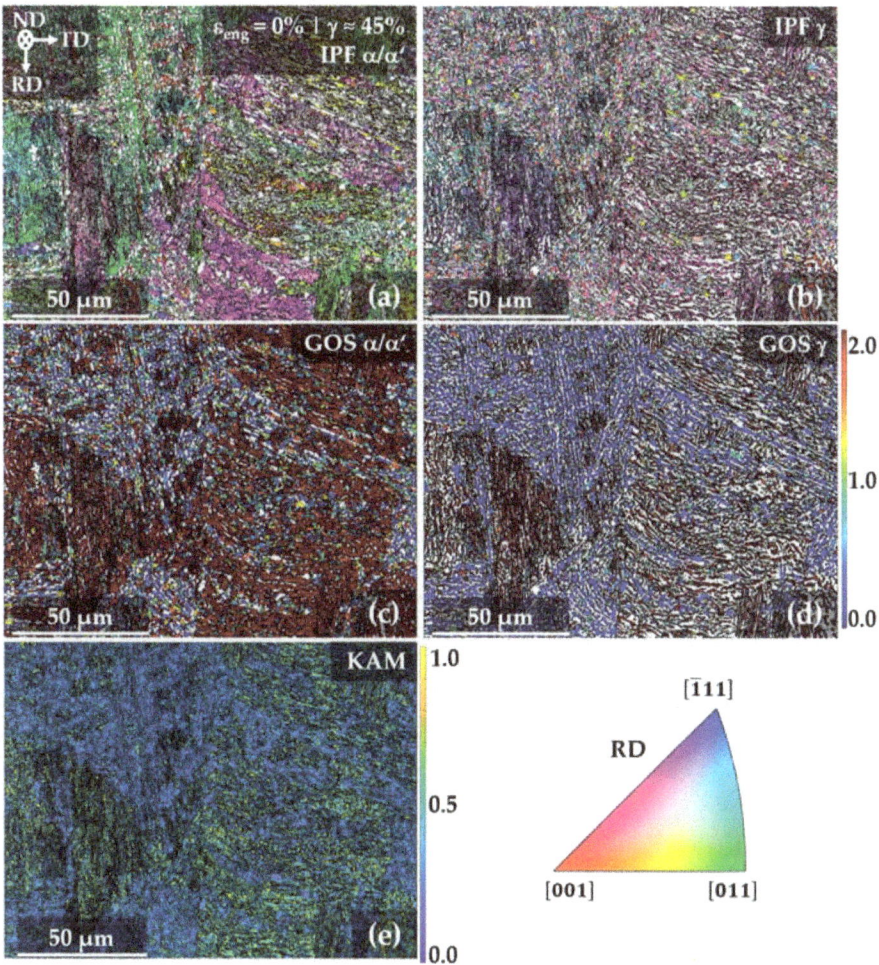

Figure 7. EBSD maps of the X6MnAl12-3 alloy prior to deformation (**a,b**) with inverse pole figure (IPF) color coding relative to the rolling direction (RD) for (**a**) α and α′ and (**b**) γ. Grain orientation spread (GOS) maps for (**c**) α and α′, (**d**) γ, and (**e**) the kernel average misorientation (KAM) map are shown. White areas in (**a–d**) correspond to (**a,c**) γ and (**b,d**) α or α′, respectively.

The EBSD IPF maps and KAM maps in Figure 8 show the microstructure of the X6MnAl12-3 alloy after uniform elongation to 2% strain. Comparing the EBSD IPF maps in Figure 8a,b showed that, similar to the undeformed state in Figure 7, the orientation of lamellar α and α′ grains did not deviate much, while the equiaxed γ grains between those lamellae were comparably randomly oriented. However, the KAM values of the γ grains were significantly higher than those of most equiaxed α grains (cf. Figure 8c,d) and also significantly higher than the KAM values of the γ grains in the undeformed state, as shown in Figure 7e.

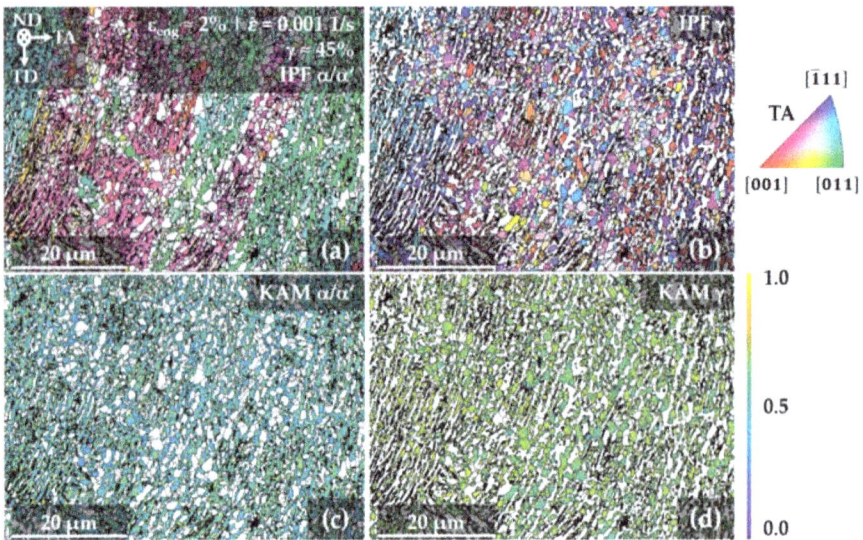

Figure 8. EBSD maps of the X6MnAl12-3 alloy at the end of the discontinuous elastic-plastic transition with (**a,b**) IPF color coding relative to the tensile axis (TA) for (**a**) α and α' and (**b**) γ. Corresponding KAM maps are shown for (**c**) α and α' and (**d**) γ. White areas in (**a–d**) correspond to (**a,c**) γ and (**b,d**) α or α', respectively.

The evolution of the microstructure and the distribution of strain between phases of the X6MnAl12-3 during further straining and after fracture for $\dot{\varepsilon}$ of 0.00001 1/s, 0.001 1/s, 1.0 1/s, and 230 1/s is shown in Figure 9 by means of EBSD IPF maps and KAM maps for α and α'. The EBSD IPF maps in Figure 9 show that the microstructure at higher elongations mainly consisted of blocks of lamellar α and α' grains, in which the orientations of α and α' were similar to each other. Comparing the KAM maps of α and α' grains after fracture for 0.00001 1/s, 1.0 1/s, and 230 1/s (cf. Figure 9b,f,h,j), the KAM values overall decreased with increasing strain rate. The KAM values of α and α' grains after elongation to fracture at 0.00001 1/s (cf. Figure 9e) and after 20% elongation at 0.001 1/s (cf. Figure 9b) were similar.

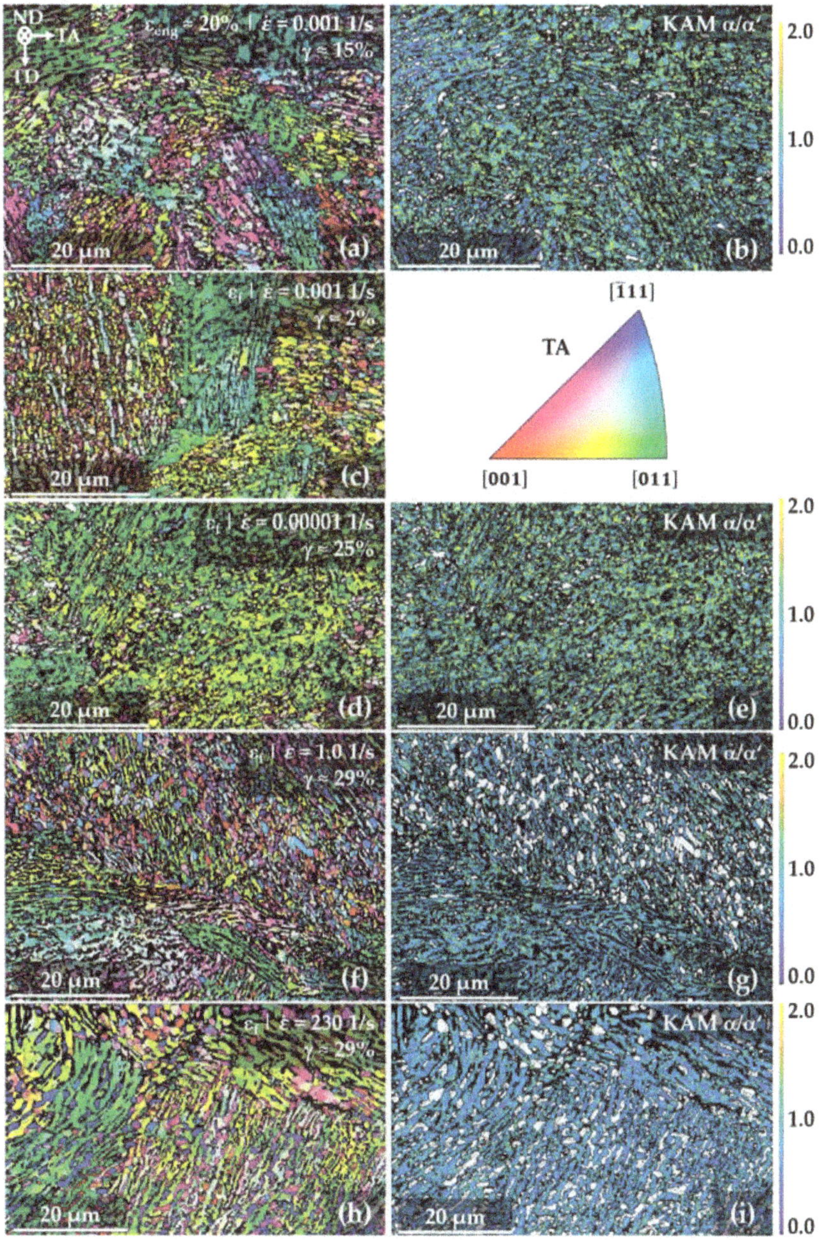

Figure 9. EBSD IPF maps for 0.001 1/s after (**a**) 20% ε_{eng} and (**c**) ε_f, as well as for the microstructure after straining until fracture at (**e**) 0.00001 1/s, (**g**) 1.0 1/s, and (**i**) 230 1/s. Corresponding EBSD maps with color coding according to KAM values of α and α' in (**b,f,h**) and (**j**) are shown. White areas in the KAM maps correspond to γ.

The in-situ DIC measurements in Figure 10 show the local strain rate, which indicates localized deformation behavior, on the surface of the tensile test specimen during the discontinuous elastic-plastic transition for the strain rates of 0.001 1/s and 0.00001 1/s. As shown in Figure 10, the discontinuous yield phenomenon was accompanied by a Lüders-band type of strain localization, which initiates in the clamping sections of the tensile test specimen and subsequently progresses towards the middle of the sample. The bands meet at the end of the Lüders strain at approximately 2.0% ε_{eng}. At higher strains, the plastic deformation proceeded homogeneously.

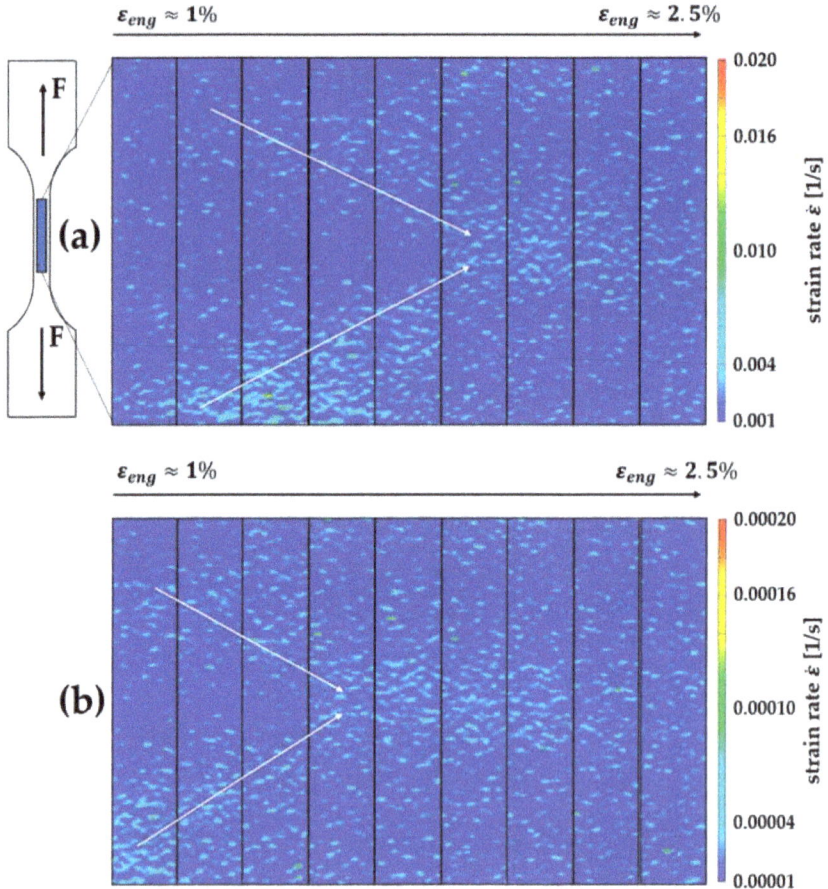

Figure 10. Visualization of local strain rate distribution in a tensile test sample during the discontinuous elastic-plastic transition using in-situ DIC during tensile testing at (**a**) 0.001 1/s and (**b**) 0.00001 1/s. White arrows highlight the propagation of localized peaks in strain rate during straining.

Figure 11 shows the strain localization behavior at high plastic strains for the specimens at 0.001 1/s and 0.00001 1/s. Comparing Figure 11a,b, tensile testing at 0.001 1/s did not lead to pronounced localized deformation behavior before necking, while two "X"-shaped areas ("X-bands") of strain localization occurred before necking during tensile testing at 0.00001 1/s. The onset of those deformation bands coincided with the spikes in the flow curve shown in Figure 6. The deformation bands propagated only in one direction and changed direction at the end of the parallel gauge length (cf. left part of Figure 11b).

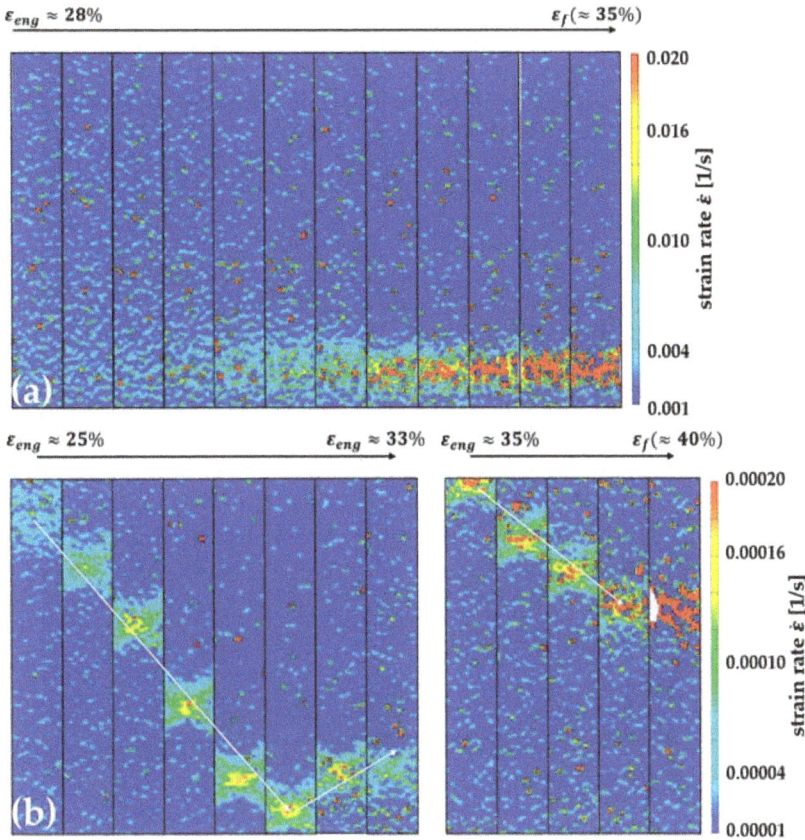

Figure 11. Visualization of the local strain rate distribution in a tensile test sample prior to necking initiation using in-situ DIC during tensile testing at (**a**) 0.001 1/s and (**b**) 0.00001 1/s. White arrows highlight the propagation of localized peaks in strain rate during straining.

4. Discussion

In this work, the strain-rate-dependent mechanical properties and strain localization behavior of an intercritically annealed medium-manganese X6MnAl12-3 steel were investigated.

IA at 675 °C for 2 h after cold rolling led to a mixed microstructure consisting of γ, α, and α'. The distinction between α and α' can be made based on the lamellar morphology and the GOS values (cf. Figure 7), for which a threshold value of 2 can be used to distinguish recrystallized grains with low internal stress (GOS below 2) and deformed grains with higher internal stress (GOS over 2) [32]. For the purpose of distinguishing between α and α' in this work, a lamellar morphology in combination with a GOS value over 2 is considered to indicate α'. However, it should be noted that recovered α' could also show GOS values below 2, as the diffusion of C out of the distorted bcc lattice could reduce the lattice distortion of α' and accordingly reduce the GOS value. Applying the GOS criterion for γ (cf. Figure 7d) reveals that the majority of γ grains, except lamellar grains of reverted γ situated between α'-lamellae, showed GOS values significantly lower than 1. Whereas the microstructure after cold rolling can be identified as fully martensitic (cf. Figure 2c), IA led to a microstructure consisting of lamellar reverted γ and α and α' grains as well as recrystallized, mostly equiaxed γ and α grains. Al and Mn were partitioned with Mn being enriched and Al being depleted in γ (cf. Figures 2d and 4a). The approximated Mn and Al contents of 14.0 wt.% and 2.5 wt.% for the calculation of the SFE using a subregular solution thermodynamic model [7] resulted in an SFE of γ at room temperature (RT) of 10 mJ/m². The low alloying content of 0.064 wt.% does not allow for an accurate determination of the actual C content in austenite via EDS. However, it can be assumed that the SFE is not significantly impacted by the partitioning of C to γ because of the very low alloying content of C compared to the significant γ phase fraction of 45% [6–8,33]. Accordingly, for the calculation of the SFE in this work, the C content is approximated to be 0.064 wt.%. This is in line with the findings of experimental approaches using atom probe tomography and EDS on a MMnS with very similar chemical composition and phase fractions after IA [22].

Overall, the high YS in the X6MnAl12-3 alloy, despite the low C content, can be explained by the presence of hard α' and the ultrafine-grained microstructure, which affects the flow stress significantly [34,35]. In the case of the microstructure of the X6MnAl12-3 alloy after IA, the critical stress required for plastic deformation is different for each α', α, and γ. Due to the low C content, the critical stress value for γ is significantly lower than for α and especially for α'. As a result, the initial plastic deformation is mainly accommodated by deformation in γ, which is pointed out by the comparison of KAM values for γ, α, and α' in Figure 8. High KAM values are an indication of the existence of geometrically necessary dislocations [36]. The different critical stress values of γ, α, and α' for plastic deformation are also the reason for the discontinuous elastic-plastic transition during quasistatic tensile testing. Due to the polynomial function that is used to smooth the oscillating force signal during the analysis of the high-speed dynamic tensile tests, it remains unclear whether the discontinuous elastic-plastic transition also occurs at higher strain rates, as a previous study has found to be the case for a multi-phase MMnS [28].

During deformation, the TRIP effect leads to a reduction in γ phase fraction, as summarized in Figure 12.

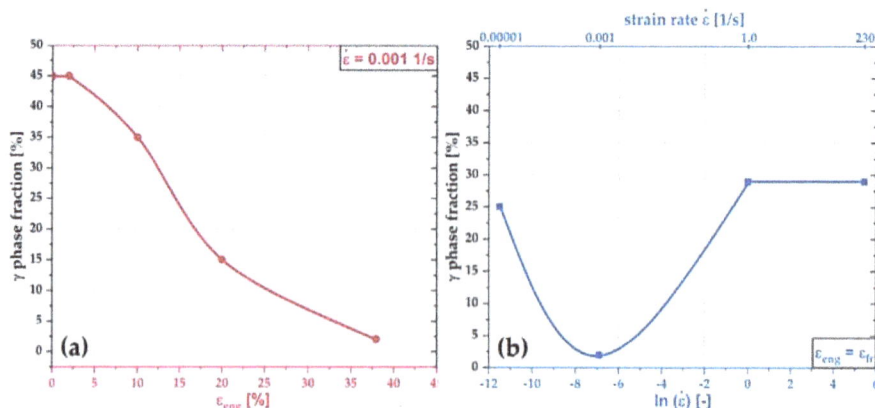

Figure 12. Summary of the evolution of γ phase fraction (**a**) during straining at 0.001 1/s and (**b**) comparison of strain-rate-dependent γ phase fractions after fracture.

The activation of the TRIP effect is determined by the stability of γ, which depends on the chemical composition, temperature and internal stress σ_γ [37]. With progressing plastic deformation, σ_γ increases and reaches a critical level for the activation of the TRIP effect. Accordingly, as shown in Figures 7 and 12, the phase fraction of γ is reduced. As stated in the previous paragraph, the flow stress of multi-phase steels like the X6MnAl12-3 MMnS is determined by the individual effective flow stress of each microstructural component γ, α, and α'. As evidenced by the discontinuous elastic-plastic transition, plastic deformation will be accommodated to different degrees by γ, α, and α', and intergranular strain accommodation is impeded [24–26]. However, at low ε̇, strain accommodation processes like dislocation generation in neighboring grains can occur over a longer period of time until the critical stress for initiation of the TRIP effect is reached. As Figures 8 and 12 show, this results in higher phase fractions of γ after fracture for a strain rate of 0.00001 1/s. This is also reflected by the higher KAM values in α and α' after deformation in Figure 9 for lower strain rates and is in agreement with previous reports of decreasing twin volume fractions with lower strain rates in high-manganese TWIP steels [38], where dislocation activity increased at lower strain rates [39]. Accordingly, the strain-rate sensitivity at low and medium $\dot{\varepsilon}$ of the tested X6MnAl12-3 MMnS on the microscale regarding the activation of secondary deformation mechanisms is high.

At $\dot{\varepsilon} \geq 1.0\,1/\text{s}$, adiabatic heating effects can cause drastic temperature increases of up to 100–150 °C in the tensile test specimen during deformation [29,36]. As such, the SFE is increased in-situ to 27–38 mJ/m^2 in γ based on the chemical composition derived from EDS [7], which is higher than the critical value of 18 mJ/m^2, above which the TRIP effect is suppressed or partially suppressed, as has been shown previously for HMnS and multi-phase MMnS [4,27–29,40–42]. Therefore, higher γ phase fractions are present after fracture (cf. Figures 8 and 12). This results in an increase in UTS with increasing strain rate >1.0 1/s (cf. Figure 3c), because strain accommodation processes between γ, α, and α' do not occur to the same extent as at lower strain rates. This is also reflected by the low KAM values of α and α' at higher strain rates (cf. Figure 9), as a higher fraction of plastic deformation is accommodated by γ at 1.0 1/s than at lower strain rates.

Following the remarks of the previous paragraphs, the impact of $\dot{\varepsilon}$ on the accommodation of plastic deformation in the X6MnAl12-3 MMnS can be described as a strain-rate-dependent strength of α and α' relative to γ ($\Delta\sigma$). At high strain rates, strain accommodation processes are partially inhibited, increasing $\Delta\sigma$. At low strain rates, sufficient time for the dislocation generation at grain boundaries of α and α' is available to accommodate the increased stress level in neighboring γ grains. Accordingly, $\Delta\sigma$ is lower than at high strain rates. As a result, the accommodation of plastic deformation is much more homogeneously distributed over γ, α, and α' at very slow strain rates. Consequently, the TRIP effect, which is activated above a critical stress value for σ_γ, is partially suppressed at very low strain rates. Figure 13 illustrates the strain-rate-dependent microstructural development.

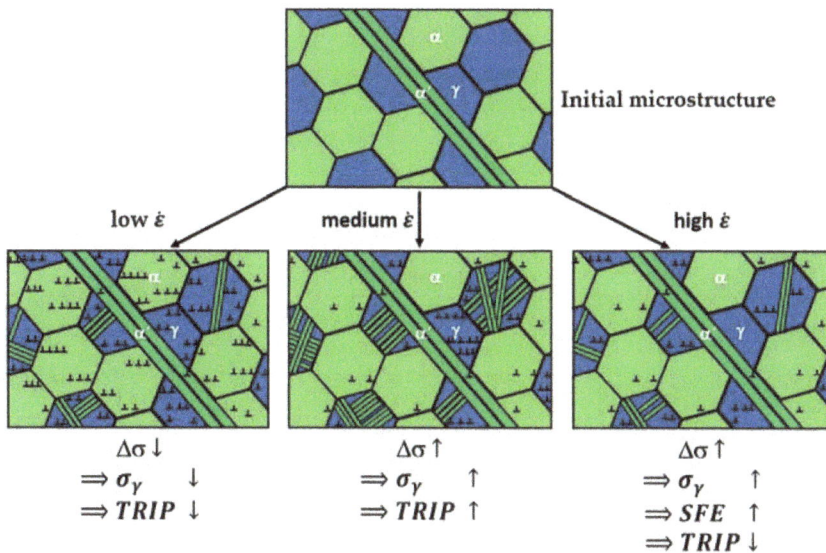

Figure 13. Impact of the strain-rate-dependent strength of α and α' relative to γ ($\Delta\sigma$) on the microstructural evolution during straining at different strain rates.

The low WHR, UE, and ε_f at 1.0 1/s compared to quasistatic $\dot{\varepsilon}$ (cf. Figure 6) can be explained by three effects. Firstly, adiabatic heating at high $\dot{\varepsilon}$ enables dynamic recovery processes in γ, which decreases the WHR [4,27]. Additionally, the adiabatic heating also increases the SFE of γ and leads to a partial suppression of the TRIP effect during plastic deformation (cf. Figures 4 and 5) [7,8,27–29]. Thirdly, the high $\Delta\sigma$ suppresses the participation of the harder α and α' phases in accommodating plastic deformation to a large extent. Increasing $\dot{\varepsilon}$ >1.0 1/s led to an increase in WHR and, accordingly, to higher UTS and ε_f (cf. Figure 3), which is typically observed in conventional TRIP steels, dual-phase (DP) steels, and high-manganese TWIP steels with positive strain-rate sensitivity [27,43,44], as well as in multi-phase MMnS [28]. However, at 1.0 1/s, this effect is not strong enough to compensate for the partial suppression of the TRIP effect, which results in a low WHR compared to quasistatic $\dot{\varepsilon}$. This minimum in UE and ε_f at medium-high $\dot{\varepsilon}$ has also been previously observed in high-manganese TWIP steels [27]. Comparing the γ phase fractions after elongation to fracture (cf. Figures 5 and 12) reveals that increasing $\dot{\varepsilon}$ >1.0 1/s did not lead to an increase in the retained γ phase fraction, as has also previously been reported in multi-phase MMnS [28]. This is because the adiabatic heating effect primarily occurs during necking so that during the initial stages of plastic deformation, the TRIP effect can still occur. Due to this, the UE does not increase significantly. Overall, the interplay of effects on a microscale leads to a negative strain-rate sensitivity in the quasistatic $\dot{\varepsilon}$ regime with regards to the macroscopic mechanical properties, especially for the UTS, UE and ε_f. At high $\dot{\varepsilon}$, the strain-rate sensitivity is slightly positive, similar to conventional TRIP and DP steels.

Although the α' phase fraction after straining to fracture at 0.00001 1/s is not as high as at 0.001 1/s (cf. Figures 8 and 12), indicating a lower activity of the TRIP effect, the WHR is similar (cf. Figure 6). Usually, the TRIP effect leads to a strong increase in WHR [1,27]. However, as the participation of the harder phases α and α' in accommodating plastic deformation increases at lower strain rates due to a lower relative $\Delta\sigma$, the necessary stress for further straining increases to a similar extent. Accordingly, the WHR in the quasistatic tensile testing regime is not impacted significantly by the strain rate.

Recently, the impact of microstructural heterogeneity and the activation of secondary deformation mechanisms has been found to be of critical importance for the initiation and propagation of deformation bands, resulting in serrated flow curves [14,45]. Following the remarks of the previous paragraphs, the occurrence of serrated flow in the X6MnAl12-3 alloy at very slow strain rates could be due to the enhanced accommodation of plastic deformation by relatively low $\Delta\sigma$, which enables the propagation of deformation bands and the occurrence of serrated flow. Comparing the flow curves of the X6MnAl12-3 alloy at 0.001 1/s and 0.00001 1/s in Figure 6 reveals that the onset of strain localization during plastic deformation at 0.00001 1/s and 0.001 1/s occurs at approximately 28% strain, as is also shown by the in-situ DIC measurements in Figure 11. However, as the propagation of stresses is inhibited by the higher $\Delta\sigma$ at 0.001 1/s, necking is initiated in contrast to the occurrence of the serrated flow at 0.00001 1/s. This also explains the higher total elongation at 0.00001 1/s.

5. Conclusions

The strain-rate-dependent deformation behavior and mechanical properties of a X6MnAl12-3 medium-manganese steel (MMnS) were investigated. Intercritical annealing at 675 °C for 2 h led to an ultrafine-grained multi-phase microstructure consisting of approximately 45% reverted austenite (γ) and 55% recrystallized ferrite (α) and recovered martensite (α'). The following conclusions can be drawn:

- The strain-rate sensitivity of the macroscopic mechanical properties is different for low quasistatic strain rates and high strain rates $\dot{\varepsilon}$, which is due to the high strain-rate sensitivity of the deformation behavior on the microscale. On the microscale, the strain-rate-dependent strength of α and α' relative to γ ($\Delta\sigma$) leads to the observed strain-rate sensitivity of the $\gamma \rightarrow \alpha'$ phase transformation (TRIP).

- At low $\dot{\varepsilon}$, $\Delta\sigma$ is relatively small, which enables the accommodation of plastic deformation by α or α' and compensates the negative impact of the partial suppression of the TRIP effect on the work-hardening rate (WHR). At very low $\dot{\varepsilon}$, serrated flow occurs, which is due to the TRIP mechanism in combination with a significant accommodation of plastic deformation by α and α'.
- At high $\dot{\varepsilon}$, plastic deformation is not accommodated by α and α' to a large extent due to a very high $\Delta\sigma$. Additionally, adiabatic heating effects partially suppress the TRIP effect in γ.
- At a medium quasistatic $\dot{\varepsilon}$ (e.g., 0.001 1/s), $\Delta\sigma$ is relatively high, and plastic deformation is primarily accommodated by γ, which induces the pronounced activation of the TRIP mechanism.

Author Contributions: Conceptualization, S.S.; formal analysis, S.S.; investigation, S.S.; project administration, W.B.; supervision, C.H. and W.B.; visualization, S.S.; writing–original draft, S.S.; writing–review & editing, S.S., C.H. and W.B.

Funding: This research was funded by the "Deutsche Forschungsgemeinschaft" (DFG) within the Sonderforschungsbereich (Collaborative Research Center) 761 "Steel–ab initio".

Conflicts of Interest: The authors declare no conflict of interest. The funders had no role in the design of the study; in the collection, analyses, or interpretation of data; in the writing of the manuscript, or in the decision to publish the results.

References

1. Lee, Y.-K.; Han, J. Current opinion in medium manganese steel. *Mater. Sci. Technol.* **2015**, *31*, 843–856. [CrossRef]
2. DeCooman, B.C. High Mn TWIP steel and medium Mn steel. In *Automotive Steels: Design, Metallurgy, Processing and Applications*; Rana, R., Singh, S.B., Eds.; Woodhead Publishing: Cambridge, UK, 2017; pp. 317–385.
3. Bouaziz, O.; Allain, S.; Scott, C.P.; Cugy, P.; Barbier, D. High manganese austenitic twinning induced plasticity steels: A review of the microstructure properties relationships. *Curr. Opin. Solid State Mater. Sci.* **2011**, *15*, 141–168. [CrossRef]
4. DeCooman, B.C.; Estrin, Y.; Kim, S.K. Twinning-induced plasticity (TWIP) steels. *Acta Mater.* **2018**, *142*, 283–362. [CrossRef]
5. Bouaziz, O. Strain-hardening of twinning-induced plasticity steels. *Scr. Mater.* **2012**, *66*, 982–985. [CrossRef]
6. Pierce, D.T.; Jiménez, J.A.; Bentley, J.; Raabe, D.; Oskay, C.; Wittig, J.E. The influence of manganese content on the stacking fault and austenite/ε-martensite interfacial energies in Fe–Mn–(Al–Si) steels investigated by experiment and theory. *Acta Mater.* **2014**, *68*, 238–253. [CrossRef]
7. Saeed-Akbari, A.; Imlau, J.; Prahl, U.; Bleck, W. Derivation and Variation in Composition-Dependent Stacking Fault Energy Maps Based on Subregular Solution Model in High-Manganese Steels. *Metall. Mater. Trans. A* **2009**, *40*, 3076–3090. [CrossRef]
8. Dumay, A.; Chateau, J.P.; Allain, S.; Migot, S.; Bouaziz, O. Influence of addition elements on the stacking-fault energy and mechanical properties of an austenitic Fe–Mn–C steel. *Mater. Sci. Eng. A* **2008**, *483–484*, 184–187. [CrossRef]
9. Bambach, M.; Conrads, L.; Daamen, M.; Güvenç, O.; Hirt, G. Enhancing the crashworthiness of high-manganese steel by strain-hardening engineering, and tailored folding by local heat-treatment. *Mater. Des.* **2016**, *110*, 157–168. [CrossRef]
10. Haase, C.; Barrales-Mora, L.A.; Roters, F.; Molodov, D.A.; Gottstein, G. Applying the texture analysis for optimizing thermomechanical treatment of high manganese twinning-induced plasticity steel. *Acta Mater.* **2014**, *80*, 327–340. [CrossRef]
11. Allain, S.; Bouaziz, O.; Lebedkina, T.; Lebyodkin, M. Relationship between relaxation mechanisms and strain aging in an austenitic FeMnC steel. *Scr. Mater.* **2011**, *64*, 741–744. [CrossRef]
12. Hoffmann, S.; Bleck, W.; Berme, B. In Situ Characterization of Deformation Behavior of Austenitic High Manganese Steels. *Steel Res. Int.* **2012**, *83*, 379–384. [CrossRef]
13. Renard, K.; Ryelandt, S.; Jacques, P.J. Characterisation of the Portevin-Le Châtelier effect affecting an austenitic TWIP steel based on digital image correlation. *Mater. Sci. Eng. A* **2010**, *527*, 2969–2977. [CrossRef]

14. Sevsek, S.; Brasche, F.; Haase, C.; Bleck, W. Combined deformation twinning and short-range ordering causes serrated flow in high-manganese steels. *Mater. Sci. Eng. A* **2019**. [CrossRef]

15. Wang, X.G.; Wang, L.; Huang, M.X. In-situ evaluation of Lüders band associated with martensitic transformation in a medium Mn transformation-induced plasticity steel. *Mater. Sci. Eng. A* **2016**, *674*, 59–63. [CrossRef]

16. Ma, Y.; Song, W.; Bleck, W. Investigation of the Microstructure Evolution in a Fe-17Mn-1.5Al-0.3C Steel via In Situ Synchrotron X-ray Diffraction during a Tensile Test. *Materials* **2017**, *10*, 1129. [CrossRef]

17. Ma, Y. Medium-manganese steels processed by austenite-reverted-transformation annealing for automotive applications. *Mater. Sci. Technol.* **2017**, *33*, 1713–1727. [CrossRef]

18. Lee, S.; DeCooman, B.C. Annealing Temperature Dependence of the Tensile Behavior of 10 pct Mn Multi-phase TWIP-TRIP Steel. *Metall. Mater. Trans. A* **2014**, *45*, 6039–6052. [CrossRef]

19. Lee, S.-J.; Lee, S.; DeCooman, B.C. Mn partitioning during the intercritical annealing of ultrafine-grained 6% Mn transformation-induced plasticity steel. *Scr. Mater.* **2011**, *64*, 649–652. [CrossRef]

20. Cai, Z.H.; Ding, H.; Misra, R.D.K.; Ying, Z.Y. Austenite stability and deformation behavior in a cold-rolled transformation-induced plasticity steel with medium manganese content. *Acta Mater.* **2015**, *84*, 229–236. [CrossRef]

21. Emo, J.; Maugis, P.; Perlade, A. Austenite growth and stability in medium Mn, medium Al Fe-C-Mn-Al steels. *Comput. Mater. Sci.* **2016**, *125*, 206–217. [CrossRef]

22. Benzing, J.T.; Kwiatkowski da Silva, A.; Morsdorf, L.; Bentley, J.; Ponge, D.; Dutta, A.; Han, J.; McBride, J.R.; Van Leer, B.; Gault, B.; Raabe, D.; Wittig, J.E. Multi-scale characterization of austenite reversion and martensite recovery in cold-rolled medium-Mn steel. *Acta Mater.* **2019**, *166*, 512–530. [CrossRef]

23. Dutta, A.; Ponge, D.; Sandlöbes, S.; Raabe, D. Strain partitioning and strain localization in medium manganese steels measured by in situ microscopic digital image correlation. *Materialia* **2019**, *5*, 100252. [CrossRef]

24. Tan, X.; Ponge, D.; Lu, W.; Xu, Y.; Yang, X.; Rao, X.; Wu, D.; Raabe, D. Carbon and strain partitioning in a quenched and partitioned steel containing ferrite. *Acta Mater.* **2019**, *165*, 561–576. [CrossRef]

25. Gibbs, P.J.; de Cooman, B.C.; Brown, D.W.; Clausen, B.; Schroth, J.G.; Merwin, M.J.; Matlock, D.K. Strain partitioning in ultra-fine grained medium-manganese transformation induced plasticity steel. *Mater. Sci. Eng. A* **2014**, *609*, 323–333. [CrossRef]

26. Latypov, M.I.; Shin, S.; DeCooman, B.C.; Kim, H.S. Micromechanical finite element analysis of strain partitioning in multiphase medium manganese TWIP+TRIP steel. *Acta Mater.* **2016**, *108*, 219–228. [CrossRef]

27. Brüx, U.; Frommeyer, G.; Grässel, O.; Meyer, L.W.; Weise, A. Development and characterization of high strength impact resistant Fe-Mn-(Al-, Si) TRIP/TWIP steels. *Steel Res. Int.* **2002**, *73*, 294–298. [CrossRef]

28. Poling, W.A. Tensile Deformation of Third Generation Advanced High Strength Sheet Steels under High Strain Rates. Ph.D. Dissertation, Colorado School of Mines, Golden, CO, USA, 2016. Available online: http://hdl.handle.net/11124/170454 (accessed on 14 March 2019).

29. Rana, R.; De Moor, E.; Speer, J.G.; Matlock, D.K. On the Importance of Adiabatic Heating on Deformation Behavior of Medium-Manganese Sheet Steels. *JOM* **2018**, *70*, 706–713. [CrossRef]

30. Bachmann, F.; Hielscher, R.; Schaeben, H. Texture Analysis with MTEX—Free and Open Source Software Toolbox. *SSP* **2010**, *160*, 63–68. [CrossRef]

31. Hielscher, R.; Schaeben, H. A novel pole figure inversion method: Specification of the MTEX algorithm. *J. Appl. Crystallogr.* **2008**, *41*, 1024–1037. [CrossRef]

32. Field, D.; Bradford, L.; Nowell, M.; Lillo, T. The role of annealing twins during recrystallization of Cu. *Acta Mater.* **2007**, *55*, 4233–4241. [CrossRef]

33. Bouaziz, O.; Zurob, H.; Chehab, B.; Embury, J.D.; Allain, S.; Huang, M. Effect of chemical composition on work hardening of Fe-Mn-C TWIP steels. *Mater. Sci. Technol.* **2011**, *27*, 707–709. [CrossRef]

34. Gutierrez-Urrutia, I.; Zaefferer, S.; Raabe, D. The effect of grain size and grain orientation on deformation twinning in a Fe–22wt.% Mn–0.6wt.% C TWIP steel. *Mater. Sci. Eng. A* **2010**, *527*, 3552–3560. [CrossRef]

35. Dini, G.; Ueji, R.; Najafizadeh, A. Grain Size Dependence of the Flow Stress of TWIP Steel. *MSF* **2010**, *654–656*, 294–297. [CrossRef]

36. Calcagnotto, M.; Ponge, D.; Demir, E.; Raabe, D. Orientation gradients and geometrically necessary dislocations in ultrafine grained dual-phase steels studied by 2D and 3D EBSD. *Mater. Sci. Eng. A* **2010**, *527*, 2738–2746. [CrossRef]

37. Twardowski, R.; Prahl, U. Microstructure based flow curve modeling of high-Mn steels with TWIP and TRIP effect. *Comput. Methods Mater. Sci.* **2012**, *12*, 130–136.
38. Meyers, M.A.; Vöhringer, O.; Lubarda, V.A. The onset of twinning in metals: A constitutive description. *Acta Mater.* **2001**, *49*, 4025–4039. [CrossRef]
39. Li, K.; Yu, B.; Misra, R.D.K.; Han, G.; Tsai, Y.T.; Shao, C.W.; Shang, C.J.; Yang, J.R.; Zhang, Z.F. Strain rate dependence on the evolution of microstructure and deformation mechanism during nanoscale deformation in low carbon-high Mn TWIP steel. *Mater. Sci. Eng. A* **2019**, *742*, 116–123. [CrossRef]
40. Saeed-Akbari, A.; Mosecker, L.; Schwedt, A.; Bleck, W. Characterization and Prediction of Flow Behavior in High-Manganese Twinning Induced Plasticity Steels: Part, I. Mechanism Maps and Work-Hardening Behavior. *Metall. Mater. Trans. A* **2012**, *43*, 1688–1704. [CrossRef]
41. Gutierrez-Urrutia, I.; Raabe, D. Grain size effect on strain hardening in twinning-induced plasticity steels. *Scr. Mater.* **2012**, *66*, 992–996. [CrossRef]
42. Gutierrez-Urrutia, I.; Zaefferer, S.; Raabe, D. Electron channeling contrast imaging of twins and dislocations in twinning-induced plasticity steels under controlled diffraction conditions in a scanning electron microscope. *Scr. Mater.* **2009**, *61*, 737–740. [CrossRef]
43. Kim, J.-H.; Kim, D.; Han, H.N.; Barlat, F.; Lee, M.-G. Strain rate dependent tensile behavior of advanced high strength steels: Experiment and constitutive modeling. *Mater. Sci. Eng. A* **2013**, *559*, 222–231. [CrossRef]
44. Curtze, S.; Kuokkala, V.-T.; Hokka, M.; Peura, P. Deformation behavior of TRIP and DP steels in tension at different temperatures over a wide range of strain rates. *Mater. Sci. Eng. A* **2009**, *507*, 124–131. [CrossRef]
45. Jo, M.C.; Choi, J.H.; Lee, H.; Zargaran, A.; Ryu, J.H.; Sohn, S.S.; Kim, N.J.; Lee, S. Effects of solute segregation on tensile properties and serration behavior in ultra-high-strength high-Mn TRIP steels. *Mater. Sci. Eng. A* **2019**, *740–741*, 16–27. [CrossRef]

![metals logo] *metals*

MDPI

Article

Deformation Behavior of a Double Soaked Medium Manganese Steel with Varied Martensite Strength

Alexandra Glover [1,*], Paul J. Gibbs [2], Cheng Liu [2], Donald W. Brown [2], Bjørn Clausen [2], John G. Speer [1] and Emmanuel De Moor [1,*]

[1] Advanced Steel Processing and Products Research Center, Colorado School of Mines, Golden, CO 80401, USA
[2] Los Alamos National Laboratory, Los Alamos, NM 87545, USA
*   Correspondence: aglover@mymail.mines.edu (A.G.); edemoor@mines.edu (E.D.M.)

Received: 15 June 2019; Accepted: 4 July 2019; Published: 7 July 2019

Abstract: The effects of athermal martensite on yielding behavior and strain partitioning during deformation is explored using in situ neutron diffraction for a 0.14C–7.14Mn medium manganese steel. Utilizing a novel heat treatment, termed double soaking, samples with similar microstructural composition and varied athermal martensite strength and microstructural characteristics, which composed the bulk of the matrix phase, were characterized. It was found that the addition of either as-quenched or tempered athermal martensite led to an improvement in mechanical properties as compared to a ferrite plus austenite medium manganese steel, although the yielding and work hardening behavior were highly dependent upon the martensite characteristics. Specifically, athermal martensite was found to promote continuous yielding and improve the work hardening rate during deformation. The results of this study are particularly relevant when considering the effect of post-processing thermal heat treatments, such as tempering or elevated temperature service environments, on the mechanical properties of medium manganese steels containing athermal martensite.

Keywords: neutron diffraction; austenite stability; medium manganese steel; double soaking; localized deformation

---

1. Introduction

Modern automotive designs, which desire increased passenger safety in conjunction with vehicle light-weighting and increased fuel efficiency, require the development of new advanced high strength steel (AHSS) grades [1–3]. The first generation of advanced high strength steels, which includes dual-phase, transformation-induced plasticity, and complex phase steels, are primarily ferrite-based. The second generation of AHSS utilize alloys with high manganese contents to generate fully austenitic microstructures such as twinning-induced plasticity (TWIP) steels. The high alloy contents needed to achieve complete austenite stabilization present processing challenges which, combined with the increased cost of these alloys, have delayed the application of TWIP steels in automotive applications. With these processing challenges in mind, the third-generation of AHSS have been proposed with the goal of generating higher strength–ductility combinations than the first generation of AHSS while using lower alloying element concentrations than the second generation of AHSS.

One of the new steel grades proposed to reach the property targets desired for the third generation of AHSS are medium manganese steels. These steels generally contain between 5–10 wt % Mn and 0.1–0.3 wt % C, and are intercritically annealed to form a microstructure of ferrite plus austenite [4–6]. With typical retained austenite volume fractions between 5 and 30 vol %, the improved mechanical properties of these steels can be attributed in part to the strain-induced transformation of austenite to martensite, which has been shown to help maintain a high rate of work hardening during deformation [4,7–11]. In intercritically annealed medium manganese steels, the resistance of austenite to strain-induced transformation has been shown to be highly dependent upon intercritical annealing heat treatment parameters [5,6,12]. This is because the intercritical annealing temperature within the ferrite plus austenite phase field determines the equilibrium austenite volume fraction and the associated C and Mn enrichment.

These intercritically annealed medium manganese steels, which contain primarily retained austenite and ferrite, may be strengthened through the substitution of ferrite with athermal martensite. Several studies of medium manganese steels which contain small volume fractions of athermal martensite have shown, in some conditions, to promote continuous yielding and improve the overall strength–ductility combination of the material [5,6,12–14]. Due to the nature of the single intercritical annealing heat treatment used in these studies, the introduction of martensite was only possible through the incomplete stabilization of austenite during intercritical annealing. This approach leads to retained austenite with reduced levels of carbon and manganese partitioning, which may not be desirable.

The newly proposed double soaking heat treatment provides a unique opportunity to study the influence of martensite in medium manganese steels [15,16]. This is because the double soaking heat treatment allows for a large volume fraction of martensite with variable strength to be generated while maintaining high volume fractions of retained austenite with significant manganese partitioning. Previous research has shown that the double soaking heat treatment is capable of generating ultimate tensile strengths above 1600 MPa in conjunction with total elongations above 13% in a 0.14C–7.14Mn steel [16]. In this study, the response of a 0.14C–7.14Mn steel to the double soaking or double soaking plus tempering heat treatment was examined using in situ neutron diffraction. This allowed for the role of martensite strength and characteristics, such as dislocation density, in yielding behavior and during plastic deformation to be studied.

2. Materials and Methods

The double soaking heat treatment is shown schematically in Figure 1. The first step, the primary soaking treatment in the intercritical region, is characterized by primary austenite (γ_P) nucleation and growth and manganese partitioning from ferrite to austenite. Next, a short secondary soaking treatment is applied at a higher temperature as either a continuous or discontinuous heat treatment. The microstructure at the secondary soaking temperature consists of primary austenite present after the initial soak (γ_P), newly formed secondary austenite (γ_s), and (potentially) some remaining ferrite (α), depending upon the selected secondary soaking temperature and time. During the secondary soaking treatment, a substantial amount of manganese diffusion from the enriched γ_P to the newly formed γ_s is not desired and may be avoided given the slow rate of manganese diffusion in austenite combined with the short secondary soaking time. Depending upon the time and temperature of the secondary soaking treatment, varying amounts of carbon diffusion are expected. Finally, the steel is quenched to room temperature, which is below the martensite start (M_s) temperature of γ_s, causing the austenite formed during the secondary soaking operation to transform into martensite. This results in a final microstructure of retained austenite, martensite, and (potentially) intercritical ferrite. An optional tempering treatment, in this instance at temperatures above the M_s, may also be added to modify martensite strength. This additional heat treatment is classified as the double soaking plus tempering (DS-T) heat treatment in this investigation.

This study explored the influence of martensite strength on austenite stability using a steel with the composition 0.14C–7.17Mn–0.21Si (wt %). This steel was cold-rolled to a final thickness of 1.4 mm and then intercritically annealed by industrial batch annealing, which constitutes the primary soaking treatment. Following the primary soaking treatment, the A_{c1} and A_{c3} temperatures were measured at 668 and 825 °C, respectively, and the microstructure was found to consist of 40 vol % austenite and 60 vol % ferrite using X-ray diffraction. The tensile properties generated following the primary soaking heat treatment are shown in Figure 2. The primary soaked material was characterized as having an ultimate tensile strength of 1038 MPa combined with a total elongation of 41%. The material underwent localized deformation upon yielding, with the yield point drop and yield point elongation characteristic of many medium manganese steels. This was followed by a region of moderate work hardening and limited post-uniform elongation.

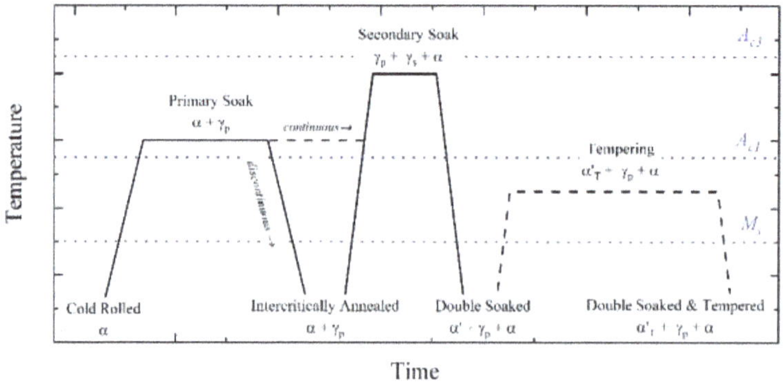

Figure 1. Schematic of the double soaking and double soaking plus tempering heat treatment. Expected changes in microstructure are indicated: Ferrite (α), primary austenite (γ_P), secondary austenite (γ_s), martensite (α'), and tempered martensite (α'_T).

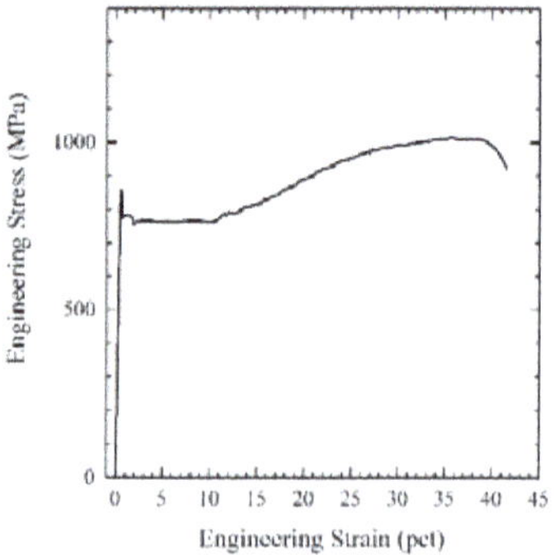

Figure 2. Room temperature, quasi-static engineering stress-strain curve for the industrially batch annealed (primary soaked) material.

In this study, two experimental heat treatments were then applied to the batch annealed (BA) material. These heat treatments were designed to generate equivalent volume fractions of martensite while varying martensite strength. The first heat treatment, called the double soaking (DS) treatment, involved the application of the secondary soaking heat treatment at 800 °C for 30 s to the BA steel [16]. The second heat treatment, identified as the double soaking plus tempering (DS-T) treatment, added both a secondary soaking heat treatment at 800 °C for 30 s followed by a quench to room temperature as well as subsequent application of a tempering treatment at 450 °C for 300 s. Both heat treatments were applied using salt pots, with an average heating rate of 80 °C/s.

Uniaxial tensile testing was conducted using ASTM E8 subsized samples machined parallel to the rolling direction with a gauge section of 25 mm. All tests were performed on a Instron 1125 (Noorwood, MA, USA) screw-driven loading frame at a constant engineering strain rate of 4.5×10^{-4} s^{-1} in a method consistent with ASTM E8 [17]. Applied load and crosshead displacement were recorded at a rate of 10 Hz. Sample deformation was monitored using 2D digital image correlation (DIC), a technique which relies on the computer vision approach to extract whole-field displacement data [18]. Images were captured using a CCD camera with a resolution of 2448 × 2048 pixels and a lens with a 75 mm focal length. The physical resolution was 22.5 μm. Images were taken at a frequency of 5 Hz. Before testing, a random pattern was printed onto each specimen surface by first spraying white paint as a background and then black paint to generate random speckles. Using the displacement vectors reported by the VIC 3D software developed by Correlated Solutions (Irmo, SC, USA) and a virtual 25.4 mm (1 in) extensometer, the macroscopic strain for each sample was characterized [19].

In situ neutron diffraction was performed on the SMARTS instrument at the Lujan Center at Los Alamos National Laboratory [20]. Phase fractions, elastic lattice strains, and diffraction peak width measurements were made during tensile deformation using two banks of detectors oriented at ± 90° to the incident beam. One detector bank collected data from the transverse direction (normal to the specimen thickness), and the other collected data from the axial direction (along the axis of maximum tension). Tensile specimens, with a 44.5 mm reduced gauge section and a 6.4 mm gauge width, were deformed incrementally at a constant engineering strain rate of 4.5×10^{-4} s^{-1}, and diffraction patterns were recorded while the sample was held at a constant displacement for approximately 1800 s [17]. The reported stress values were taken as the final value measured during each diffraction hold, and the magnitude of stress relaxation measured during each diffraction hold varied between 20 and 60 MPa following macroscopic yielding for both heat treatment conditions. Sample deformation was monitored using 2D DIC during the in situ neutron diffraction. Due to the increased length of the interrupted neutron diffraction tests, the sampling interval was reduced to 0.0034 Hz. Using the displacement vectors reported by the VIC 3D software developed by Correlated Solutions and a virtual 25.4 mm (1 in) extensometer, the macroscopic strain for each sample was determined [19].

Two phases were considered in the analysis of the diffraction data—body centered cubic (BCC) ferrite/martensite (α/α'), and face centered cubic (FCC) austenite (γ). The alpha prime (α') martensite fraction in the starting microstructure was estimated from dilatometric scans, as martensite tetragonality was insufficient to differentiate body centered tetragonal (BCT) martensite from BCC ferrite in the diffraction data [21]. Austenite fractions were determined at each deformation step using the Rietveld analysis function within the GSAS software package [22].

Single peak fitting using the Rawplot subroutine of GSAS was used to measure changes in interplanar spacing at each deformation step where diffraction data were collected. Elastic lattice strain (ε_{hkl}), a measurement of elastic strain partitioning between phases, was calculated using Equation (1):

$$\varepsilon_{hkl} = \left(d^{\sigma}_{hkl} - d^{0}_{hkl} \right) / d^{0}_{hkl} \qquad (1)$$

where d^{σ}_{hkl} is the interplanar spacing for a set of lattice plane normals, parallel to the diffraction vector measured under an applied stress, and d^{0}_{hkl} is the interplanar spacing in the stress-free condition. Stress-free interplanar spacings were measured by extrapolating the linear-elastic region of the true stress-lattice strain curve to a zero-load condition, which should have removed any effects of sample/fixture unbending. The reported elastic lattice strain values were all taken from the axial detector bank.

3. Results

The initial phase fractions for the specimens with the DS and DS-T heat treatments are shown in Table 1 along with the heat treatment parameters applied to each sample. The austenite phase fractions were measured during the initial diffraction measurement in the axial direction at a stress of 40 MPa based upon full Rietveld full pattern refinements. The athermal martensite phase fractions were estimated through the application of the lever rule to dilatometric data in a method consistent with Kang et al. [23]. Ferrite was assumed to compose the remaining balance of the microstructure, as no other phases were identified in the diffraction patterns of the unstrained samples.

Table 1. Heat treatment parameters and resulting phases fractions.

Sample ID	Heat Treatment Heat Treatment Parameters	γ vol %	α' vol %	α vol %
DS	800 °C, 30 s	28	$60\left(\alpha'_F\right)$	12
DS-T	800 °C, 30 s + 450 °C, 300 s	29	$60\left(\alpha'_T\right)$	11

Engineering stress-strain curves for the DS and DS-T conditions are shown in Figure 3a,c, respectively. The DS condition exhibited continuous yielding, followed by a region with a high rate of work hardening, and no post-uniform elongation. This resulted in an ultimate tensile strength of 1710 MPa and total elongation of 12.8% for the DS sample tested in uniaxial tension. The corresponding quasi-static and interrupted neutron diffraction true stress-true strain tensile curves are shown in Figure 3b,d. The neutron diffraction curve, which shows the true stress-true strain location of each incremental step where diffraction data were collected, had reduced true stress values as compared to the corresponding true strain value on the quasi-static curve. Additionally, the total elongation of the neutron diffraction sample was reduced as compared to the sample tested under quasi-static conditions in uniaxial tension. Overall, the tensile curve characteristics of interest, including the high rate of work hardening and continuous yielding behavior, were maintained in the sample that was incrementally deformed during neutron diffraction.

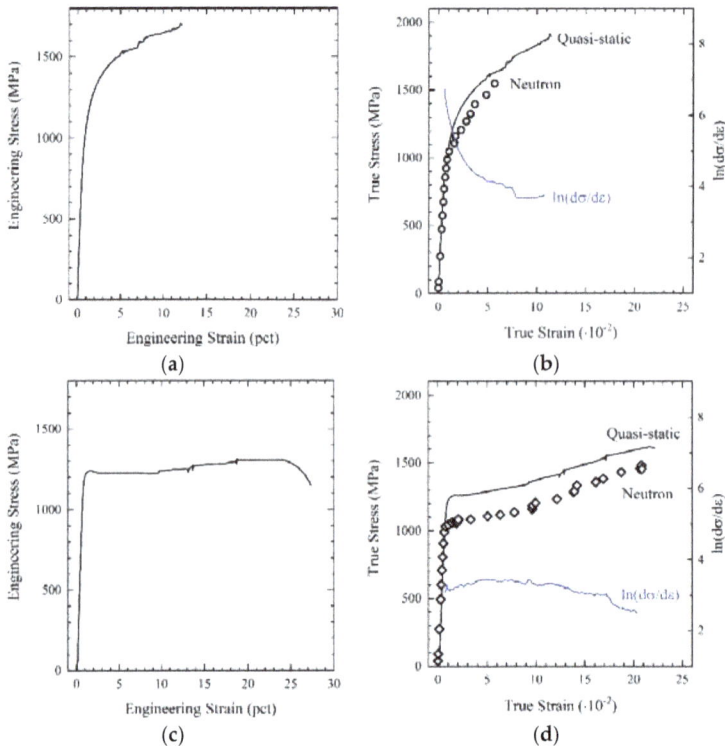

Figure 3. Room temperature engineering and true stress-strain curves for the (**a,b**) double soaked (DS) and (**c,d**) double soaked and tempered (DS-T) samples tested in both quasi-static and in situ neutron diffraction testing conditions along with the instantaneous work hardening rate, $\ln(d\sigma/d\varepsilon)$.

Representative indexed diffraction patterns, measured at a range of true stress values in the axial orientation for the DS and DS-T heat treatment conditions, are shown in Figure 4a,b respectively. The phase fractions of only two phases—BCC ferrite/martensite and FCC austenite—could be quantified. Small diffraction peaks from the {101} orientation of hexagonal close packed (HCP) epsilon martensite (ε) were evident in the diffraction patterns for both DS and DS-T samples at increased stress levels, although the intensity was not sufficient to allow the epsilon martensite volume fractions to be estimated.

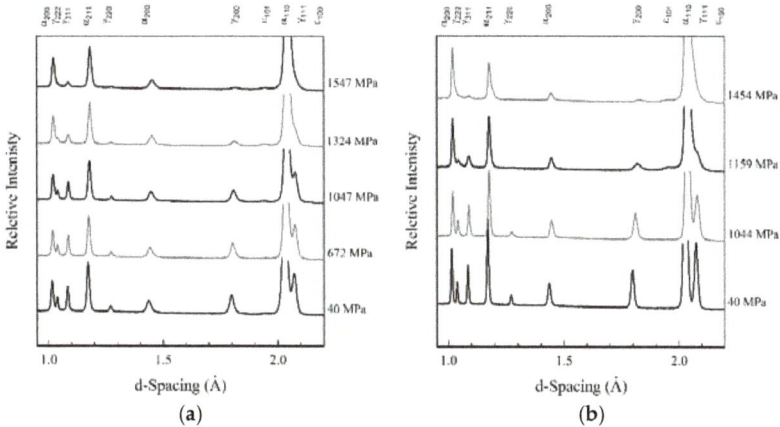

Figure 4. Representative in situ neutron diffraction data during uniaxial deformation for the (**a**) double soaked and (**b**) double soaked plus tempered heat treatment conditions. True stresses at each diffraction condition are indicated.

Using the diffraction patterns collected during each deformation step, the austenite fractions were calculated for both the DS and DS-T samples. The calculated austenite fractions are plotted as a function of true stress in Figure 5, and, in all cases, the calculated uncertainty was less than ±0.12 vol %. In the unstrained condition, the DS sample, shown in Figure 5a, contained 28 vol % austenite. Upon macroscopic yielding, the austenite transformation occurred at a nearly constant rate with increasing stress. The final austenite volume fraction measured before fracture was 7.5 vol %. In the DS-T condition, the strain-induced transformation of austenite occurred primarily during the yield point elongation (YPE), during which approximately 12 vol % of austenite transformed. In the region of limited work hardening that follows YPE, the strain-induced transformation of austenite slows significantly with only 4 vol % austenite transformed before sample failure. In both the DS and DS-T heat treatment conditions retained austenite volume fractions decreased by less than 1 vol % before macroscopic yielding was observed. This may indicate that a small fraction of stress-assisted austenite transformation occurred. Due to the small volume fraction (potentially) transformed by the stress-assisted mechanism, the resulting impact on tensile properties was likely insignificant.

Figure 6 shows the evolution of elastic lattice strains for the {211} orientation of the BCC phases (α/α') and the {311} orientation of FCC austenite as a function of macroscopic true strain measured in the axial orientation. By applying an isostrain assumption ($\sigma = E \cdot \varepsilon$) the elastic lattice strain values shown in Figure 6 can be assumed to be proportional to an equivalent stress [24]. The elastic lattice strain values calculated for the DS condition are shown in Figure 6a, and the values for the DS-T condition are shown in Figure 6c along with the austenite volume fractions calculated at each deformation step. In Figure 6b,d, the elastic lattice strains and austenite volume fractions measured at low macroscopic true strain values are magnified for both the DS and DS-T sample conditions. In all cases, error bars were omitted, as they were smaller than the data points. Additionally, the calculated linear fit for the linear elastic region of the γ and α elastic lattice strains are plotted, as they are helpful in identifying the order in which individual phases yield. In all plots of elastic lattice strain, the {211} orientation of the BCC phases (α/α') and the {311} orientation of austenite are displayed, as they have been shown to have the most linear lattice strain response, thus making them the best orientations to use when characterizing the macroscopic stresses or strains for a sample tested in uniaxial tension with limited texture [25].

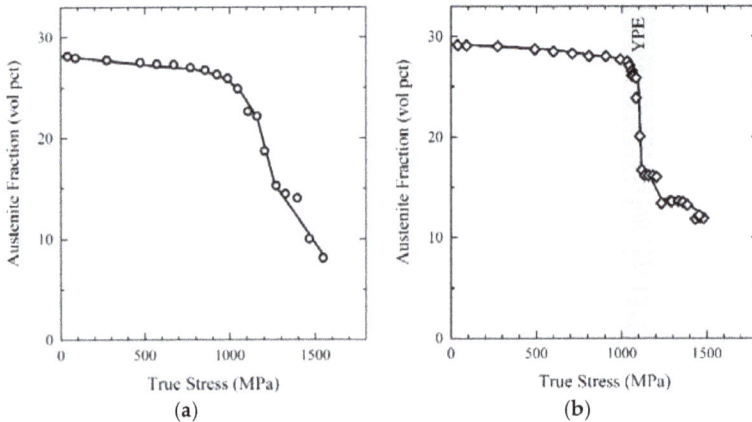

Figure 5. Austenite volume fractions as a function of macroscopic true stress for the (**a**) double soaked (DS) and (**b**) double soaked and tempered (DS-T) heat treatment conditions.

In the DS condition, three distinct deformation stages can be identified from the elastic lattice strain data, which is plotted as a function of macroscopic true strain in Figure 6b. In stage I, linear elastic (reversible) deformation occurred in both α/α' and γ. Stage II was initiated at a macroscopic true strain value of 0.0044 and a corresponding estimated macroscopic true stress of 626 MPa when $\gamma_{\{311\}}$ deviated from linear elastic loading, as shown in Figure 6b. This deviation from reversible elastic loading indicates the initiation of plastic deformation in γ. Based upon the corresponding austenite volume fractions plotted in Figure 6b, the transformation of significant volume fractions of austenite has not yet occurred. Finally, stage III was initiated when yielding occurred in the α/α' matrix at a macroscopic true strain of 0.0056 and corresponding macroscopic true stress of 768 MPa. At the onset of stage III, significant volume fractions of austenite were retained, although the strain-induced transformation of austenite began as indicated by the decrease in austenite fractions with increasing macroscopic true strain, clearly seen in Figure 6a. With an increasing macroscopic true strain, the elastic lattice strain (which is directly proportional to stress) in the BCC phases continued to increase until the neutron diffraction sample fractured at a macroscopic true strain of 0.056.

In the elastic lattice strain data for the DS-T sample, shown in Figure 6c, three stages of deformation can also be identified. Following the reversible elastic deformation of α/α' and γ in stage I, stage II was initiated by yielding in austenite at a macroscopic true strain of 0.0034 and corresponding estimated macroscopic true stress of 624 MPa, as indicated in Figure 6d. Following the initiation of stage II, stress in both γ and α/α' continued to increase as shown by the increasing elastic lattice strains in both phases. During stage II, austenite fractions remained nearly constant, indicating that the strain-induced transformation of austenite was not yet occurring. The yielding of the α/α' phase, at a macroscopic true strain of 0.0041 and corresponding macroscopic true stress of 718 MPa, initiated stage III. During stage III, stress levels in austenite were relatively constant and austenite fractions began to decrease. The most rapid rate of austenite transformation occurred during the YPE, as identified in the macroscopic true-stress, true-strain curve, after which the transformation rate slowed significantly. The corresponding elastic lattice strain for the α/α' phase increased slightly throughout stage III with increasing macroscopic true strain, indicating increasing stress in the α/α' phase.

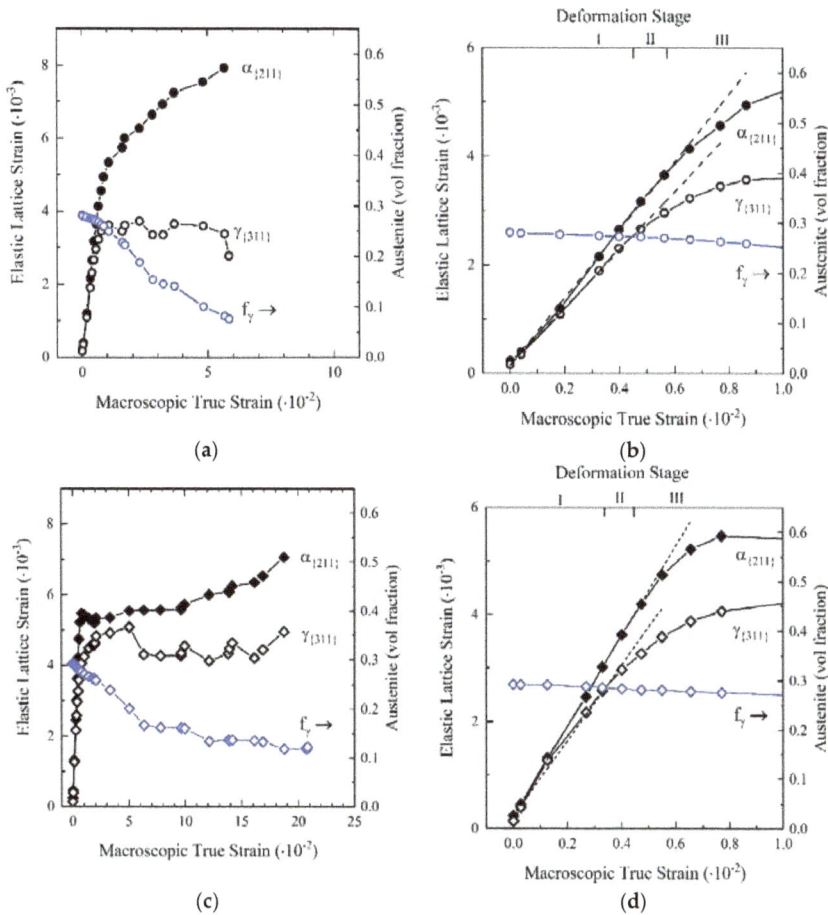

Figure 6. Elastic lattice strain as a function of macroscopic true strain for the {211} orientation of the BCC phases (α/α′) and {311} orientation of austenite for the (**a**,**b**) DS and (**c**,**d**) DS-T heat treatment conditions.

4. Discussion

The heat treatments presented in this work were selected to generate microstructures containing the same amount of retained austenite with similar strength while varying the strength and characteristics of the matrix phase. These modifications were achieved through the application of the double soaking treatment with an added tempering step. This modified the strength and mobile dislocation density along with other microstructural characteristics of the athermal martensite, which composed the bulk for the matrix phase. These differences were seen to significantly influence true stress-true strain curves (Figure 3)—the yielding behavior and evolution of the incremental work hardening rate during deformation, in particular.

4.1. Role of Martensite in Yielding Behavior

The results of this study suggest that the yielding behavior of medium manganese steels is highly dependent upon the characteristics of the matrix phase. Significant changes in yielding behavior were observed between the sample containing as-quenched athermal martensite (DS heat treatment) and the sample containing tempered athermal martensite (DS-T heat treatment). Specifically, during the deformation of the initial double soaked microstructure, which contained large fractions of as-quenched athermal martensite, continuous yielding was observed. With the addition of a tempering treatment, which was likely accompanied by a reduction in dislocation density and precipitation of carbides [26–29], the discontinuous yielding traditionally seen in medium manganese steels was observed in the macroscopic true-stress, true-strain curve. Additionally, the load drop and YPE-like behavior observed in the $\alpha_{(211)}$ elastic lattice strain data shown in Figure 6c would be consistent with the pinning of mobile dislocations by carbon segregation and carbide precipitation in martensite during tempering, as proposed by Krauss and Swarr [27].

The results of this study support the findings of Steineder et al. and De Cooman et al., both of whom suggested that the introduction of a large volume fraction of mobile dislocations due to the formation of athermal martensite was effective at preventing discontinuous yielding in medium manganese steels [14,30]. The prevention of discontinuous yielding through an increase in mobile dislocation density has also been shown effective in studies which utilized pre-straining to increase mobile dislocation density [31]. Other proposed mechanisms for the elimination of discontinuous yielding in medium manganese steels, such as the modification of austenite strength and volume fractions or the sequential deformation of individual phases during yielding, were not found to influence the yielding behavior of the medium manganese steel used in this study, as these factors were held constant between the two sample conditions examined [13,32,33].

4.2. Role of Martensite during Plastic Deformation

The role of martensite during plastic deformation is evident through a comparison of the macroscopic tensile properties shown in Figure 3 and the elastic lattice strains shown in Figure 6. The DS condition, with a matrix of primarily as-quenched athermal martensite, maintains a high rate of work hardening following macroscopic yielding. Through an examination of the elastic lattice strain data for the corresponding heat treatment, it is evident that the rapidly increasing elastic lattice strains in the α/α' phase in stage III due to work hardening and the dynamic replacement of austenite with martensite is responsible for the high work hardening rate seen in the macroscopic true stress-true strain curve. The corresponding elastic lattice strains for the γ phase remain nearly constant during stage III, as the stress in austenite is limited by the transformation criteria. Overall, the elastic lattice strain data for the DS condition indicate that the high rate of work hardening can be attributed to work hardening in the α/α' matrix and the strain-induced transformation of austenite to martensite.

In the DS-T condition, where the matrix was primarily composed of tempered athermal martensite, the macroscopic true stress-true strain curve shows that the ultimate tensile strength and work hardening rate decreased as compared to the DS condition. The elastic lattice strain data and austenite fractions shown as a function of true strain in Figure 6 are helpful in understanding the contribution of martensite to deformation behavior following yielding. The similar increase in stress partitioned to the α/α' phase during stage III can be attributed to two factors. The first factor to consider is the limited work hardening response of tempered martensite, which composes the majority phase in the matrix. Studies of low-carbon martensite tempered at intermediate temperatures (300–450 °C) for similar times have shown that martensite tempered in this regime generally produces an extremely limited work hardening response during deformation [27–29,34,35]. This low rate of work hardening has been attributed to the replacement of the cellular dislocation structure found in as-quenched martensite, with a non-cellular distribution of dislocations which by-pass carbides formed during tempering, leading to a limited dislocation interaction and accumulation [27,34]. Secondly, the reduced rate of austenite transformation following the macroscopic YPE limits the fraction of martensite added to the matrix. This reduces the overall strength of the material at high macroscopic strains, as lower strength austenite has not been transformed to higher strength martensite. Overall, the reduced work hardening response of the tempered athermal martensite composing the majority of the matrix and the reduced rate of austenite transformation following YPE lead to the overall reduced work hardening rate seen in the macroscopic true stress-true strain curve.

Metals **2019**, *9*, 761

5. Conclusions

The role of martensite characteristics and strength in a 0.14C–7.14Mn steel was explored using a novel double soaking heat treatment. The double soaking process utilizes a two-step heat treatment where the primary soak generates a significant volume fraction of retained austenite stabilized by carbon and manganese partitioning, and the secondary soak substitutes athermal martensite for ferrite in the matrix. In this study, the double soaking heat treatment was applied to generate microstructures with equivalent volume fractions of austenite and martensite while varying martensite strength. Using in situ neutron diffraction to study the role of athermal martensite in yielding and plastic deformation, it was shown that:

1. The inclusion of large volume fractions of as-quenched athermal martensite in the microstructure of medium manganese steels promotes continuous yielding in both the FCC and BCC/BCT phases and in the overall macroscopic tensile curve. Following tempering, discontinuous yielding was found to occur in both the BCC/BCT phase and in the overall macroscopic tensile curve. The shift from continuous to discontinuous yielding is believed to be linked to a reduction in mobile dislocation density in the BCC/BCT phase as a result of the tempering heat treatment.
2. The work hardening rate of a medium-manganese steel is highly dependent upon the properties of the matrix. An initial matrix composed primarily of as-quenched athermal martensite is effective in producing a sustained high rate of work hardening, primarily due to work hardening in the martensitic phase and the dynamic strain-induced transformation of austenite to martensite.
3. The substitution of ferrite with martensite (as-quenched or tempered) in the initial microstructure of a medium manganese steel improves the strength–ductility product of the material.

Author Contributions: Conceptualization, formal analysis, and writing—original draft preparation A.G.; conceptualization, methodology, and writing—review and editing, E.D.M., J.G.S., and P.J.G.; methodology, formal analysis, resources, and data curation, C.L., D.W.B., B.C.

Funding: The sponsors of the Advanced Steel Processing and Products Research Center (ASPPRC) at the Colorado School of Mines are gratefully acknowledged. This work benefited from the use of the Lujan Neutron Scattering Center at LANSCE, LANL. Los Alamos National Laboratory is operated by Triad National Security, LLC, for the National Nuclear Security Administration of U.S. Department of Energy (Contract No. 89233218CNA000001).

Conflicts of Interest: The authors declare no conflict of interest. The funders had no role in the design of the study; in the collection, analyses, or interpretation of data; in the writing of the manuscript, or in the decision to publish the results.

References

1. Matlock, D.; Speer, J.G. Design Considerations for the Next Generation of Advanced High Strength Sheet Steels. In Proceedings of the Third International Conference on Advanced Structural Steels, Gyeongju, Korea, 22–24 August 2006; pp. 774–781.
2. De Moor, E.; Gibbs, P.J.; Speer, J.G.; Matlock, D.K. Strategies for third-generation advanced high-strength steel development. *AIST Trans.* **2010**, *7*, 133–144.
3. Matlock, D.K.; Speer, J.G. Third generation of AHSS: Microstructure design concepts. In *Microstructure and Texture in Steels*; Springer: London, UK, 2009; pp. 185–205. ISBN 184882453X.
4. Miller, R.L. Ultrafine-grained microstructures and mechanical properties of alloy steels. *Metall. Trans.* **1972**, *3*, 905–912. [CrossRef]
5. Merwin, M.J. Low-carbon manganese TRIP steels. *Mater. Sci. Forum* **2007**, *539–543*, 4327–4332. [CrossRef]
6. Gibbs, P.J.; De Moor, E.; Merwin, M.J.; Clausen, B.; Speer, J.G.; Matlock, D.K. Austenite stability effects on tensile behavior of manganese-enriched-austenite transformation-induced plasticity steel. *Metall. Mater. Trans. A* **2011**, *42*, 3691–3702. [CrossRef]
7. Patel, J.R.; Cohen, M. Criterion for the action of applied stress in the martensitic transformation. *Acta Metall.* **1953**, *1*, 531–538. [CrossRef]
8. Olson, G.B.; Azrin, M. Transformation Behavior of TRIP steels. *Metall. Mater. Trans. A* **1978**, *9A*, 713–721. [CrossRef]

9. Olson, G.B.; Cohen, M. Stress assisted isothermal martensitic transformation: Application to TRIP steels. *Metall. Trans. A* **1982**, *13A*, 1907–1914. [CrossRef]
10. Stringfellow, R.G.; Parks, D.M.; Olson, G.B. A constitutive model for transformation plasticity accompanying strain-induced martensitic transformations in metastable austenitic steels. *Acta Metall. Mater.* **1992**, *40*, 1703–1716. [CrossRef]
11. Haidemenopoulos, G.N.; Aravas, N.; Bellas, I. Kinetics of strain-induced transformation of dispersed austenite in low-alloy TRIP steels. *Mater. Sci. Eng. A* **2014**, *615*, 416–423. [CrossRef]
12. Steineder, K.; Schneider, R.; Krizan, D.; Béal, C.; Sommitsch, C. Comparative investigation of phase transformation behavior as a function of annealing temperature and cooling rate of two medium-Mn steels. *Steel Res. Int.* **2015**, *86*, 1179–1186. [CrossRef]
13. Gibbs, P.J.; De Cooman, B.C.; Brown, D.W.; Clausen, B.; Schroth, J.G.; Merwin, M.J.; Matlock, D.K. Strain partitioning in ultra-fine grained medium-manganese transformation induced plasticity steel. *Mater. Sci. Eng. A* **2014**, *609*, 323–333. [CrossRef]
14. Steineder, K.; Krizan, D.; Schneider, R.; Béal, C.; Sommitsch, C. On the microstructural characteristics influencing the yielding behavior of ultra-fine grained medium-Mn steels. *Acta Mater.* **2017**, *139*, 39–50. [CrossRef]
15. De Moor, E.; Speer, J.G.; Matlock, D.K. Heat Treating Opportunities for Medium Manganese Steels. In Proceedings of the First International Conference on Automobile Steel & the 3rd International Conference on High Manganese Steels, Chengdu, China, 15–18 Novermber 2016; Luo, H., Ed.; Metallurgical Industry Press: Beijing, China, 2016; pp. 182–185.
16. Glover, A.G.; Speer, J.G.; De Moor, E. Double Soaking of a 0.14C-7.14Mn Steel. In Proceedings of the International Symposium on New Developments in Advanced High-Strength Sheet Steels, Keystone, CO, USA, 30 May–2 June 2017.
17. ASTM E8 / E8M-13. *Standard Test Method for Tension Testing of Metallic Materials*; ASTM International: West Conshohocken, PA, USA, 2015; p. 29.
18. Sutton, M.A.; McNeill, S.R.; Helm, J.D.; Chao, Y.J. Advances in Two-Dimensional and Three-Dimensional Computer Vision. In *Photomechanics, Topics in Applied Physics*; Rastogi, P.K., Ed.; Springer: Berlin, Germany, 2000.
19. Solutions, C. *Vic 3D v7, Reference Manual*; Correlated Solutions: Columbia, SC, USA, 2018.
20. Bourke, M.A.M.; Dunand, D.C.; Ustundag, E. SMARTS—a spectrometer for strain measurement in engineering. *Appl. Phys. A* **2002**, *1709*, 1707–1709. [CrossRef]
21. Huang, J.; Poole, W.J.; Militzer, M. Austenite Formation during Intercritical Annealing. *Metall. Mater. Trans. A* **2004**, *35*, 3363–3375. [CrossRef]
22. Rietveld, H.M. A profile refinement method for nuclear and magnetic structures. *J. Appl. Crystallogr.* **1969**, *2*, 65–71. [CrossRef]
23. Kang, J.Y.; Park, S.J.; Suh, D.W.; Han, H.N. Estimation of phase fraction in dual phase steel using microscopic characterizations and dilatometric analysis. *Mater. Charact.* **2013**, *84*, 205–215. [CrossRef]
24. Bao, G.; Hutchinson, J.W.; McMeeking, R.M. The flow stress of dual-phase, non-hardening solids. *Mech. Mater.* **1991**, *12*, 85–94. [CrossRef]
25. Clausen, B.; Lorentzen, T.; Leffers, T. Self consistent modelling of the plastic deformatiton of fcc polycrystals and its implicaitons for diffraction measurements of internal stresses. *Acta Mater.* **1998**, *46*, 3087–3098. [CrossRef]
26. Speich, G.R.; Leslie, W.C. Tempering of Steel. *Metall. Mater. Trans. A* **1972**, *3*, 1043–1054. [CrossRef]
27. Swarr, T.; Krauss, G. The effect of structure on the deformation of as-quenched and tempered martensite in an Fe-0.2 pct C alloy. *Metall. Mater. Trans. A* **1976**, *7*, 41–47. [CrossRef]
28. Massardier, V.; Goune, M.; Fabregue, D.; Selouane, A.; Douillard, T.; Bouaziz, O. Evolution of microstructure and strength during the ultra-fast tempering of Fe-Mn-C martensitic steels. *J. Mater. Sci.* **2014**, *49*, 7782–7796. [CrossRef]
29. Krauss, G. Tempering of lath martensite in low and medium carbon steels: Assessment and challenges. *Steel Res. Int.* **2017**, *88*, 1–18. [CrossRef]
30. De Cooman, B.C.; Gibbs, P.J.; Lee, S.; Matlock, D.K. Transmission electron microscopy analysis of yielding in ultrafine-grained medium Mn transformation-induced plasticity steel. *Metall. Mater. Trans. A* **2013**, *44*, 2563–2572. [CrossRef]

31. Li, Z.C.; Misra, R.D.K.; Ding, H.; Li, H.P.; Cai, Z.H. The significant impact of pre-strain on the structure-mechanical properties relationship in cold-rolled medium manganese TRIP steel. *Mater. Sci. Eng. A* **2018**, *712*, 206–213. [CrossRef]
32. Luo, H.; Dong, H.; Huang, M. Effect of intercritical annealing on the luders strains of medium Mn transformation-inducted plasticity steels. *Mater. Des.* **2015**, *83*, 42–48. [CrossRef]
33. Emadoddin, E.; Akbarzadeh, A.; Daneshi, G.H. Correlation betweeen luder strain and retained austenite in TRIP-assisted cold rolled sheet steels. *Mater. Sci. Eng. A* **2007**, *447*, 174–179. [CrossRef]
34. Materkowski, J.; Krauss, G. Tempered martensite embrittlement in SAE 4340 steel. *Metall. Trans. A* **1979**, *10*, 1643–1651. [CrossRef]
35. Winter, P.L.; Woodward, R.L. Effect of tempering temperature on the work-hardening rate of five HSLA steels. *Metall. Trans. A* **1986**, *17 A*, 307–313. [CrossRef]

metals

MDPI

Article

Development of a Cr-Ni-V-N Medium Manganese Steel with Balanced Mechanical and Corrosion Properties

Tarek Allam [1,2,*], Xiaofei Guo [1], Simon Sevsek [1], Marta Lipińska-Chwałek [3,4],
Atef Hamada [5], Essam Ahmed [2] and Wolfgang Bleck [1]

[1]   Steel Institute (IEHK), RWTH Aachen University, 52056 Aachen, Germany;
      xiaofei.guo@iehk.rwth-aachen.de (X.G.); simon.sevsek@iehk.rwth-aachen.de (S.S.);
      bleck@iehk.rwth-aachen.de (W.B.)
[2]   Metallurgical and Materials Engineering Department, Suez University, 43528 Suez, Egypt;
      essam.ahmed@suezuniv.edu.eg
[3]   Central Facility for Electron Microscopy, RWTH Aachen University, 52074 Aachen, Germany;
      m.lipinska@fz-juelich.de
[4]   Forschungszentrum Jülich GmbH, Ernst Ruska-Centre for Microscopy and Spectroscopy with
      Electrons (ER-C) and Peter Grünberg Institute, Microstructure Research (PGI-5), D-52425 Jülich, Germany
[5]   Egypt-Japan University of Science and Technology E-JUST, New Borg El-Arab City, 21934 Alexandria, Egypt;
      atef.hamada@ejust.edu.eg
*   Correspondence: tarek.allam@iehk.rwth-aachen.de

Received: 15 May 2019; Accepted: 19 June 2019; Published: 22 June 2019

Abstract: A novel medium manganese (MMn) steel with additions of Cr (18%), Ni (5%), V (1%), and N (0.3%) was developed in order to provide an enhanced corrosion resistance along with a superior strength–ductility balance. The laboratory melted ingots were hot rolled, cold rolled, and finally annealed at 1000 °C for 3 min. The recrystallized single-phase austenitic microstructure consisted of ultrafine grains (~1.3 μm) with a substantial amount of Cr- and V-based precipitates in a bimodal particle size distribution (100–400 nm and <20 nm). The properties of the newly developed austenitic MMn steel X20CrNiMnVN18-5-10 were compared with the standard austenitic stainless steel X5CrNi18-8 and with the austenitic twinning-induced plasticity (TWIP) steel X60MnAl17-1. With a total elongation of 45%, the MMn steel showed an increase in yield strength by 300 MPa and in tensile strength by 150 MPa in comparison to both benchmark steels. No deformation twins were observed even after fracture for the MMn steel, which emphasizes the role of the grain size and precipitation-induced change in the austenite stability in controlling the deformation mechanism. The potentio-dynamic polarization measurements in 5% NaCl revealed a very low current density value of 7.2×10^{-4} mA/cm^2 compared to that of TWIP steel X60MnAl17-1 of 8.2×10^{-3} mA/cm^2, but it was relatively higher than that of stainless steel X5CrNi18-8 of 2.0×10^{-4} mA/cm^2. This work demonstrates that the enhanced mechanical properties of the developed MMn steel are tailored by maintaining an ultrafine grain microstructure with a significant amount of nanoprecipitates, while the high corrosion resistance in 5% NaCl solution is attributed to the high Cr and N contents as well as to the ultrafine grain size.

Keywords: MMn steel X20CrNiMnVN18-5-10; V alloying; corrosion resistance; precipitations; ultrafine grains

---

1. Introduction

High manganese steels (HMnSs) with the twinning-induced plasticity (TWIP) effect exhibit an excellent combination of high ultimate tensile strength, large uniform elongation, and high work hardening capacity, which makes them a competitive candidate for a wide range of uses in automotive industry, liquefied natural gas (LNG)-shipbuilding, and the oil and gas industry [1,2]. However, some related material and technological limitations restrict their industrial applications. A processing route consisting of cold rolling followed by a recovery-annealing heat treatment, during which the dislocation density is reduced while previously introduced deformation twins are thermally stable, has been proposed as a solution for the low yield strength of TWIP steels [3,4]. Moreover, the use of microalloying elements has been approached under consideration of the stacking fault energy (SFE), which determines the austenite stability and controls the activation of the secondary deformation mechanism, such as the TWIP effect or deformation induced phase transformation (TRIP—transformation-induced plasticity) in addition to dislocation slip [5]. Due to the high dissolution rate of Mn, TWIP steels exhibit a relatively poor wet corrosion resistance [6,7]. In this regard, different single or combined additions of Cr, Cu, Si, and Al were investigated to enhance the corrosion behavior [8–10]. In a different way to enhance the corrosion resistance and maintain the attractive mechanical properties of TWIP steels, the so-called Fe-Mn-Cr-N TWIP steels have been proposed [11] as a cost-effective substitute for both conventional stainless steel X5CrNi18-8 and HMnS [12]. In that respect, the change in the austenite stability due to N- and Cr- addition has been investigated to design an Fe-Cr-Mn-N system with a stable austenitic microstructure before and after deformation, which can satisfy the corrosion and mechanical requirements of industrial applications [13]. Another barrier to the use of HMnS in different applications is their susceptibility to delayed fracture due to hydrogen embrittlement (HE). Among the various methodologies that have been suggested to alleviate HE in HMnS are Al-addition [14–16], grain refinement [17,18], or the introduction of hydrogen traps, e.g., k-carbides [19]. However, the role of precipitates in improving hydrogen embrittlement is still a matter of discussion [20–22]. Recently, medium manganese steels (MMnSs) with a duplex microstructure have received significant attention due to their excellent mechanical properties and reduced production costs compared to HMnS [23–27].

In light of progress made towards attaining enhanced mechanical and corrosion behavior for TWIP steels and avoiding their material and technological limitations, we designed a new alloying concept with a medium Mn content. This concept is based on the demonstrated excellent corrosion resistance of conventional Fe-Cr-Ni stainless steels and the unique mechanical behavior of TWIP steels, taking into account the austenite stability by adjusting SFE within the TWIP range [28]. In addition, a high V content (1 wt.%) was considered to introduce a significant amount of precipitates to control the grain size. In addition, the impacts of precipitation-strengthening on corrosion resistance, austenite stability, and the activation of additional deformation mechanisms were investigated.

2. Materials and Methods

The alloying concept was designed to develop an austenitic-microstructure MMn steel that can exhibit a high corrosion resistance and a superior strength–ductility balance. Accordingly, C, Mn, Ni, and N were optimized to stabilize the austenite. Cr was added to achieve the envisaged high corrosion resistance, besides the role of N in increasing the pitting resistance. V alloying is considered to increase the mechanical properties by precipitation strengthening and grain refining mechanisms. The chemical composition of the developed MMn steel X20CrNiMnVN18-5-10 is listed in Table 1. A 50 kg heat was ingot-casted and homogenized at 1200 °C for 4 h. Subsequently, the homogenized block was subjected to hot and cold rolling processes to a final thickness of 1.5 mm with a 62.5% cold reduction. The specimens required for different investigations were manufactured before the microstructure was adjusted in a recrystallization annealing treatment at 1000 °C for 3 min in a salt bath. The annealing treatment was designed based on the thermodynamic calculations and precipitation kinetics using Thermo-Calc Software TCFE Steels/Fe-alloys version 9 and MatCalc version 6.02, respectively.

Table 1. Chemical composition in wt.% of the developed medium manganese (MMn) steel as well as the applied benchmark steels.

Elements	C	Si	Mn	Al	Cr	Ni	V	N
MMn X20CrNiMnVN18-5-10	0.17	0.43	10.4	-	17.7	4.7	0.9	0.26
BenchmarkX5CrNi18-8	0.04	0.49	1.2	-	18.0	8.3	-	0.060
Benchmark X60MnAl17-1	0.60	0.06	16.8	1.1	0.6	0.2	0.049	0.008

The developed microstructure was investigated by means of a Zeiss Gemini scanning electron microscopy (SEM) (Carl Zeiss Microscopy GmbH, Jena, Germany) equipped with an energy dispersive X-ray spectrometer (EDS) (Oxford X-Max50, Oxford Instruments, Abingdon, UK) at acceleration voltages of 15 kV and working distances within 10 to 20 mm. The metallographic samples were prepared through the standard preparation route, including cutting and mechanical grinding up to grit 1200 followed by mechanical polishing using 3 μm and 1 μm diamond suspension on a Struers Abrapol-2 (Struers GmbH, Willich, Germany). V2A etching solution was applied to reveal the developed microstructure. The samples used for electron backscatter diffraction (EBSD) measurements were additionally electro-polished at room temperature for 20 s at 22 V using an electrolyte consisting of 700 mL ethanol, 100 mL butyl glycol, and 78 mL 60%-perchloric acid. The EBSD mappings were captured by the EDAX-TSL Hikari detector and analyzed by the OIM-Data Collection-V7.3 software (AMETEK-EDAX Inc., USA). EBSD scans were recorded with a step size of 150 nm applying an acceleration voltage of 15 to 20 kV and a probe current of approximately 30 nA. Evolution of the deformation mechanism was investigated using ex-situ EBSD measurements at the center of the parallel gauge length of tensile samples, which were pulled to different elongation strains. Post-processing of the EBSD measurements was conducted using the HKL Channel 5 software (version 5.12j, Oxford Instruments, Abingdon, UK) and the MATLAB-based MTEX toolbox [29,30] and included the removal of wild spikes and a careful noise reduction, which took at least 5 neighboring data points into account. The precipitation state after recrystallization was identified on electron transparent specimens prepared with the carbon extraction replica method. Conventional bright field imaging of the extracted precipitates was performed using the FEI Tecnai F20 transmission electron microscope (TEM) [31] operated at 200 kV. For detailed characterization of the size distribution and chemical composition of the precipitates, a high angle annular dark field (HAADF) and energy dispersive X-ray (EDX) spectral imaging in a high-resolution Cs probe corrected scanning transmission electron microscope (STEM) FEI Titan G2 80–200 STEM [32] operated at 200 kV was applied. The size of the precipitates was determined from HAADF STEM micrographs of an arbitrarily selected areas of extraction replicas (each 6 μm × 6 μm in size), analyzed with the aid of image processing software, Image J®.

The mechanical properties, namely yield strength (YS), ultimate tensile strength (UTS), and total elongation (El), were evaluated by means of quasi-static tensile testing. The tensile tests were conducted at room temperature with a strain rate of 10^{-3} s^{-1} using a universal tensile testing machine of type Z100 (Zwick/Roell GmbH & Co. KG, Germany) on the recrystallized annealed A30-specimens of 6 mm width and 30 mm parallel gauge length. The strain and force were measured by a videoXtens extensometer attached to the specimen and an Xforce load cell, respectively. Corrosion properties of the MMn steel were evaluated by dynamic polarization measurement in 5% NaCl with a controlled scanning rate of 0.2 mV/s. The corrosion potential, E_{corr}, and corrosion rate, i_{corr}, at the open current potential were determined on the polarization curves by extrapolating the Tafel plots. Mechanical and corrosion values of the developed austenitic MMn steel were compared with two benchmarks, i.e., austenitic stainless steel X5CrNi18-8 and austenitic TWIP steel X60MnAl17-1 (listed in Table 1). The benchmark steels were tested by the same methods used for testing the developed MMn steel in this study.

3. Results

3.1. Thermodynamic and Kinetics Calculations

The recrystallization annealing treatment of the developed MMn steel was designed based on both the thermodynamic calculations of equilibrium phases and the precipitation kinetics. Figure 1a shows the fraction of equilibrium phases based on the actual chemical composition of the MMn steel. It is clear that the austenite phase (FCC_A1) along with carbides ($M_{23}C_6$-type) and nitrides (MN-type) are the expected equilibrium phases by applying an annealing treatment at a temperature of 1000 °C. Figure 1b shows the kinetic of the possible precipitates formed due to annealing at 1000 °C. The kinetics results indicate the formation of the HCP_A3#2 phase (M_2N-type) that starts to saturate within 3 min, while the carbides ($M_{23}C_6$-type) do not tend to form at this temperature.

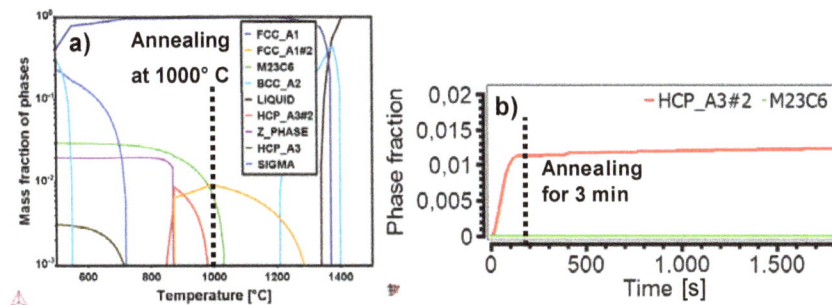

Figure 1. Thermodynamic calculations and precipitation kinetics. (**a**) Mass fraction of equilibrium phases diagram vs. temperature showing that austenite (FCC_A1), carbides ($M_{23}C_6$-type), and nitrides FCC_A1#2 (MN-type) are the equilibrium phases at 1000 °C (Thermo-Calc TCFE 9, MOBFE4). (**b**) Precipitation kinetics at 1000 °C showing the formation of HCP_A3#2 (M_2N-type) which comes up to saturation within 3 min (MatCalc version 6.02).

3.2. Microstructure

The thermomechanical processing of the MMn steel resulted in a fine-grained and recrystallized austenitic microstructure containing a considerable amount of precipitates as observed from the SEM image shown in Figure 2a. For the relatively large particles (<150 nm), the EDX-point analysis was applied to reveal their approximate alloying contents. The EDX-spectra represented in Figure 2b shows that the relatively large particles resemble Cr,Fe-rich carbides. The SEM observations and particle analyses are in accordance with the Thermo-Calc equilibrium calculations (Figure 1a). However, the expected precipitation kinetics conducted by MatCalc (Figure 1b) contradicts the formation of Cr-rich carbides during annealing at 1000 °C for 3 min. Instead, it supports the formation of V-rich nitrides. TEM and STEM/EDX investigations confirmed the presence of Cr,Fe-rich carbides (Cr:Fe atomic ratio of 4:1) and Cr,V-rich nitro-carbides (V:Cr atomic ratio of 1:2) as well as V,Cr-rich nitro-carbides (V:Cr atomic ratio of 5:1) with particle sizes of 100 to 150 nm, as indicated in the STEM HAADF image in Figure 2c. Moreover, V,Cr-rich nitro-carbides (V:Cr atomic ratio 5:1) were identified with a relatively larger particle size of ~200 to 400 nm. Additionally, very fine V,Cr-rich nitrides (V:Cr atomic ratio 5:1) with a particle size of <20 nm were observed (in accordance with Thermo-Calc calculations) as depicted in the inset of Figure 2c.

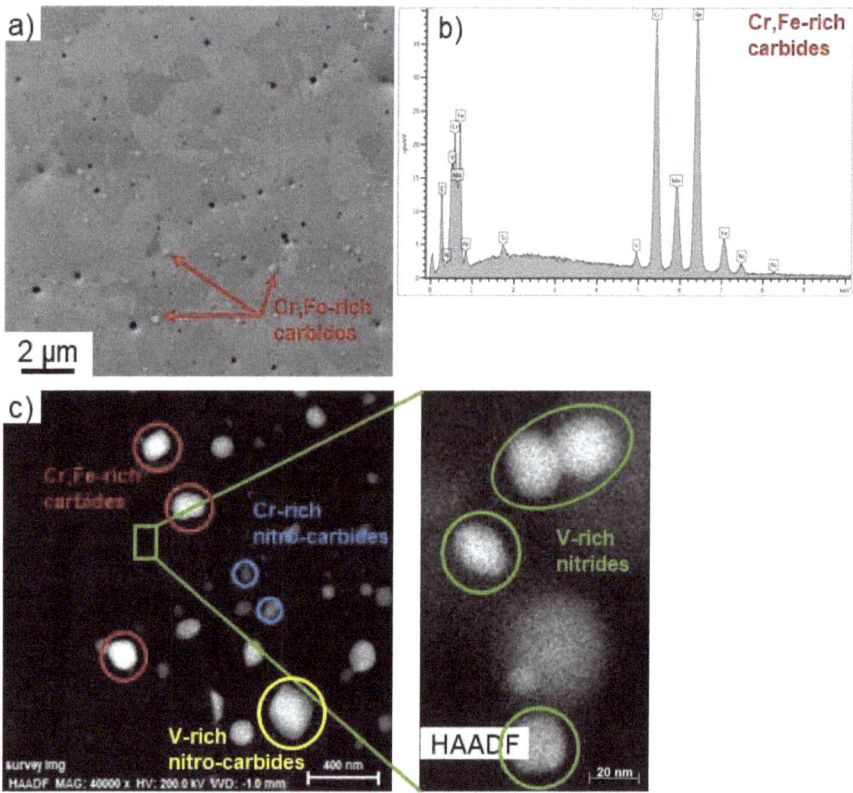

Figure 2. (**a**) SEM micrograph showing the adjusted microstructure after annealing of cold-rolled specimens at 1000 °C for 3 min. In a), the grain boundaries of recrystallized austenitic microstructure are not visible, however, a considerable amount of precipitates can be observed. (**b**) Representative EDX-point analyses for some visible coarse particles. (**c**) STEM HAADF images representing different precipitates with different sizes in the nano-scale. The precipitates were identified with high resolution STEM EDX spectra imaging.

Figure 3 represents the analyses of the EBSD measurements for the MMn steel with and without deformation. The orientation and phase maps in the as-annealed state (in Figure 3a) show that the applied annealing treatment at 1000 °C for 3 minutes resulted in an ultrafine recrystallized austenitic microstructure without a pronounced texture. The average grain size of the adjusted microstructure is around 1.3 µm according to the grain size distribution shown in Figure 3b. The developed austenite microstructure remains mechanically stable after straining to 20% elongation in a tensile irrupted test without any indication for either transformation-induced plasticity (TRIP) nor TWIP effects. However, the specimen pulled until the fracture (with a total elongation of ~48%) showed low amounts of less than 1% ε-martensite (in red) and approximately 2% α'-martensite (in green) at dispersed regions. Deformation twins were not observed at any deformation stage under the applied resolution during EBSD measurements. The corresponding phase maps recorded in the direction of tensile axis for both deformed states, i.e., at 20% elongation and at fracture strain, are shown Figure 3c,d, respectively.

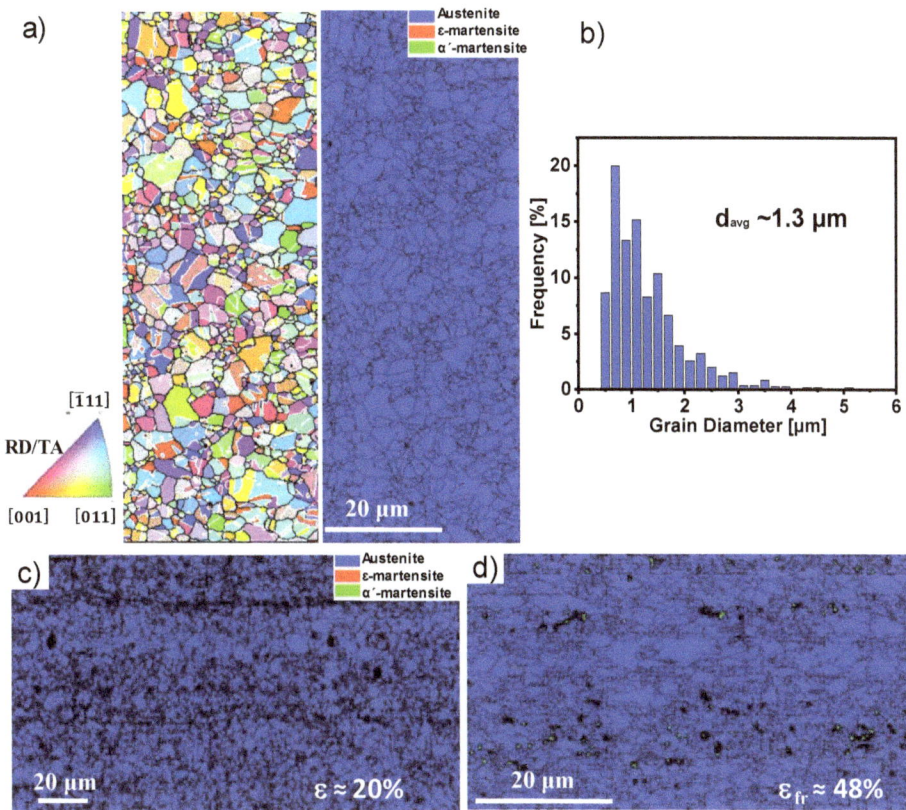

Figure 3. Electron backscatter mappings of the MMn steel annealed at 1000 °C for 3 min. (**a**) The orientation map and the spatial distributions of phases for the developed microstructure without deformation. RD and TA denote the rolling direction and tensile axis, respectively. (**b**) Grain size distribution and average grain size determined by the line-intercept method considering the annealing twins. (**c**) and (**d**) The spatial distribution of phases (austenite in blue, ε-martensite in red, α'-martensite in green, and black not indexed) formed at 20% elongation and at fracture strain, respectively.

3.3. Mechanical Properties

Quasi-static tensile properties, namely YS, UTS, and total El, for the MMn steel were compared to benchmark stainless steel X5CrNi18-8 and TWIP steel X60MnAl17-1 as depicted in Figure 4. According to the engineering stress–strain curves and the corresponding inset bar-chart in Figure 4a, the MMn steel shows higher YS and UTS values and a relatively lower total El that is still reasonably high for cold formability applications. The applied annealing treatment led to YS and UTS values of ~600 and 975 MPa, respectively, along with a total El of more than 45%. This demonstrates a jump in YS and UTS of approximately 300 and more than 150 MPa, respectively, compared to their counterparts of benchmark steels. Figure 4b represents the corresponding work hardening rate (WHR) calculated from the tensile test data. It is worthy to note that besides the higher YS and UTS values of the MMn steel, the WHR is also higher than both benchmark steels up to true strain values of ~0.16 and 0.25, respectively. The MMn steel exhibits an initial WHR (measured at 0.02) of ~3.4 GPa, which is higher than those of other benchmark steels. This initial high WHR decreases gradually and reached the same WHR of X5CrNi18-8 and X60MnAl17-1 benchmark steels at true stress values of ~1.1 and 1.2 GPa, respectively. Moreover, the WHR decreases at a slightly lower rate after around >0.18 true strain (>20% elongation) as can be realized from its slope illustrated by black dotted lines.

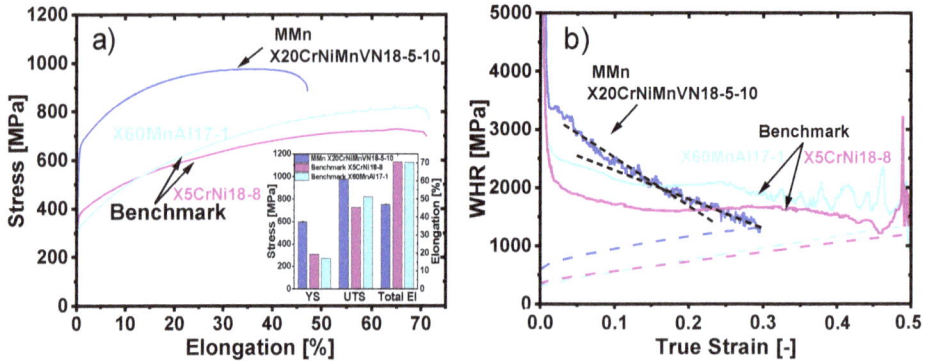

Figure 4. Tensile properties of the MMn steel X20CrNiMnVN18-5-10 in comparison to the benchmark stainless steel X5CrNi18-8 and TWIP steel X60MnAl17-1. (**a**) Engineering stress–strain curves with a bar chart showing the achievable strength and ductility levels. (**b**) The work hardening rate (WHR) curves.

3.4. Corrosion Properties

The corrosion properties of MMn steel were evaluated and compared to the benchmark steels. Figure 5a displays the results of the dynamic polarization measurements, which were carried out to evaluate the corrosion properties in terms of the open current potential (E_{corr}) and corrosion current density (i_{corr}) of the MMn steel compared to the benchmark steels. Obviously, the MMn steel exhibits a distinctively higher positive E_{corr} value of −299 mV_{SCE} than that observed for TWIP steel X60MnAl17-1 of −658 mV_{SCE}, while stainless steel X5CrNi18-8 shows the highest positive E_{corr} value of −144 mV_{SC}. Moreover, the i_{corr} (determined by Tafel lines calculations) shows a very low value of 7.2×10^{-4} mA/cm^2 compared to TWIP steel X60MnAl17-1 with a value of 8.2×10^{-3} mA/cm^2, but it is slightly higher than that of stainless steel X5CrNi18-8 with a value of 2.0×10^{-4} mA/cm^2. The bar chart in Figure 5b points out the corrosion current ratio (CCR) of each tested grade with respect to the benchmark stainless steel X5CrNi18-8. The calculated CCR values demonstrate that the corrosion behavior of MMn steel is superior to the TWIP steel X60MnAl17-1, while it is lower but still could be compared to the stainless steel X5CrNi18-8.

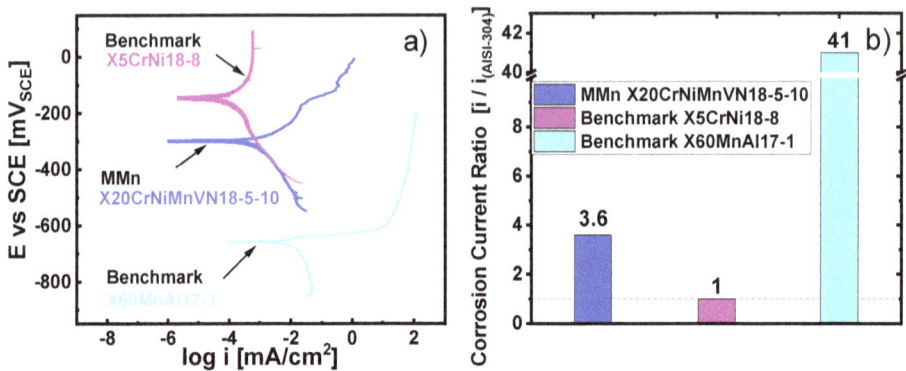

Figure 5. Dynamic polarization measurements in 5% NaCl showing the corrosion properties of the MMn steel compared to the benchmark steels. (**a**) Recorded polarization curves. (**b**) The corrosion current ratio with respect to the benchmark stainless steel X5CrNi18-8.

4. Discussion

4.1. Interplay among Different Strengthening Effects Induced by V Alloying

The developed ultrafine grain microstructure containing large Cr,Fe-rich carbides, Cr,V-rich nitro-carbides, V,Cr-rich nitro-carbides, and very fine V,Cr-rich nitrides shows superior strength–ductility balance, which emphasizes the synergetic effect of the nanoprecipitates and the ultrafine grains on controlling the strength level and hardening mechanism. The substantial increase in YS of MMn steel compared to the benchmark steels is attributed to the Ashby–Orowan effect for precipitation-strengthening, Hall–Petch effect for grain-size strengthening, solid–solution, and dislocation strengthening induced by V alloying. Since the formed precipitates are of different sizes, they will contribute differently to the increment in yield strength. According to the Ashby–Orowan relationship [33], the increment in yield strength increases as the size of particles decreases. Based on the size distribution analysis of the formed precipitates, >80% of the whole particles are smaller than 50 nm, which emphasizes the major role of the precipitation-strengthening mechanism.

There are additional effects caused by V alloying that act in a close interplay with nanoprecipitates on improving the YS of the MMn steel. As observed from the grain size distribution in Figure 3b, the annealing treatment at 1000 °C for 3 min resulted in a recrystallized and ultrafine grained microstructure that contributes to the increase in yield strength, which is commonly referred to as the Hall–Petch effect [34,35]. In addition to Ashby–Orowan and Hall–Petch effects, solid–solution strengthening due to the dissolved V and N solutes contributes to the YS value of the MMn steel. As can be noted from Figure 6, the thermodynamic equilibrium austenite at the applied annealing temperature still contains around 0.2 and 0.1 wt.% of dissolved V and N solutes, respectively. Norström [36] demonstrated a significant increase in yield strength of N-alloyed stainless steel due to the solid solution-hardening effect. Moreover, Werner [37] established that the increment in yield strength due to solid–solution strengthening of N-alloyed stainless steel was considerably higher than that achieved by grain refinement at the same amount of N.

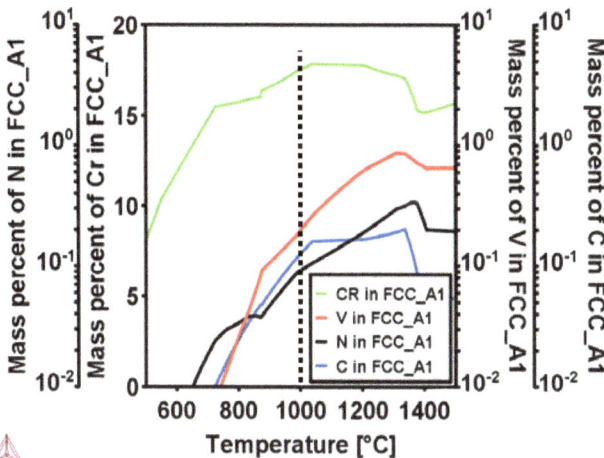

Figure 6. The change in equilibrium solubility in the mass percent of C, N, Cr, and V in austenite with temperature for the MMn steel.

Furthermore, the formation of precipitates is considered to be responsible for the suppression of the complete dislocation annihilation during annealing and hence increasing the yield strength via the dislocation hardening effect. In the same context for V-alloyed MMn steel, Hu et al. [38] reported a small increment in yield strength due to grain refinement, however, they emphasized the role of

VC precipitates and dislocation hardening in increasing the yield strength. The observed increase in yield strength due to V alloying is in agreement with several studies [39–42] that reported solely the pronounced effect of V precipitates; however, the current study emphasizes additional effects, such as grain–boundary, solid–solution, and dislocation strengthening.

4.2. Strain Hardening Behavior

Although the SFE for the MMn steel was adjusted in the TWIP range, it shows a high WHR at early deformation stages with a gradual decrease until fracture, which is quite different from the common WHR of conventional TWIP steels. Despite the gradual decrease in its WHR, the MMn steel maintains higher values of WHR than those of benchmark steels until around 20% elongation, thus allowing for a superior UTS value as observed in Figure 4. However, the gradual decrease in its WHR facilitates the early strain localization and results in a relatively lower but still comparable total elongation to the benchmark steels.

The high WHR at early deformation stages can be attributed to the grain refinement effect. However, the suppression of deformation twin formation results in a continuous drop in WHR. The austenite to martensite transformation is considered the reason for the slight decrease in the drop rate of WHR at later deformation stages (higher than 20% elongation) as observed from the EBSD maps in Figure 3d. Several studies emphasized the profound effect of grain refinement on the restriction of deformation twins. Ueji et al. [43] studied the twinning behavior of different high Mn austenitic steels with various mean grain sizes (1.8, 7.2, and 49.6 μm) and reported that the deformation twinning is strongly inhibited by grain refinement. Rahman et al. [44] reported the increase in twin nucleation stress with decreasing grain size. Lee et al. [45] attributed the increase in twinning stress to the suppression of dislocation activity and movement of partial dislocations by interaction with a high dislocation density in fine grain sized specimens. In the same regard, Gutierrez-Urrutia et al. [46,47] ascribed the increase in twinning stress resistance to the activation of multiple slip propagation, which is a perquisite to deformation twin formation. In general, the work hardening rate is related to the grain size, dislocation density, and the active deformation mechanism defined mostly by SFE [48–51]. Since the microstructure of the MMn steel contains a considerable amount of precipitates, the local chemical changes in the precipitates surroundings, especially C and N, will lead eventually to a decrease in the SFE and hence the possibility of austenite to martensite transformation as observed in Figure 3d. This has also been reported by Yen et al. [42].

4.3. Enhanced Corrosion Resistance

The potentiodynamic polarization measurements of the MMn steel demonstrate an enhanced corrosion resistance, which is superior to that of TWIP steel X60MnAl17-1 but still lower than that of stainless steel X5CrNi18-8. This enhanced corrosion resistance over the TWIP steel can be ascribed to the high Cr content. The thermodynamic equilibrium Cr content in austenite does not decrease significantly by the formation of precipitates at 1000 °C as shown in Figure 6. Nevertheless, the formed precipitates allow for the creation of corrosion cells that contribute detrimentally to corrosion resistance and passivation behavior of the MMn steel. Therefore, the lower corrosion resistance and the apparent increase in i_{corr} compared to stainless steel X5CrNi18-8 can be accounted for the formed precipitates. Although the corrosion current ratio of the MMn steel with respect to X5CrNi18-8 is 3.6, it was expected to be higher than this ratio due to precipitation, since Yan et al. [52] found that the precipitates of chromium-rich carbides in 316 L stainless steel resulted in severe pitting corrosion. However, in the current study, it seems that the grain refinement plays a significant role in enhancing the pitting resistance, in addition to the role of N. The pitting resistance equivalent number (PREN), which is a predictive measurement for the resistance of stainless steels to localized pitting corrosion based on their chemical composition [53], increases with N addition as indicated in Equation (1). The PREN number is 21.4 for the MMn steel, while 19.8 for the stainless steel X5CrNi18-8. Moreover, Hamada et al. [54] reported a significant improvement in the corrosion resistance of 301 LN as a result of grain refinement. It is well established that the segregation of impurities or the formation of chromium nitrides on the grain boundaries of stainless steel is more severe in a coarse-grained structure than in an ultrafine-grained structure, which is more homogeneous and has a smoother structure. Furthermore, the compactness and stability of the passive film formed on nano/submicron-grained structure is significantly increased [55,56]:

$$PREN = 1 \times \%Cr + 3.3 \times \%Mo + 16 \times \%N. \tag{1}$$

The current study demonstrates that V alloying resulted in a synergetic effect of precipitates and ultrafine grain size on both mechanical properties and corrosion behavior for the MMn steel. Indeed, the designed alloy concept provides the possibility to overcome some HMnS limitations, such as low yield strength values, poor corrosion resistance, and processing problems due to the high Mn content. Furthermore, the preliminary investigations on its hydrogen embrittlement behavior showed a significant improvement compared to conventional TWIP steels, even those Al-alloyed. However, a proper adjustment of the type and amount of precipitates could further enhance the corrosion resistance mechanical behavior of the MMn. This requires deeper understanding for the nature of the interaction between precipitates and other micro- and nanostructure defects.

5. Conclusions

The main aim of the present study was to develop a novel austenitic MMn steel X20CrNiMnVN18-5-10 that exhibits an enhanced corrosion resistance and a superior strength–ductility balance. The alloying additions were optimized to attain an ultrafine austenitic microstructure with a significant amount of nanoprecipitates. Alloying with V (1 wt.%) and N (0.3 wt.%) has been considered for precipitation and grain size control. The corrosion resistance and mechanical properties of the MMn steel were investigated and compared to their counterparts of two benchmark steels, i.e., austenitic stainless steel X5CrNi18-8 and austenitic TWIP steel X60MnAl17-1. Based on the findings, the following conclusions can be drawn:

1. Recrystallization annealing treatment at 1000 °C for 3 min resulted in an ultrafine austenitic microstructure with an average grain size of ~1.3 μm containing a considerable amount of V- and Cr-based precipitates in a bimodal particle size distribution (100–400 nm and <20 nm). More than 80% of the precipitates are smaller than 50 nm.
2. The alloying concept (especially the high V and N contents) results in a high yield strength of ~600 MPa via an interplay among different mechanisms, namely the Ashby–Orowan effect, Hall–Petch effect, solid–solution, and dislocation strengthening.
3. Although the work hardening rate (WHR) at early deformation stages is very high (3.4 GPa), the suppression of deformation twin formation by ultrafine grains until fracture resulted in a continuous drop in WHR. The slope of the WHR decreases at high strains (>20%) due to the austenite to martensite transformation.
4. The enhanced corrosion resistance of the newly developed MMn steel is attributed to the high Cr and N contents even after precipitation. It seems that the ultrafine grain microstructure plays an important role in improving the corrosion resistance despite the detrimental effect of precipitates.

Author Contributions: T.A. designed, performed, analyzed, and interpreted the experimental data, such as heat treatment, SEM, tensile tests, and mechanical properties. X.G. performed the corrosion tests, precipitate size measurements, and analyzed the data. M.L.-C. performed the TEM and STEM/EDX investigations and processed the data. S.S. performed the EBSD measurements and analyzed the data. A.H. and E.A. designed and produced the material. W.B. contributes to ideas and intensive discussions. T.A. wrote the original draft. All authors contributed to interpretation of the results and writing the final version of the manuscript.

Acknowledgments: A.H. and E.A. gratefully acknowledge the STDF-Egypt for supporting the melt production. T.A. expresses his gratitude to the DAAD for the personal financial support. DFG within the Collaborative Research Center (SFB) 761 is appreciated for supporting the processing and testing the developed steel grade.

Conflicts of Interest: The authors declare no conflict of interest.

References

1. Grässel, O.; Krüger, L.; Frommeyer, G.; Meyer, L.W. High strength Fe–Mn–(Al, Si) TRIP/TWIP steels development—Properties—Application. *Int. J. Plast.* **2000**, *16*, 1391–1409. [CrossRef]
2. de Cooman, B.C.; Estrin, Y.; Kim, S.K. Twinning-induced plasticity (TWIP) steels. *Acta Mater.* **2018**, *142*, 283–362. [CrossRef]
3. Bouaziz, O.; Allain, S.; Scott, C.P.; Cugy, P.; Barbier, D. High manganese austenitic twinning induced plasticity steels: A review of the microstructure properties relationships. *Curr. Opin. Solid State Mater. Sci.* **2011**, *15*, 141–168. [CrossRef]
4. Haase, C.; Ingendahl, T.; Güvenç, O.; Bambach, M.; Bleck, W.; Molodov, D.A.; Barrales-Mora, L.A. On the applicability of recovery-annealed Twinning-Induced Plasticity steels: Potential and limitations. *Mater. Sci. Eng. A* **2016**, *649*, 74–84. [CrossRef]
5. Dumay, A.; Chateau, J.-P.; Allain, S.; Migot, S.; Bouaziz, O. Influence of addition elements on the stacking-fault energy and mechanical properties of an austenitic Fe–Mn–C steel. *Mater. Sci. Eng. A* **2008**, *2008*, 184–187. [CrossRef]
6. Grajcar, A. Corrosion resistance of high-Mn austenitic steels for the automotive industry. In *Corrosion Resistance*; InTech: Rijeka, Croatia, 2012.
7. Kannan, M.B.; Raman, R.S.; Khoddam, S. Comparative studies on the corrosion properties of a Fe–Mn–Al–Si steel and an interstitial-free steel. *Corros. Sci.* **2008**, *50*, 2879–2884. [CrossRef]
8. Zhang, Y.S.; Zhu, X.M. Electrochemical polarization and passive film analysis of austenitic Fe–Mn–Al steels in aqueous solutions. *Corros. Sci.* **1999**, *41*, 1817–1833. [CrossRef]
9. Zhang, Y.S.; Zhu, X.M.; Liu, M.; Che, R.X. Effects of anodic passivation on the constitution, stability and resistance to corrosion of passive film formed on an Fe-24Mn-4Al-5Cr alloy. *Appl. Surf. Sci.* **2004**, *222*, 89–101. [CrossRef]
10. Dieudonné, T.; Marchetti, L.; Wery, M.; Miserque, F.; Tabarant, M.; Chêne, J.; Allely, C.; Cugy, P.; Scott, C.P. Role of copper and aluminum on the corrosion behavior of austenitic Fe–Mn–C TWIP steels in aqueous solutions and the related hydrogen absorption. *Corros. Sci.* **2014**, *83*, 234–244. [CrossRef]
11. Toor, I.-U.-H.; Hyun, P.J.; Kwon, H.S. Development of high Mn–N duplex stainless steel for automobile structural components. *Corros. Sci.* **2008**, *50*, 404–410. [CrossRef]
12. Di Schino, A.; Kenny, J.M.; Mecozzi, M.G.; Barteri, M. Development of high nitrogen, low nickel, 18%Cr austenitic stainless steels. *J. Mater. Sci.* **2000**, *35*, 4803–4808. [CrossRef]
13. Mosecker, L.; Saeed-Akbari, A. Nitrogen in chromium–manganese stainless steels: A review on the evaluation of stacking fault energy by computational thermodynamics. *Sci. Technol. Adv. Mater.* **2013**, *14*, 33001. [CrossRef] [PubMed]
14. de Cooman, B.C.; Chin, K.-G.; Kim, J. High Mn TWIP Steels for Automotive Applications. In *New Trends and Developments in Automotive System Engineering*; Chiaberge, M., Ed.; InTech: Bratislava, Slovakia, 2011; ISBN 978-953-307-517-4.
15. Kim, Y.; Kang, N.; Park, Y.; Choi, I.; Kim, G.; Kim, S.; Cho, K. Effects of the strain induced martensite transformation on the delayed fracture for Al-added TWIP steel. *J. Korean Inst. Met. Mater.* **2008**, *46*, 780–787.
16. Koyama, M.; Akiyama, E.; Tsuzaki, K. Hydrogen Embrittlement in Al-added Twinning-induced Plasticity Steels Evaluated by Tensile Tests during Hydrogen Charging. *ISIJ Int.* **2012**, *52*, 2283–2287. [CrossRef]
17. Zan, N.; Ding, H.; Guo, X.; Tang, Z.; Bleck, W. Effects of grain size on hydrogen embrittlement in a Fe-22Mn-0.6C TWIP steel. *Int. J. Hydrogen Energy* **2015**, *40*, 10687–10696. [CrossRef]
18. Park, I.-J.; Lee, S.-M.; Jeon, H.-H.; Lee, Y.-K. The advantage of grain refinement in the hydrogen embrittlement of Fe–18Mn–0.6C twinning-induced plasticity steel. *Corros. Sci.* **2015**, *93*, 63–69. [CrossRef]
19. Timmerscheidt, T.; Dey, P.; Bogdanovski, D.; von Appen, J.; Hickel, T.; Neugebauer, J.; Dronskowski, R. The Role of κ-Carbides as Hydrogen Traps in High-Mn Steels. *Metals* **2017**, *7*, 264. [CrossRef]
20. Ooi, S.W.; Ramjaun, T.I.; Hulme-Smith, C.; Morana, R.; Drakopoulos, M.; Bhadeshia, H.K.D.H. Designing steel to resist hydrogen embrittlement Part 2—Precipitate characterisation. *Mater. Sci. Technol.* **2018**, *34*, 1747–1758. [CrossRef]

21. Zhang, Z.; Moore, K.L.; McMahon, G.; Morana, R.; Preuss, M. On the role of precipitates in hydrogen trapping and hydrogen embrittlement of a nickel-based superalloy. *Corros. Sci.* **2019**, *146*, 58–69. [CrossRef]
22. Takahashi, J.; Kawakami, K.; Kobayashi, Y. Origin of hydrogen trapping site in vanadium carbide precipitation strengthening steel. *Acta Mater.* **2018**, *153*, 193–204. [CrossRef]
23. Suh, D.W.; Ryu, J.H.; Joo, M.S.; Yang, H.S.; Lee, K.; Bhadeshia, H.K.D.H. Medium-Alloy Manganese-Rich Transformation-Induced Plasticity Steels. *Metall. Mater. Trans. A* **2013**, *44*, 286–293. [CrossRef]
24. Lee, Y.-K.; Han, J. Current opinion in medium manganese steel. *Mater. Sci. Technol.* **2014**, *31*, 843–856. [CrossRef]
25. Sun, B.; Ding, R.; Brodusch, N.; Chen, H.; Guo, B.; Fazeli, F.; Ponge, D.; Gauvin, R.; Yue, S. Improving the ductility of ultrahigh-strength medium Mn steels via introducing pre-existed austenite acting as a "reservoir" for Mn atoms. *Mater. Sci. Eng. A* **2019**, *749*, 235–240. [CrossRef]
26. Sun, B.; Fazeli, F.; Scott, C.; Brodusch, N.; Gauvin, R.; Yue, S. The influence of silicon additions on the deformation behavior of austenite-ferrite duplex medium manganese steels. *Acta Mater.* **2018**, *148*, 249–262. [CrossRef]
27. Kaar, S.; Schneider, R.; Krizan, D.; Béal, C.; Sommitsch, C. Influence of the Quenching and Partitioning Process on the Transformation Kinetics and Hardness in a Lean Medium Manganese TRIP Steel. *Metals* **2019**, *9*, 353. [CrossRef]
28. Hamada, A.; Juuti, T.; Khosravifard, A.; Kisko, A.; Karjalainen, P.; Porter, D.; Kömi, J. Effect of silicon on the hot deformation behavior of microalloyed TWIP-type stainless steels. *Mater. Des.* **2018**, *154*, 117–129. [CrossRef]
29. Hielscher, R.; Schaeben, H. A novel pole figure inversion method: Specification of the MTEX algorithm. *J. Appl. Cryst.* **2008**, *41*, 1024–1037. [CrossRef]
30. Bachmann, F.; Hielscher, R.; Schaeben, H. Texture Analysis with MTEX—Free and Open Source Software Toolbox. *SSP* **2010**, *160*, 63–68. [CrossRef]
31. Luysberg, M.; Heggen, M.; Tillmann, K. FEI Tecnai G2 F20. *JLSRF* **2016**, *2*. [CrossRef]
32. Kovács, A.; Schierholz, R.; Tillmann, K. FEI Titan G2 80-200 CREWLEY. *JLSRF* **2016**, *2*. [CrossRef]
33. Gladman, T. Precipitation hardening in metals. *Mater. Sci. Technol.* **1999**, *15*, 30–36. [CrossRef]
34. Hansen, N. Hall–Petch relation and boundary strengthening. *Scr. Mater.* **2004**, *51*, 801–806. [CrossRef]
35. Hall, E.O. The Deformation and Ageing of Mild Steel: III Discussion of Results. *Proc. Phys. Soc. Sect. B* **1951**, *64*, 747–753. [CrossRef]
36. Norström, L.-Å. The influence of nitrogen and grain size on yield strength in Type AISI 316L austenitic stainless steel. *Met. Sci.* **1977**, *11*, 208–212. [CrossRef]
37. Werner, E. Solid solution and grain size hardening of nitrogen-alloyed austenitic steels. *Mater. Sci. Eng. A* **1988**, *101*, 93–98. [CrossRef]
38. Hu, B.; He, B.B.; Cheng, G.J.; Yen, H.W.; Luo, H.W.; Huang, M.X. Super-High-Strength and Formable Medium Mn Steel Manufactured by Warm Rolling Process. 23 March 2019. Available online: https://ssrn.com/abstract=3358845 (accessed on 26 April 2019).
39. Yazawa, Y.; Furuhara, T.; Maki, T. Effect of matrix recrystallization on morphology, crystallography and coarsening behavior of vanadium carbide in austenite. *Acta Mater.* **2004**, *52*, 3727–3736. [CrossRef]
40. Atasoy, Ö.A.; Özbaysal, K.; Inal, O.T. Precipitation of vanadium carbides in 0.8% C-13% Mn-1% V austenitic steel. *J. Mater. Sci.* **1989**, *24*, 1393–1398. [CrossRef]
41. Sohn, S.S.; Song, H.; Jo, M.C.; Song, T.; Kim, H.S.; Lee, S. Novel 1.5 GPa-strength with 50%-ductility by transformation-induced plasticity of non-recrystallized austenite in duplex steels. *Sci. Rep.* **2017**, *7*, 1255. [CrossRef]
42. Yen, H.-W.; Huang, M.; Scott, C.P.; Yang, J.-R. Interactions between deformation-induced defects and carbides in a vanadium-containing TWIP steel. *Scr. Mater.* **2012**, *66*, 1018–1023. [CrossRef]
43. Ueji, R.; Tsuchida, N.; Terada, D.; Tsuji, N.; Tanaka, Y.; Takemura, A.; Kunishige, K. Tensile properties and twinning behavior of high manganese austenitic steel with fine-grained structure. *Scr. Mater.* **2008**, *59*, 963–966. [CrossRef]
44. Rahman, K.M.; Vorontsov, V.A.; Dye, D. The effect of grain size on the twin initiation stress in a TWIP steel. *Acta Mater.* **2015**, *89*, 247–257. [CrossRef]

45. Lee, S.-I.; Lee, S.-Y.; Han, J.; Hwang, B. Deformation behavior and tensile properties of an austenitic Fe-24Mn-4Cr-0.5C high-manganese steel: Effect of grain size. *Mater. Sci. Eng. A* **2019**, *742*, 334–343. [CrossRef]

46. Gutierrez-Urrutia, I.; Zaefferer, S.; Raabe, D. The effect of grain size and grain orientation on deformation twinning in a Fe–22 wt.% Mn–0.6 wt.% C TWIP steel. *Mater. Sci. Eng. A* **2010**, *527*, 3552–3560. [CrossRef]

47. Gutierrez-Urrutia, I.; Raabe, D. Grain size effect on strain hardening in twinning-induced plasticity steels. *Scr. Mater.* **2012**, *66*, 992–996. [CrossRef]

48. Bouaziz, O.; Allain, S.; Scott, C. Effect of grain and twin boundaries on the hardening mechanisms of twinning-induced plasticity steels. *Scr. Mater.* **2008**, *58*, 484–487. [CrossRef]

49. Idrissi, H.; Renard, K.; Ryelandt, L.; Schryvers, D.; Jacques, P.J. On the mechanism of twin formation in Fe–Mn–C TWIP steels. *Acta Mater.* **2010**, *58*, 2464–2476. [CrossRef]

50. Saeed-Akbari, A.; Mosecker, L.; Schwedt, A.; Bleck, W. Characterization and Prediction of Flow Behavior in High-Manganese Twinning Induced Plasticity Steels: Part I. Mechanism Maps and Work-Hardening Behavior. *Metall. Mater. Trans. A* **2012**, *43*, 1688–1704. [CrossRef]

51. Renard, K.; Jacques, P.J. On the relationship between work hardening and twinning rate in TWIP steels. *Mater. Sci. Eng. A* **2012**, *542*, 8–14. [CrossRef]

52. Yan, S.; Shi, Y.; Liu, J.; Ni, C. Effect of laser mode on microstructure and corrosion resistance of 316L stainless steel weld joint. *Opt. Laser Technol.* **2019**, *113*, 428–436. [CrossRef]

53. Okada, T.; Hashino, T. A contribution to the kinetic theory of pitting corrosion. *Corros. Sci.* **1977**, *17*, 671–689. [CrossRef]

54. Hamada, A.S.; Karjalainen, L.P.; El-Zeky, M.A. Effect of anodic passivation on the corrosion behaviour of Fe-Mn-Al steels in 3.5% NaCl. In *Passivation of Metals and Semiconductors, and Properties of Thin Oxide Layers*; Elsevier: Amsterdam, The Netherlands, 2006; pp. 77–82.

55. Hamada, A.S.; Karjalainen, L.P.; Somani, M.C. Electrochemical corrosion behaviour of a novel submicron-grained austenitic stainless steel in an acidic NaCl solution. *Mater. Sci. Eng. A* **2006**, *431*, 211–217. [CrossRef]

56. Jinlong, L.; Hongyun, L. Comparison of corrosion properties of passive films formed on phase reversion induced nano/ultrafine-grained 321 stainless steel. *Appl. Surf. Sci.* **2013**, *280*, 124–131. [CrossRef]

![metals logo] *metals*

MDPI

Article

Influence of Microstructural Morphology on Hydrogen Embrittlement in a Medium-Mn Steel Fe-12Mn-3Al-0.05C

Xiao Shen [1], Wenwen Song [1,*], Simon Sevsek [1], Yan Ma [1], Claas Hüter [2], Robert Spatschek [2] and Wolfgang Bleck [1]

[1] Steel Institute (IEHK), RWTH Aachen University, Intzestraße 1, 52072 Aachen, Germany
[2] IEK-2, Forschungszentrum Jülich, Wilhelm-Johnen-Straße, 52425 Jülich, Germany
*   Correspondence: wenwen.song@iehk.rwth-aachen.de; Tel.: +49-241-8095-815

Received: 24 June 2019; Accepted: 21 August 2019; Published: 24 August 2019

Abstract: The ultrafine-grained (UFG) duplex microstructure of medium-Mn steel consists of a considerable amount of austenite and ferrite/martensite, achieving an extraordinary balance of mechanical properties and alloying cost. In the present work, two heat treatment routes were performed on a cold-rolled medium-Mn steel Fe-12Mn-3Al-0.05C (wt.%) to achieve comparable mechanical properties with different microstructural morphologies. One heat treatment was merely austenite-reverted-transformation (ART) annealing and the other one was a successive combination of austenitization (AUS) and ART annealing. The distinct responses to hydrogen ingression were characterized and discussed. The UFG martensite colonies produced by the AUS + ART process were found to be detrimental to ductility regardless of the amount of hydrogen, which is likely attributed to the reduced lattice bonding strength according to the H-enhanced decohesion (HEDE) mechanism. With an increase in the hydrogen amount, the mixed microstructure (granular + lamellar) in the ART specimen revealed a clear embrittlement transition with the possible contribution of HEDE and H-enhanced localized plasticity (HELP) mechanisms.

Keywords: medium-Mn steel; austenite-reversed-transformation; retained austenite; hydrogen embrittlement; ultrafine-grained microstructure; strain-hardening behavior

1. Introduction

A tremendous number of studies have focused on medium-Mn steels (MMnS) due to their excellent combination of mechanical properties and production cost as well as the understanding of complex deformation mechanisms and microstructure-property relationships [1–8]. By austenite-reverted-transformation (ART) annealing, medium-Mn steels often display ultrafine-grained (UFG) austenite-ferrite/martensite duplex microstructures [5,6]. The reverted austenite contributes to complex deformation mechanisms, enhancing the mechanical performance in MMnS. Deformation-induced martensitic phase transformation, the transformation-induced-plasticity (TRIP) effect, was reported to eliminate the localized deformation behavior in UFG microstructures and impede dislocation motion, resulting in considerably high ductility and work-hardening rate [1,2]. More recently, deformation-induced twinning was found in MMnS with a manganese content of 6 to 12 wt% after the α′-martensite formation, which was referred to as a medium-Mn twinning-induced-plasticity (TWIP) + TRIP steel [3,4]. Tailoring the mechanical properties by modifying the microstructure has also been intensively studied, involving the influences of intercritical annealing temperature [9–11], intercritical annealing time [10,12] and alloying elements [13–15].

Hydrogen embrittlement (HE) resistance is an important criterion for modern high-strength steels applied in the automobile industry, energy industry, aerospace industry and chemistry industry. The ferrite/martensite phase is considered to be vulnerable to hydrogen ingression, which contributes to the severe level of HE [16–20]. The austenite, as a critical phase in MMnS to improve mechanical performance, was reported to trap the hydrogen atoms effectively and leave the absorbed hydrogen in a more activated state, resulting in brittle fracture by the localized TRIP effect [16,17,19]. Han et al. [18] investigated the HE behavior in cold-rolled and hot-rolled Fe-7Mn-0.1C steel, correlating the contributions of well-established HE mechanisms, such as hydrogen-enhanced decohesion (HEDE) [21–24] and hydrogen-enhanced localized plasticity (HELP) [21,22,25,26] with microstructural features. The lamellar microstructure was found to be sensitive to hydrogen ingression compared to the granular one [18]. Most recently, Jeong et al. [20] argued that the lath-type austenite had a beneficial effect on relieving the hydrogen susceptibility compared with the equiaxed one due to the obstructed migration of hydrogen, which was contrary to the results of Han et al. [18].

The interaction between hydrogen and microstructural features has been studied in different steel grades [27–29]. However, little systematic work has been done to investigate the influences of microstructural morphology on HE susceptibility in MMnS, excluding the divergences in austenite fraction and mechanical properties. In the present work, by designing heat treatment, two different microstructures were produced and revealed comparable mechanical properties and austenite fraction. The effects of microstructural features on HE were investigated from the perspectives of hydrogen absorption and hydrogen-induced fracture modes. With a comprehensive understanding of the hydrogen–microstructure relation, we could design high-performance MMnS with superior HE resistance that can be applied for industry in the near future.

2. Materials and Methods

The chemical composition of the investigated medium-Mn steel Fe-12Mn-3Al-0.05C is given in Table 1. The 80 kg laboratory-melt alloy was cast into a 140 mm × 140 mm × 500 mm ingot in an ALD vacuum induction furnace (ALD Vacuum Tech. GmbH, Hanau, Germany), followed by hot forging to a bar with a cross section of 160 mm × 40 mm at 1150 °C. The thickness of the bar was finally reduced from 40 mm to 2.5 mm by hot rolling with reheating between passes. The material was subsequently homogenized at 1100 °C for 2 h to dissolve micro-segregations, followed by water quenching. The material was subsequently austenitized at 850 °C for 10 min and cold rolled with a reduction in thickness of 50% to the final sheet thickness of 1.25 mm. Two different heat treatment routes were selected to generate different microstructures with comparable mechanical properties by screening tests. A step of austenitization was applied to alter the initial microstructure, which was decisive to an as-annealed microstructure [7]. The cold-rolled specimen was austenitized at 800 °C for 20 min and then ART annealed at 650 °C for 15 min and termed AUS + ART. The other group of specimens were ART annealed at 675 °C for 2 h and denoted as ART. Figure 1 schematically shows the thermal cycles.

For hydrogen charging, the specimens were ground with SiC paper up to 2000 grit and polished with 3 μm diamond suspension. Specimens were subsequently hydrogen-charged for 2, 8 and 24 h, using the cathodic hydrogen-charging method. The specimens were subject to slow strain rate tensile tests (SSRT) immediately after hydrogen charging, using the tensile machine produced by Fritz Fackert KG (Moers, Germany) with a maximum load of 30 kN. The displacement speed of the crosshead was fixed at 2.5×10^{-5} mm/s, corresponding to an initial strain rate of 10^{-6} s^{-1}. The fractured specimens were stored in a liquid nitrogen atmosphere before hydrogen measurement. The hydrogen content was measured by thermal desorption analysis (TDA) at the Welding and Joining Institute (ISF) of the RWTH Aachen University. The temperature range for measurement was from room temperature to 800 °C with a heating rate of 20 °C/min. The application of an external diffusible hydrogen module enabled the detection and analysis of both interstitial hydrogen and trapped hydrogen.

Table 1. Chemical composition of the investigated material as determined by wet chemical analysis.

Element	C	Si	Mn	P	S	Al	Fe
wt.%	0.064	0.2	11.7	0.006	0.003	2.9	Bal.

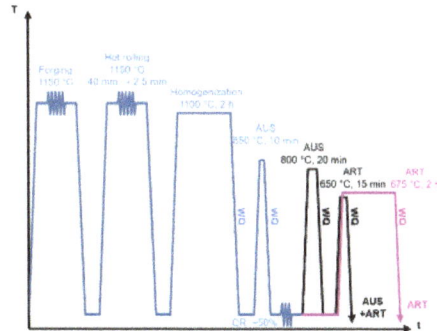

Figure 1. Schematic illustration of thermal cycles on cold-rolled Fe-12Mn-3Al-0.05C steel after cold rolling (WQ denotes water cooling; CR denotes cold rolling).

The inspection of fracture surfaces was carried out using a ZEISS DSM 982 scanning electron microscope (SEM) (Carl Zeiss Microscopy GmbH, Jena, Germany) with an acceleration voltage of 15 kV. For electron backscatter diffraction (EBSD) measurements, the specimens were ground with SiC paper up to 4000 grit and subsequently polished using 3 μm, 1 μm diamond suspension. Electropolishing was then performed at 28 V for 20 s using an electrolyte containing 700 mL ethanol (C_2H_5OH), 100 mL butyl glycol ($C_6H_{14}O_2$), and 78 mL perchloric acid (60%, $HClO_4$). EBSD measurements were performed with an acceleration voltage of 10 kV and a step size of 50 nm. The working distance was between 18 mm and 24 mm. A post-processing routine employing the HKL Channel 5 software (version 5.12j, Oxford Instruments, Abingdon-on-Thames, UK) was utilized, as well as the MATLAB-based MTEX toolbox [30,31].

Quantitative analysis on the amount of austenite and its mechanical stability was carried out using synchrotron X-ray diffraction (SYXRD) data, performed at beamline P02.1, PETRA III in Deutsches Elektronen-Synchrotron (DESY) center (Bahrenfeld, Hamburg, Germany). The measurement was operated with high beamline energy of 60 keV and a fixed wavelength of ~0.207 Å. The collected two-dimensional diffraction patterns were converted into intensity versus 2θ data by Fit2D software [32] and further analyzed using Rietveld refinement [33] by Material Analysis Using Diffraction (MAUD) software V 2.8.

3. Results

3.1. Microstructure

3.1.1. Electron Backscatter Diffraction (EBSD) Analysis of Microstructure

EBSD measurements were employed to characterize the phase distribution, the orientation distribution, and the misorientation. Figure 2a shows the forescatter diode (FSD) image of the ART specimen to highlight the UFG microstructure. Figure 2b shows that the austenite (blue area) had two main locations, between the martensite laths (green area) and in the globular-grain region. The red arrows mark the region with fine-grained globular structure, while the white arrows indicate the lamellar structure. The length of the lamellae ranged from 10 to 20 μm with a lamellar spacing of about 0.8 μm and the grain size of the fine-grained globular structure was around 1.5 μm. The inverse pole figure (IPF) of the ART specimen (Figure 2c) highlighted that the majority of deformed lamellar martensite showed orientations close to the <112>‖ND-fiber (ND: normal direction), which denoted the retainment of cold-deformed microstructure. The grain orientation spread (GOS) figure (Figure 2d) shows that the globular austenite grains, as identified by the phase map, displayed low GOS values below 1, which is usual for recrystallized grains [34].

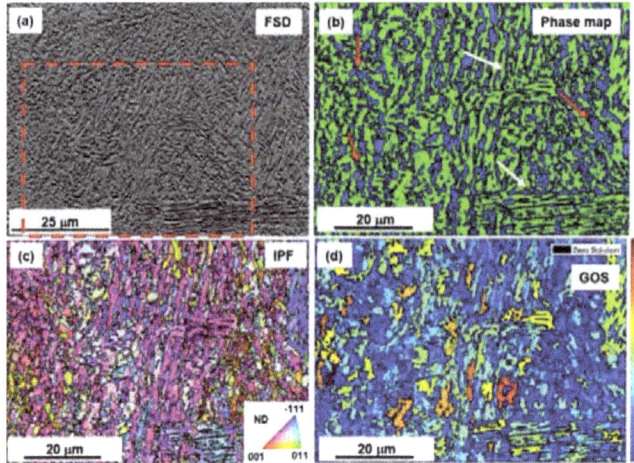

Figure 2. Electron backscatter diffraction measurement of initial microstructure in austenite-reverted-transformation annealed specimen, including (**a**) forescatter diode image, (**b**) phase distribution map, (**c**) inverse pole figure and (**d**) grain orientation spread image. The dashed box in (**a**) denotes the analyzed region for (**b**–**d**); red arrows in (**b**) indicates globular grains; white arrows marks the band-shaped lamellar grains.

Figure 3a shows the FSD image of the AUS + ART specimen, with an indication of a further analysed area. Figure 3b exhibits the initial microstructure of AUS + ART specimen, which consisted of an ultrafine-grained lamellar structure and globular structure. The red arrows mark the region with ultrafine-grained globular structure, while the white arrows indicate the ultrafine-grained lamellar structure. During the step of austenitization, the deformed microstructure was fully eliminated and replaced by quenched martensite. The size of martensite packages ranged from 1 to 5 μm, the size of globular grains went down to 1 μm after subsequent ART annealing as a result of the competition between austenite reversion and martensite recrystallization. Unlike the ART specimen, AUS + ART specimen did not show a preferred orientation but revealed random orientation in IPF (Figure 3c). As shown in the GOS figure (Figure 3d), the ultrafine-grained martensite colonies manifested a low grain orientation spread, which indicates that they were recrystallized during intercritical annealing. The volume fraction of austenite determined from the EBSD measurement lacked accuracy due to the limited step size and spatial resolution. The explicit quantitative analyses of the microstructure is given in the following section.

3.1.2. Quantitative Analysis of Microstructure by Synchrotron X-ray Diffraction (SYXRD)

SYXRD enables quantitative analysis of mechanical stability and volume fraction of austenite. Interrupted tensile tests were employed with a termination strain of 5%, 15% and 25%. The SYXRD measurements were performed on the uniform elongation zone next to necking areas. Figure 4a,b show the SYXRD profile of specimens with increasing deformation degree. The volume fractions of austenite within the deformed and undeformed microstructure were determined from fitting results by MAUD software. In the initial microstructure, ART specimen and AUS + ART specimen revealed a comparable amount of austenite, with a slight difference of ~2%. During deformation, austenite transformed into martensite due to relatively low mechanical stability, resulting in strain-hardening behavior. At the end of the deformation (engineering strain from 25% to 30.1%), the transformation amount of austenite in AUS + ART was still high, which dropped from 40.4% to 36.4%. Conversely, an extremely small amount of austenite transformation from 40.4% to 40.2% was observed in ART specimen with an engineering strain from 25% to 33.1%, which suggests that the TRIP effect rarely contributed to deformation at this stage.

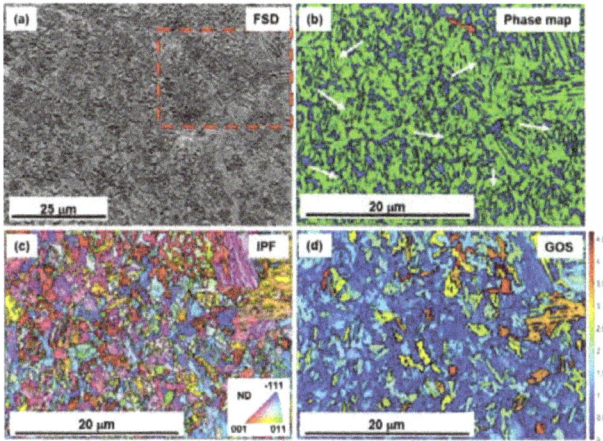

Figure 3. Electron backscatter diffraction measurement of initial microstructure in AUS + ART annealed specimen, including (**a**) forescatter diode image, (**b**) phase distribution map, (**c**) inverse pole figure and (**d**) grain orientation spread image. The dashed box in (**a**) denotes the analyzed region for (**b**–**d**); red arrows in (**b**) indicates globular grains; white arrows marks the band-shaped lamellar grains.

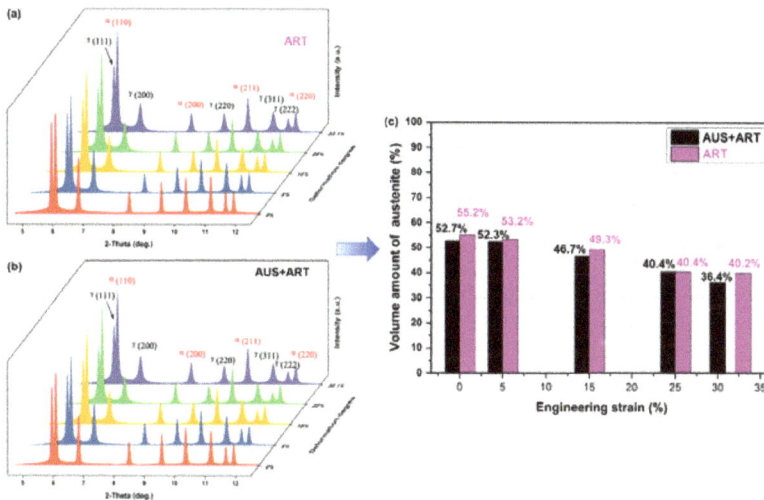

Figure 4. Synchrotron X-ray diffraction profile of (**a**) austenite-reverted-transformation specimen; (**b**) AUS + ART specimen with increasing deformation degree. (**c**) volume fraction of austenite in samples in the as-annealed condition as well as taken from interrupted tensile tests at different strains.

3.2. Mechanical Properties

3.2.1. Tensile Properties

Figure 5a shows the engineering stress-engineering strain diagrams of the medium-Mn steel Fe-12Mn-3Al-0.05C after heat treatment ART and AUS + ART. The quasi-static tensile tests were conducted at room temperature at a consistent strain rate of 10^{-3} s^{-1}. The mechanical properties of ART and AUS + ART specimens were comparable. The ART specimen revealed a slightly higher total elongation of 33.1% and a relatively lower ultimate tensile strength (UTS) of 811 MPa. The yield strength was 680 MPa. AUS + ART specimen exhibited an improved yield strength of 701 MPa and an enhanced UTS of 891 MPa with a total elongation of just over 30%.

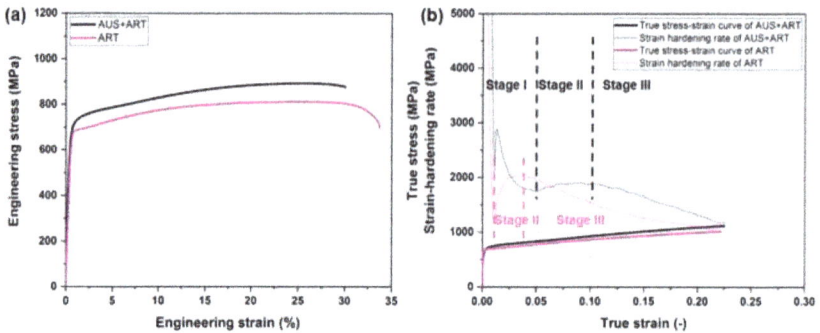

Figure 5. (a) Engineering stress-engineering strain curves; (b) true stress-true strain curves and strain-hardening curves of the medium-Mn steel Fe-12Mn-3Al-0.05C after heat treatment austenite-reverted-transformation (ART) and AUS + ART.

Figure 5b illustrates true stress-true strain curves and the respective strain-hardening curves of the medium-Mn steel Fe-12Mn-3Al-0.05C after heat treatment ART and AUS + ART. Three stages can be distinguished clearly in the curves according to their distinct characteristics. In stage I, the strain-hardening rate dropped from an extremely high value. The ART specimen terminated stage I earlier than AUS + ART specimen. During stage II, the strain-hardening rates increased during plastic deformation before the strain-hardening rates decreased in stage III until the UTS point. The end values of stage III were almost the same in two specimens, around 1000 MPa. In general, the elevated strain-hardening rate contributed to the enhanced strength in AUS + ART specimen.

3.2.2. Mechanical Degradation Due to Hydrogen Ingression

To evaluate the hydrogen embrittlement susceptibility, ART specimen and AUS + ART specimen underwent an ex situ slow strain rate tensile (SSRT) test until fracture at different states, namely, uncharged, 2 h H-charged, 8 h H-charged, 24 h H-charged. The strain rate was set at 1×10^{-6} s^{-1} to ensure that the fracture behavior was predominated by the delayed hydrogen-induced cracks [35]. The elastic part was calibrated and modified with a defined slope, in order to shorten the elastic region and make the results comparable. Figure 6 exhibits the mechanical properties before and after hydrogen ingression in two groups of specimens. The HE susceptibility can be evaluated by the degradation of mechanical properties. On the one hand, the hydrogen content manifested little influence on strength in both groups of specimens. Yield strengths remained unchanged while the UTS points were not reached before hydrogen-induced brittle fracture. On the other hand, the distinct responses of ductility to hydrogen ingression were observed. The ART specimen experienced a clear ductile-to-brittle transition with increasing hydrogen uptake. With an increase in hydrogen charging time, the strain losses became greater. Differently, the presence of even a small amount of hydrogen in the AUS + ART specimen led to catastrophic failure. Upon charging with hydrogen, the specimen fractured at a very early stage with a limited total elongation of around 4%, regardless of the hydrogen content.

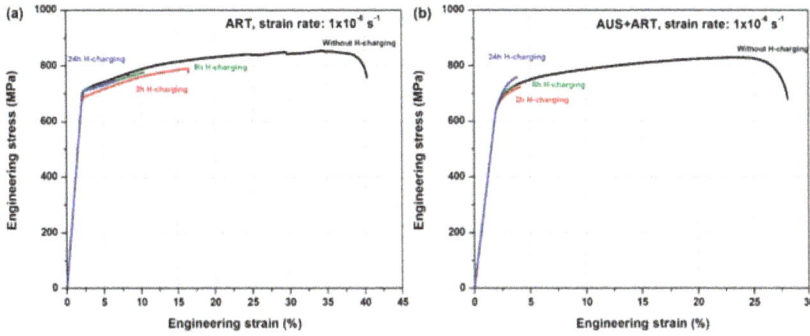

Figure 6. Evaluation of mechanical degradation by ex-situ slow strain rate tensile test at a strain rate of 10^{-6} s^{-1} in (**a**) ART specimen; and (**b**) AUS + ART specimen.

3.3. Hydrogen Uptake

For a comprehensive understanding of hydrogen uptake in H-charged medium-Mn steel Fe-12Mn-3Al-0.05C, the thermal desorption rates are plotted in Figure 7. For uncharged specimens, the hydrogen desorption rates remained low from room temperature to 500 °C, and started to increase subsequently until 800 °C, which mostly referred to the hydrogen constrained at irreversible traps with high desorption energy. These hydrogen atoms were introduced during the steel making process. After 2 h of hydrogen charging, both ART specimen and AUS + ART specimen showed an immediate increase in diffusive hydrogen content, containing 3.07 ppm and 2.42 ppm hydrogen, respectively. The contents of diffusive hydrogen increased dramatically with the extended charging time. The hydrogen desorption rate curve always experienced a decreased slope around 170 °C and a top peak around 350 °C to 400 °C. These two features corresponded to a desorption peak of the martensite phase and austenite phase, respectively [18,36]. The hydrogen concentration saw a significant difference when the H-charging time was extended to 24 h. The hydrogen concentration in AUS + ART specimen was elevated although they consisted of a similar constitution.

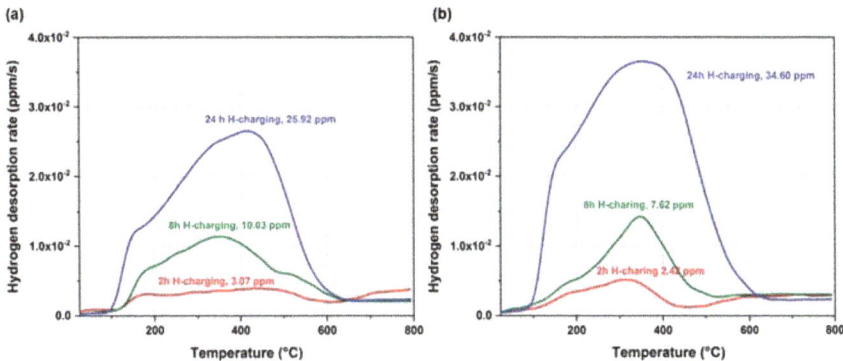

Figure 7. Hydrogen desorption rate as a function of temperature in (**a**) austenite-reverted-transformation (ART) specimen; and (**b**) AUS + ART specimen.

3.4. Fractography

In order to understand the influence of hydrogen ingression on the fracture mode, the photographs of fracture surfaces were taken by SEM. Figure 8 reveals the fracture surface at the edge region (close to the lateral surface) of ART specimen at different hydrogen charging states. Compared with the middle regions of the fracture surface, edge regions revealed more apparent characteristics of hydrogen-induced failure and crack initiation due to the limited penetration depth. The difference was remarkable by comparing the fracture features, which can account for the transition from ductile to brittle. The specimen without hydrogen charging (Figure 8a) showed fully ductile dimples at the edge regions. Due to the non-uniaxial stress state, dimples next to the surface were sheared. The presence of 3.07 ppm hydrogen content (Figure 8b) led to mainly quasi-cleavage fracture, which refers to brittle failure on non-cleavage planes. 8-h hydrogen charging induced moderate hydrogen content of 10.03 ppm, producing pronounced quasi-cleavage and a large area of flat facets on the fracture surfaces (Figure 8c). After hydrogen charging for 24 h (Figure 8d), 25.92 ppm hydrogen content resulted in intergranular cleavage regions and rugged facets. The ductile to brittle transition features were clear by comparing the fracture features at the edge regions, which could account for the increase of strain loss with increasing hydrogen charging time.

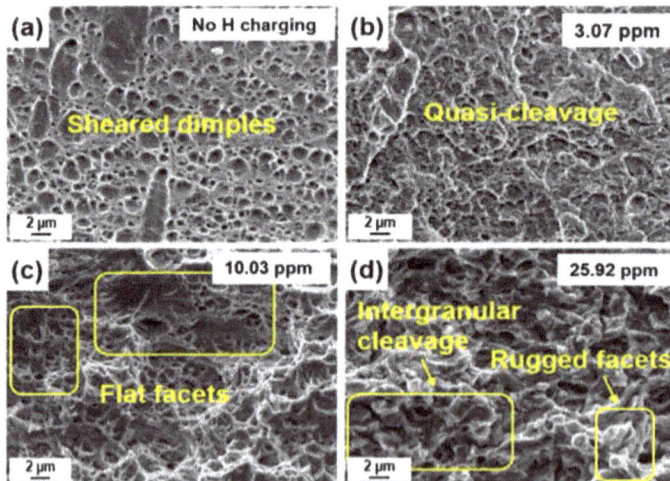

Figure 8. Fracture surfaces at the edge region of austenite-reverted-transformation annealed specimens undergone slow strain rate tensile until fracture. (**a**) No H-charged state; (**b**) H-charged for 2 h with 3.07 ppm; (**c**) H-charged for 8 h with 10.03 ppm. (**d**) H-charged for 24 h with 25.92 ppm.

Hydrogen ingression in AUS + ART specimens always led to more than 85% strain loss, regardless of the hydrogen content. Figure 9 displayed the fracture surfaces at the edge region of AUS + ART specimen. The edge region of the uncharged specimen revealed a fully ductile fracture (Figure 9a). The elongated dimples were also caused by a non-uniaxial stress state near the surface. Recalling from the mechanical degradation that was evaluated by SSRT, the fracture surface at the edge region was assumed to be consistent regardless of the hydrogen content because they all showed extremely early fracture (~4% total elongation). From Figure 9b–d, representing hydrogen charging for 2 h, 8 h, 24 h, respectively, the fracture surfaces at the edge region were composed of quasi-cleavage and a large area of flat facets. The fracture surfaces revealed typical features after hydrogen charging. A detailed interpretation of operative fracture mechanisms is discussed in the following section.

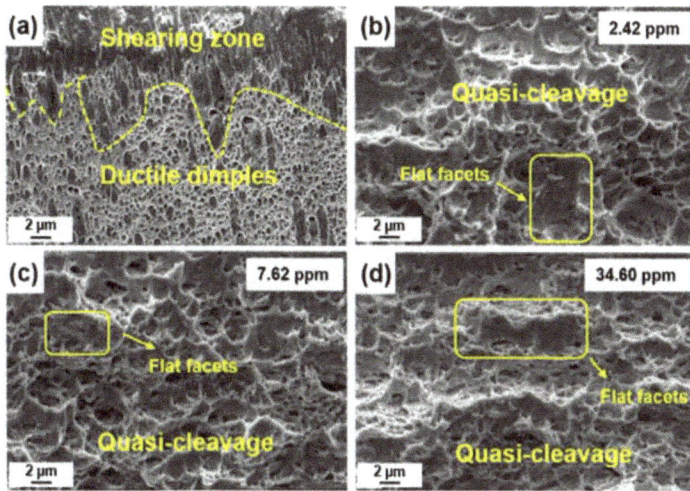

Figure 9. Fracture surfaces at the edge region of AUS + ART annealed specimens experienced slow strain rate tensile. (**a**) No H-charged state; (**b**) H-charged for 2 h with 2.42 ppm; (**c**) H-charged for 8 h with 7.62 ppm. (**d**) H-charged for 24 h with 34.60 ppm.

4. Discussion

4.1. Microstructure–Mechanical Properties Correlation

The ART specimen revealed an austenite-martensite/ferrite duplex microstructure. Austenite films were located between large martensite laths, while fine-grained globular austenite and martensite grains were also observed (Figure 2). The AUS+ART specimen revealed mainly ultrafine-grained martensite colonies with the decoration of thin austenite films, globular austenite grains and globular martensite grains (Figure 3). The microstructure features and mechanical properties are summarized in Table 2. The effective grain size was used to describe the size of the martensite package, inherited from prior austenite grains. The enhanced strength of AUS + ART specimen was mainly attributed to the strong strain-hardening effect (Figure 5b) and the Hall–Petch effect induced by the large density of interfaces. During deformation-induced martensitic transformation, dynamic Hall–Petch effect takes place when introducing new interfaces, such as phase boundaries and lath boundaries, leading to the elevated strain-hardening rates [37]. These interfaces, in turn, act as obstacles to dislocation motion, increasing the dislocation generation rate in order for materials to accommodate plastic deformation. Therefore, the superior ductility ascribes to the delay of necking with the aid of the TRIP effect and a significant accommodation of plastic deformation by generating mobile dislocations [37].

Table 2. Summary of microstructural features and mechanical properties (A_{20}: total elongation).

Heat Treatment	UTS/MPa	A_{20}/%	Effective Grain Size/µm		Austenite Fraction/%
			α' Package	Globular Grains	
ART	811	30.1	10–20	~1.5	55.2
AUS + ART	891	33.1	1–5	~1	53.1

4.2. Influences of Microstrucal Morphology on Hydrogen Embrittlement in Fe-12Mn-3Al-0.05C Steels

ART specimen and AUS + ART specimen showed different responses to hydrogen embrittlement, resulting in different failure behaviors. The prediction of the hydrogen embrittlement mechanisms focused on two main aspects, hydrogen absorption and fracture mode.

Considering the similar constitution of initial microstructure and the same hydrogen charging conditions, the divergence in hydrogen absorption might be attributed to the distinct microstructural morphologies. The very early fracture in all AUS + ART specimens with different hydrogen contents (Figure 6b) manifests that the hydrogen amount is not the critical controlling factor. Correlating the hydrogen desorption curve with the microstructure (as shown in Figure 10), the increased hydrogen content and catastrophic failure may be correlated with the large interface density in AUS + ART.

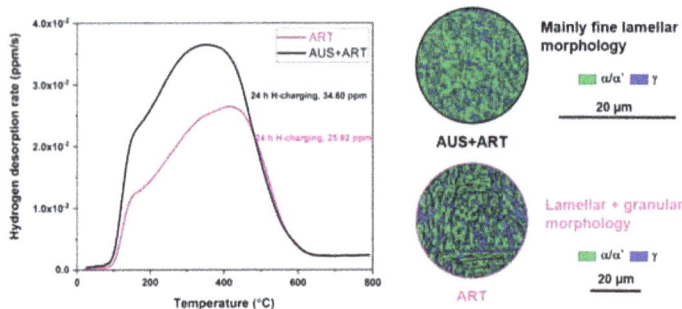

Figure 10. Hydrogen desorption rate as a function of temperature and the corresponding microstructural features.

The microstructure of AUS + ART consisted mainly of fine martensite colonies with a package size of 3 µm, which provided a large number of interfaces. These interfaces included the martensite lath boundaries, prior austenite grain boundaries and phase boundaries. During hydrogen charging, hydrogen atoms tend to trap at these interfaces. A large number of interfaces provides more trapping sites for hydrogen atoms. Meanwhile, the interface may act as the fast diffusion path for hydrogen transportation. The increased hydrogen concentration in AUS + ART specimen might be attributed to the trapping and fast diffusion behaviors of hydrogen atoms.

The distinct fracture characteristics in ART and AUS + ART specimens result from the contributions of different hydrogen embrittlement mechanisms. In ART specimen, film-like austenite grains with a large aspect ratio are still able to transform even after strong hydrogen ingression (warm-color region in Figure 11a), which is probably related to the hydrogen-enhanced localized plasticity (HELP) mechanism. The austenite film revealed a high GOS value within grains. Therefore, rugged facets in Figure 8d can be observed due to the repeated onset of a localized TRIP effect and the HEDE mechanism [18]. The TRIP effect in Mn-containing advanced high strength steels (AHSS) is considered to be vulnerable to HE, due to the localized deformation and cracking initiation, which is consistent with the published results [16,17,19]. In the highly hydrogen-affected specimen (hydrogen-charging for 24 h), the HEDE mechanism possibly became the predominant effect, reducing the cohesive bonding strength of the lattice [18,24]. Thus, the clear intergranular cleavage as directly observed in Figure 8d. Additionally, the crack propagation was reported to be impeded in the granular region, because cracks have to change the direction of propagation frequently when entering the phase in the vicinity [18].

The hydrogen trapped at interfaces was widely considered to decrease the bonding strength according to HEDE mechanism. The strong diffusion and accumulation of hydrogen atoms at interfaces in the AUS + ART specimen were assumed to be responsible for the premature fracture. The large number of interfacial defects in the UFG microstructure directly gives more chances for incubation of cracking. The reduced bonding strength may also facilitate the nucleation and emission of dislocation according to the absorption-induced dislocation emission (AIDE) mechanism [21,38], leaving small and shallow dimples at the fracture surfaces. Localized micro-strain (shown in the red region in Figure 11b) drives dislocations to interact with voids and initiate the cracks. The large area of interfaces provides a significant chance to form cracks in AUS + ART specimen, resulting in poor hydrogen embrittlement resistance.

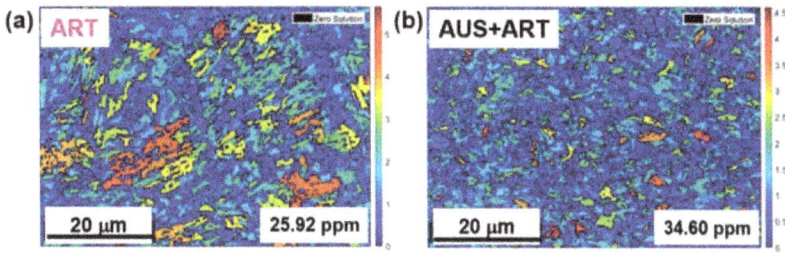

Figure 11. Grain orientation spread figures of (**a**) austenite-reverted-transformation (ART) and (**b**) AUS + ART specimens undergone slow strain rate tensile test after 24-h hydrogen charging.

4.3. Thermodynamic Assessment

To complement the experimental observations, we considered qualitatively conceivable relevant failure scenarios. While kinetic features, specifically interface diffusion, can be expected to play an important role, our focus here was on thermodynamic aspects. Concretely, we assumed that the failure originated from the precipitation of hydrogen-rich phases directly at a grain boundary (GB). We considered the dependence of the formation of such hydrogen-rich phases on the microstructural influences that resulted from heat treatment (ART or AUS + ART). The decisive quantity was the GB distribution in the different microstructures, which vary depending on the type of heat treatment. To relate the difference in GB distribution to divergencies in the formation of hydrogen-rich phases, we connected the local elastic modulus of the GBs to the modification of the local chemical equilibrium at the GBs. We referred to our previous work, where we considered the modification of the hydrogen solubility in iron due to elastic effects [39], which can result in localized nano-hydride formation [40]. As shown in Equation (68) in [39], the shifted solubility limit of hydrogen can be described as a result of a modified elastic relaxation of the coherency stresses close to and at interfaces, which are caused by the precipitation of the hydrogen-rich phase. While precipitates close to a free surface experience less elastic response to the coherency stresses, a rigid surface represents the opposite case, exhibiting a reduced elastic relaxation of the coherency stresses close to or at the interface. The resulting reduction of the hydrogen solubility limit can be up to two orders of magnitude in comparison to bulk locations at room temperature in the case of free surfaces.

To connect these fundamental findings with our experimental observations, we refer to the investigation of the local elastic properties of GBs in iron published in [41]. Though limited to symmetrical GBs, we used the presented data on the local Young's modulus of various grain boundaries as representatives of high angle and low angle grain boundaries. We noted that the range of local Young's modulus in the GB plane was 130–176 MPa in the GB, the minimum associated withe the lowest tilt angle (approx. 39 degrees), the maximum related to the highest tilt angle (approx. 109 degrees). Consequently, we can expect that a higher fraction of low angle GBs, associated with lower local Young's modulus, results in a substantially reduced hydrogen solubility, therefore supporting hydride formation and failure. After making this connection between GB character and local Young's modulus, we needed to relate the GB distribution to the hydrogen-induced failure of the specimens. We referred to [42], where the distribution of low and high angle grain boundaries was reported for TWIP steels, depending on the average grain size. Only for the UFG case, low angle grain boundaries, did we expect from our considerations to support precipitation of hydrogen-rich phases and not be suppressed. Consequently, the hydrogen failure in the UFG sample, specifically its independence on the hydrogen ingression, can be explained also from a thermodynamic perspective. The higher fraction of low angle grain boundaries in AUS + ART specimen resulted in a higher fraction of sites prone to the formation of hydrogen-rich phases, as the relaxation of the associated coherency stresses increased. The elevated hydrogen concentration in AUS + ART specimen was attributed to the fast hydrogen ingression and precipitation of brittle hydrogen-rich phases from a thermodynamic perspective, leading to premature failure.

5. Conclusions

In the current work, the mechanical properties and hydrogen embrittlement susceptibility were investigated in a cold-rolled medium-Mn Fe-12Mn-3Al-0.05C steel. Important conclusions from the present study were summarized as follows:

(1) A combination of austenitization annealing (AUS) and austenite-reversed transformation (ART) produced comparable mechanical properties (UTS = 891 MPa, Y.S. = 701 MPa, total elongation = 30.1%) as that in a routine where the ART annealing was applied immediately after cold rolling.

(2) The ultrafine-grained martensite colonies provided a large number of interfaces (prior austenite boundaries and lath boundaries) for hydrogen trapping, which increased the hydrogen ingression.

(3) ART specimen revealed a clear ductile-brittle transition with increasing hydrogen concentration. Hydrogen embrittlement is considered to be predominated by concurrent contribution of HEDE and HELP mechanisms.

(4) AUS + ART specimen exhibited extremely high hydrogen susceptibility of the ductility regardless of hydrogen concentration. The brittle failure in H-charged samples was attributed to the HEDE mechanism in the UFG microstructure with a large number of interfaces, and to possible contribution by the AIDE mechanism.

(5) Consideration of thermodynamic factors suggest that the failure discrepancy in the UFG and non-UFG specimens was likely to be related to the facilitation of hydrogen-rich phase precipitation by interfacial defects.

Author Contributions: W.S and X.S. designed the experiments. Y.M. and X.S. performed and analyzed most of the experiments (e.g., SEM, SYXRD, SSRT, TDA). S.S. performed the EBSD measurement and analyzed the data. C.H. and R.S. contributed to the theoretical interpretation in the perspective of thermodynamics. W.B. and W.S. supervised this work and contributed to intensive discussions. All authors contributed to the interpretation of the results and the writing of the final version of the manuscript.

Funding: This research work and APC were funded by the Deutsche Forschungsgemeinschaft (DFG) within the Collaborative Research Center (SFB) 761 "Steel—*ab initio*: quantum mechanics guided design of new Fe-based materials".

Acknowledgments: The synchrotron X-ray diffraction measurements were carried out at beamline P02.1 of PETRA III at DESY, a member of the Helmholtz Association (HGF), which is gratefully acknowledged. A sincere thanks also goes to colleagues at Welding and Joining Institute (ISF), RWTH Aachen university for the help with performing the hydrogen desorption experiments.

Conflicts of Interest: The authors declare no conflict of interest.

References

1. Miller, R.L. Ultrafine-grained microstructures and mechanical properties of alloy steels. *Metall. Trans.* **1972**, *3*, 905–912. [CrossRef]
2. Han, J.; Lee, S.J.; Lee, C.Y.; Lee, S.; Jo, S.Y.; Lee, Y.K. The size effect of initial martensite constituents on the microstructure and tensile properties of intercritically annealed Fe–9Mn–0.05C steel. *Mater. Sci. Eng. A* **2015**, *633*, 9–16. [CrossRef]
3. Lee, S.W.; De Cooman, B.C. Tensile behavior of intercritically annealed 10 pct Mn multi-phase steel. *Metall. Mater. Trans. A.* **2013**, *45*, 709–716. [CrossRef]
4. Lee, S.W.; Woo, W.C.; De Cooman, B.C. Analysis of the tensile behavior of 12 pct Mn multi-phase (α + γ) TWIP + TRIP steel by neutron diffraction. *Metall. Mater. Trans. A.* **2016**, *47A*, 2125–2140. [CrossRef]
5. Lee, S.J.; Lee, S.W.; De Cooman, B.C. Mn partitioning during the intercritical annealing of ultrafine-grained 6% Mn transformation-induced plasticity steel. *Scripta Mater.* **2011**, *64*, 649–652. [CrossRef]

6.   Luo, H.; Shi, J.; Wang, C.; Cao, W.; Sun, X.; Dong, H. Experimental and numerical analysis on formation of stable austenite during the intercritical annealing of 5Mn steel. *Acta Mater.* **2011**, *59*, 4002–4014. [CrossRef]
7.   Lee, Y.K.; Han, J. Current opinion in medium manganese steel. *Mater. Sci. Tech.* **2015**, *31*, 843–856. [CrossRef]
8.   Ma, Y. Medium-manganese steels processed by austenite-reverted-transformation annealing for automotive application. *Mater. Sci. Tech.* **2017**, *33*, 1713–1727. [CrossRef]
9.   Ma, Y.; Song, W.; Zhou, S.; Schwedt, A.; Bleck, W. Influence of intercritical annealing temperature on microstructure and mechanical properties of a cold-rolled medium-Mn steel. *Metals* **2018**, *8*, 357. [CrossRef]
10.   Haupt, M.; Dutta, A.; Ponge, D.; Sandlöbes, S.; Nellessen, M.; Hirt, G. Influence of intercritical annealing on microstructure and mechanical properties of a medium manganese steel. *Procedia Engineer.* **2017**, *207*, 1803–1808. [CrossRef]
11.   Edmonds, D.V.; He, K.; Rizzo, F.C.; De Cooman, B.C.; Matlock, D.K.; Speer, J.G. Quenching and partitioning martensite—A novel steel heat treatment. *Mater. Sci. Eng. A.* **2006**, *438*, 25–34. [CrossRef]
12.   Arlazarov, A.; Gouné, M.; Bouaziz, O.; Hazotte, A.; Kegel, F. Effect of intercritical annealing time on microstructure and mechanical behavior of advanced medium Mn steels. *Mater. Sci. Forum.* **2012**, *706*, 2693–2698. [CrossRef]
13.   Furukawa, T.; Huang, H.; Matsumura, O. Effects of carbon content on mechanical properties of 5%Mn steels exhibiting transformation induced plasticity. *Mater. Sci. Tech.* **1994**, *10*, 964–970. [CrossRef]
14.   De Moor, E.; Matlock, D.K.; Speer, J.G.; Merwin, M.J. Austenite stabilization through manganese enrichment. *Scripta. Mater.* **2011**, *64*, 185–188. [CrossRef]
15.   Furukawa, T. Dependence of strength–ductility characteristics on thermal history in low-carbon, 5 wt-%Mn steels. *Mater. Sci. Tech.* **1989**, *5*, 465–470. [CrossRef]
16.   Ryu, J.H.; Chun, Y.S.; Lee, C.S.; Bhadeshia, H.K.D.H.; Suh, D.W. Effect of deformation on hydrogen trapping and effusion in TRIP-assisted steel. *Acta Mater.* **2012**, *60*, 4085–4092. [CrossRef]
17.   Wang, M.; Tasan, C.C.; Koyama, M.; Ponge, D.; Raabe, D. Enhancing hydrogen embrittlement resistance of lath martensite by Introducing nano-films of interlath austenite. *Metall. Mater. Trans. A.* **2015**, *46A*, 3797–3802. [CrossRef]
18.   Han, J.; Nam, J.H.; Lee, Y.K. The mechanism of hydrogen embrittlement in intercritically annealed medium Mn TRIP steel. *Acta Mater.* **2016**, *113*, 1–10. [CrossRef]
19.   Zhang, Y.; Hui, W.; Zhao, X.; Wang, C.; Cao, W.; Dong, H. Effect of reverted austenite fraction on hydrogen embrittlement of TRIP-aided medium Mn steel (0.1C-0.5Mn). *Eng. Fail. Anal.* **2019**, *97*, 605–616. [CrossRef]
20.   Jeong, I.; Ryu, K.M.; Lee, D.G.; Jung, Y.; Lee, K.; Lee, J.S.; Suh, D.W. Austenite morphology and resistance to hydrogen embrittlement in medium Mn transformation-induced plasticity steel. *Scripta Mater.* **2019**, *169*, 52–56. [CrossRef]
21.   Lynch, S. Hydrogen embrittlement phenomena and mechanisms. *Corros. Rev.* **2012**, *30*, 105–123. [CrossRef]
22.   Robertson, I.M.; Sofronis, P.; Nagao, A.; Martin, M.L.; Wang, S.; Gross, D.W.; Nygren, K.E. Hydrogen embrittlement understood. *Metall. Mater. Trans. A* **2015**, *46A*, 2323–2341. [CrossRef]
23.   Troiano, A.R. The role of hydrogen and other interstitials in the mechanical behavior of metals. *Metallogr. Microstruct. Anal.* **2016**, *5*, 557–569. [CrossRef]
24.   Oriani, R.A. A mechanistic theory of hydrogen embrittlement of steels. *Berich. Bunsen. Gesell.* **1972**, *76*, 848–857.
25.   Beachem, C.D. A new model for hydrogen-assisted cracking (hydrogen "embrittlement"). *Metall. Mater. Trans. B.* **1972**, *3*, 441–455. [CrossRef]
26.   Robertson, I.M. The effect of hydrogen on dislocation dynamics. *Eng. Fract. Mech.* **2001**, *68*, 671–692. [CrossRef]
27.   Du, X.; Cao, W.; Wang, C.; Li, S.; Zhao, J.; Sun, Y. Effect of microstructures and inclusions on hydrogen-induced cracking and blistering of A537 steel. *Mater. Sci. Eng. A.* **2015**, *642*, 181–186. [CrossRef]
28.   Mohtadi-Bonad, M.A.; Ghesmati-kucheki, H. Important factors on the failure of pipeline steels with focus on hydrogen induced cracks and improvement of their resistance: Review paper. *Met. Mater. Int.* **2019**, *1*, 1–26.
29.   Mohtadi-Bonad, M.A.; Eskandari, M.; Szpunar, J.A. Effect of arisen dislocation density and texture components during cold rolling and annealing treatments on hydrogen induced cracking susceptibility in pipeline steel. *J. Mater. Res.* **2016**, *31*, 3390–3400. [CrossRef]
30.   Bachmann, F.; Hielscher, R.; Schaeben, H. Texture Analysis with MTEX—Free and Open Source Software Toolbox. *SSP* **2010**, *160*, 63–68. [CrossRef]

31. Hielscher, R.; Schaeben, H. A novel pole figure inversion method: Specification of the MTEX algorithm. *J. Appl. Crystallogr.* **2008**, *41*, 1024–1037. [CrossRef]
32. Hammersley, A.P.; Svensson, S.O.; Hanfland, M.; Fitch, A.N.; Hausermann, D. Two-dimensional detector software: From real detector to idealized image or two-theta scan. *High Pressure Res.* **1996**, *14*, 235–248. [CrossRef]
33. Young, R.A. *The Rietveld Method*, 4th ed.; Oxford University Press Inc.: New York, NY, USA, 2002.
34. Field, D.P.; Bradford, L.T.; Nowell, M.M.; Lillo, T.M. The role of annealing twins during recrystallization of Cu. *Acta Mater.* **2007**, *55*, 4233–4241. [CrossRef]
35. Henthorne, M. The slow strain rate stress corrosion cracking test—A 50 year retrospective. *Corrosion* **2016**, *72*, 1488–1518. [CrossRef]
36. Park, Y.; Maroef, I.; Landau, A.; Olson, D. Retained austenite as a hydrogen trap in steel welds. *Weld. J.* **2002**, *71*, 27–35.
37. Sevsek, S.; Haase, C.; Bleck, W. Strain-rate-dependent deformation behavior and mechanical properties of a multi-phase medium-manganese steel. *Metals* **2019**, *9*, 344. [CrossRef]
38. Lynch, S.P. Metallographic contributions to understanding mechanisms of environmentally assisted cracking. *Metallography* **1989**, *23*, 147–171. [CrossRef]
39. Spatschek, R.; Gobbi, G.; Hütter, C.; Chakrabarty, A.; Aydin, U.; Brinckmann, S.; Neugebauer, J. Scale bridging description of coherent phase equilibria in the presence of surfaces and interfaces. *Phys. Rev. B* **2016**, *94*, 134106. [CrossRef]
40. Pezold, J.; Lymperakis, L. Neugebauer, Hydrogen-enhanced local plasticity at dilute bulk H concentrations: The role of H-H interactions and the formation of local hydrides. *Acta Mater.* **2011**, *59*, 2969–2980. [CrossRef]
41. Bhattacharya, S.; Tanaka, S.; Shiihara, Y.; Kohyama, M. Ab initio perspective of the (110) symmetrical tilt grain boundaries in bcc Fe: application of local energy and local stress. *J. Mater. Sci.* **2014**, *49*, 3980–3995. [CrossRef]
42. Bai, Y.; Momotani, Y.; Chen, M.C.; Shibata, A.; Tsuji, N. Effect of grain refinement on hydrogen embrittlement behaviors of high-Mn TWIP steel. *Mater. Sci. Eng. A* **2016**, *651*, 935–944. [CrossRef]

MDPI

St. Alban-Anlage 66

4052 Basel

Switzerland

Tel. +41 61 683 77 34

Fax +41 61 302 89 18

www.mdpi.com

Metals Editorial Office

E-mail: metals@mdpi.com

www.mdpi.com/journal/metals

www.ingramcontent.com/pod-product-compliance
Lightning Source LLC
Chambersburg PA
CBHW051847210326
41597CB00033B/5807